NEUROSCIENCE RESEARCH PROGRESS

MEMORY CONSOLIDATION

Neuroscience Research Progress

Additional books in this series can be found on Nova's website under the Series tab.

Additional e-books in this series can be found on Nova's website under the e-book tab.

NEUROSCIENCE RESEARCH PROGRESS

MEMORY CONSOLIDATION

MANABU SAKAKIBARA
AND
ETSURO ITO
EDITORS

nova publishers
New York

Library of Congress Cataloging-in-Publication Data

ISBN: 978-1-63482-596-2
Library of Congress Control Number: 2015936786

Published by Nova Science Publishers, Inc. † New York

CONTENTS

PREFACE

"Memory Consolidation" is defined as the stabilizing processes of a memory trace after the initial acquisition. This is considered to be the process of encoding or of storage of memory. Memory consolidation has been studied as a form of modern science since the 19th century and still gives many fascinating subjects to researchers in the neuroscience and psychology fields today. The book of Hermann Ebbinghaus, entitled "MEMORY: A Contribution to Experimental Psychology (1885)", is read over (for example, see "MEMORY, From Mind to Molecules" 2nd edition, by L. R. Squire and E. R. Kandel, 2009), and the book of Donald O. Hebb, entitled "The Organization of Behavior: A Neuropsychological Theory (1949)", should be always act as reference as well.

Memory consolidation has been discussed at the cellular level since the 1950s. This concept involves the dependency of memory-related gene transcription and translation. This led to the protein-synthesis hypothesis of memory formation being proposed. Katz and Halstead (1950) described it as "an essential feature in the genesis of the memory trace is the formation, as a result of individual experience, of geometrically ordered protein molecules in the neurons of the cerebrum. These new ordered molecules then function as template molecules and thus now have the same relationship to the cell as do the origin genes". Following this suggestion, the experiments with protein synthesis inhibitor showed that the intracerebral injection of the inhibitor into animals led to amnesia. This result demonstrated that the maintenance of memory depends on the continuing synthesis of proteins.

In the 1970s, the concepts of short-term memory (STM) and long-term memory (LTM) emerged using an invertebrate model, the sea hare *Aplysia*. Kandel and his colleagues showed that a protein-synthesis inhibitor, anisomycin, did not prevent an electrophysiological correlate of short-term habituation or sensitization in an isolated ganglion of *Aplysia*. Shortly thereafter, anisomycin was also found to be an effective amnestic agent in vertebrates, proving the protein-synthesis hypothesis in the formation of LTM. That is, the use of anisomycin enabled researchers to distinguish between STM and LTM. Further, it was shown that protein must be synthesized soon after training if LTM were to be formed while STM did not require protein synthesis.

Now, memory is divided into two or three (sometimes more) critical time periods from a temporal dynamic processing point of view. Generally, STM lasts minutes and LTM can from last days and weeks to an entire lifetime. Some researchers may posit an intermediate-term memory (ITM) stage. ITM is thought to be a labile stage between STM and LTM. STM occurs within seconds, ITM ranges from minutes to a few hours, and LTM is lasts from hours to days to years. The boundaries between STM, ITM and LTM are fuzzy and can vary from situation to situation. The mechanisms of ITM formation, particularly in regards to protein-synthesis dependency or kinase-specific inhibition, seem subjective to different animals.

We have focused on the frontier studies for memory consolidation in this book. It is divided into two parts: one shows studies using invertebrates, the other showed studies regarding mammalian brains (including human trials). In Part 1, Shidara and Oka describe the variety of time scales in memory of the nematode *C. elegans*. Okada introduces the plasticity of mushroom body neurons in the honeybee *Apis*. Aonuma, Sakura and Kurabayashi shed light on the internal state of fight losers in the cricket *Gryllus*. Matsuo explains the olfactory memory of the slug *Limax*. Sakura and Watanabe show that the olfactory oscillation plays a crucial role in odor perception in *Limax* memory.

Following to the studies of these invertebrate animals, the next four chapters introduce the recent advantages of the studies using the pond snail *Lymnaea*. The chapters (one by Sakakibara and the other by Ito, Yamagishi and Ikemoto) reveal the food-avoidance conditioning from the behavior to the molecule level while Kemenes shows food-reward conditioning in *Lymnaea*. On the other hand, Lukowiak describes the forgetfulness in aerial respiratory conditioning.

In Part 2, memory consolidation is explained mainly in human-related subjects. Iemolo, De Risi and De Leonibus discuss the role of dopamine in the theoretical framework. Pardilla-Delgado, Mattingly, Alger, Chambers, Cunningham and Payne pointed out the role of sleep in memory consolidation. Fabio, Gulla' and Errante illustrated the mechanisms of memory consolidation for stimuli and objects with emotive contents. Sieb indicated the encoding and consolidation of conscious (declarative) memory.

All these new progresses in the studies of memory consolidation clearly require further approaches in multi-disciplinary Cross phyla fashion, such as from behavior to molecules. We hope this book will encourage young investigators to participate in the study field of memory consolidation.

Manabu Sakakibara
Tokai University
Professor, Neurobiological Engineering/Life Science
317 Nishino Numazu 410-0321, Shizuoka, Japan
E-mail: msakaki@chime.ocn.ne.jp

Etsuro Ito
Professor, Tokushima Bunri University
E-mail: eito@kph.bunri-u.ac.jp

December, 2014

In: Memory Consolidation
Editors: Manabu Sakakibara and Etsuro Ito
ISBN: 978-1-63482-596-2

Chapter 1

How Variety of Time Scales in Memory Could *Caenorhabditis elegans*, a Simple and Compact Neural Circuit Organism, Form?

Hisashi Shidara and Kotaro Oka*
Center for Biosciences and Informatics, School of Fundamental Sciences and Technology, Keio University, Hiyoshi, Kohoku-ku, Yokohama, Kanagawa, Japan

Abstract

Caenorhabditis elegans is one of the novel model organisms, which has been studied on over 50 years. *C. elegans* has just 302 neurons and simple neural circuits, of which connectivity has already been revealed. In addition, tractable gene manipulation thanks to its genome source has made *C. elegans* the more ideal model organism. Since early times, many kinds of assay have elucidated that *C. elegans* forms short-term and long-term memory of various sensory modal stimuli such as mechanostimulus, temperature, odorants and gustatory cues. In addition, recent studies showed memory of worms transmitted through multiple generations existed. Like this, although this model animal is very simple, it exhibits various types of time scales in learning and memory with compact neural circuits. In this chapter, first we introduce learning and memory of *C. elegans* in each sensory modality and then discuss time scales of obtaining and retaining memory. Finally, we show perspective of learning and memory researches in *C. elegans*, indicating powerful tools recently developed.

Keywords: *Ceanorhabditis elegans*, *C. elegans*, learning and memory, imaging, behavior

* Corresponding Author address; E-mail:oka@bio.keio.ac.jp.

INTRODUCTION

Ceanorhabditis elegans is a 1 mm, free-living non-parasitic nematode, which was introduced by Sydney Brenner in 1963 as a powerful model organism. Its characters such as short life cycle (< 3 days), transparent body, self-reproduction and simple structure have made the creature more attractive for scientists. Especially, the remarkable feature that *C. elegans* constructed by only 302 neurons of about 959 somatic cells contributes to neuroscience. The electron microscopy revealed wiring and connectivity of *C. elegans* nervous system (White, Southgate, Thomson, & Brenner, 1986). The nervous system has 639 chemical synapses, 890 gap junctions and 1410 neuromuscular junctions (Varshney, Chen, Paniagua, Hall, & Chklovskii, 2011), and it is divided mainly into sensory neurons, interneurons and motor neurons. Despite of the small nervous system, *C. elegans* responds to a variety of sensory modalities including mechanosensation, therosensation, chemosensation and so on with specific sensory neurons and facilitates its behavior (Bargmann, 2006).

Sensation to various sensory modalities is revealed by many kinds of behavioral assays. The observation of behavior, combined with gene manipulation and laser ablation, elucidated neurons and neural circuits including sensation, which indicated molecular and cellular mechanisms of the sensation. For example, chmotaxis assay in which worms moved to a preference substance on agar plates showed which neurons are included in chemosensaion (Bargmann, Hartwieg, & Horvitz, 1993; Bargmann & Horvitz, 1991). Learning and memory in *C. elegans* are also indicated by such behavioral assay. In the early years, thermotaxis assay, in which worm moved on plates of linear temperature gradients suggested memory of *C. elegans* (Hedgecock & Russell, 1975). In addition, in 1990, Rankin et al. precisely discussed learning and memory in *C. elegans* for the first time by checking plasticity of behavior to tapping stimulation (Rankin, Beck, & Chiba, 1990). Subsequently, olfactory and gustatory learning are clarified (Colbert & Bargmann, 1995; Wen et al., 1997). Recently, optical methods are more valuable to identify functions of neurons and neural circuits (Piggott, Liu, Feng, Wescott, & Xu, 2011).

Previous studies are performed with different assays to understand specific features which researchers are interested in. These results provide us various kinds of learning and memory in *C. elegans*. Interestingly, small neural circuits are responsible for such learning and memory. Through the ages, neural circuits of *C. elegans* were considered to be regarded as 'small-world' networks (Watts & Strogatz, 1998). Some researches supported compactness of the neural circuits, (Funabashi et al., 2001; Kawamura et al., 2001; Morita et al., 2001), thus indicating that some neurons could be overlapping to memorize experiences from other sensory modalities. Also, time courses of memory are varied each condition. These researches indicated how variety of memory processing *C. elegans* has, although it has a simple nervous system. In this chapter, we introduced learning and memory of *C. elegans* in each sensory modality, especially comparing them in aspects of conditions and time courses. Our purpose in this chapter also suggests new perspectives of learning and memory researches in *C. elegans*.

MECHANOSENSATION

C. elegans responds to various types of mechanical stimulation such as light and harsh touch, and vibration (Bounoutas & Chalfie, 2007). *C. elegans* moves sinuously forward and backward regulating dorsal-ventral muscles collaboratively. When its head is touched gently with eyelash, the worm changes its direction backward; when its tail is touched, it moves forward. Another mechanical stimulation, tap or vibration to a side of the agar plate where worms move freely also induced the change of their direction. In 1990, for the first time, learning and memory *in C. elegans* were precisely proposed by indicating habituation of taps induced withdrawal response (Rankin et al., 1990). Animals stimulated by taps swim backward for some distance and change their direction moving forward. The consecutive tapping, which decreases thire backward distance, is the index of plasticity. In the experiment, 40 trains at 10 sec interstimulus interval (ISI), where a train consisted of 6 taps, decreased worms' movement. On the other hand, the response to taps was recovered from the decrement by intense electrical shocks. Habituation is defined as a decrease in responses with stimulation repeated while the decrease recovers by dishabituation stimulation. Therefore, the result shows that *C. elegans* has potential for both habituation and dishabituation (Rankin et al., 1990). This is an example of short-term memory. In the same experiment, long-term memory, which retention of learning over 24 hr, was also investigated. Animals that received 5 blocks of 20 trains of taps at 10 sec every hour showed a reduction to the tap stimulation even the next day. Thus, *C. elegans* is capable of over 24 hr retention of habituation memory (Rankin et al., 1990).

Then, to understand features of the habituation precisely, experiments with different conditions were performed. In short-term habituation, ISI has an important role of its rate and recovery (Rankin & Broster, 1992). Compared with 6 sec ISI in 30 tap stimuli, 2 sec ISI in 30 tap stimuli caused more rapid habituation, but after 10 min the response returned to the initial level. On the other hand, 6 sec ISI habituation process was slow and did not reach to the same level of 2 sec. However, the 6 sec ISI induced habituation lasted more than 30 min. Meanwhile, processes of long-term habituation were also described in detail. To retain long-term memory, cellular processes are accompanied by gene expression. In *C. elegans* as well as other organisms, heat shock disruption caused failures to obtain long-term memory (Beck & Rankin, 1995). When a heat shock (32℃, 45 min) was presented in a 1 hr rest during training, neither before nor after training, memory consolidation was disrupted. The research also showed that the brief heat shock (32℃, 15 min) to particular term from 15 to 30 min in the 1 hr rest during training disrupted the retention of memory, indicating the narrow crucial term for memory consolidation. The brief heat shock effectively blocking long-term memory meant that the first 30 min in the 1 hr rest is necessary for memory consolidation. In fact, the over 40 min is required for long-term retention of training (Rose, Kaun, & Rankin, 2002). In addition, conformation of long-term memory is required for appropriate training condition (Beck & Rankin, 1997). When distributed training (60 sec ISI) was performed, long-term habituation was retained, although massive training and training (10 sec ISI) did not cause.

Behavior induced by mechanosensation was driven by small compact neural circuits. The identification of neurons in the head and tail touch circuits was performed with laser ablation (Chalfie & Sulston, 1981; Chalfie et al., 1985). The laser ablation and the assaying touch response indicate that *C. elegans* senses gentle touch with 5 mechanosensory neurons

ALML/R, AVM and PLML/R. ALML/R and AVM located at anterior, whose structure are microtubule cells, are required for the head touch perception; PLML/R located at posterior for the tail touch perception (Chalfie et al., 1985). In addition, four interneurons, AVD, AVA, AVB and PVC, which are categorized into command interneurons, are necessary for touch sensation. Commonly, these mechanosensory circuits are divided into two types; the anterior sensory motor circuits to integrate inputs from the touch response at anterior and the posterior sensory motor circuits to integrate inputs at posterior. The balance of the two circuits exerts light touch induced behavior. The withdraw response is also mediated by the same neural circuits for light touch response as well as three other neurons, PVDL/R sensory neurons and DVA interneuron (Wicks & Rankin, 1995). These additional neurons are thought to have a functional role to integrate the anterior and posterior inputs.

For cellular and molecular roles of short-term habituation, *eat-4*, the *C. elegans* homologue of the mammalian glutamate vesicular transporter (VGLUT1) is the first candidate gene, which is expressed in ALM, AVM and PLM (Lee, Sawin, Chalfie, Horvitz, & Avery, 1999; Rankin & Wicks, 2000). Although *eat-4* mutant, loss-of-function worm, showed a normal response to a single tap, the mutant indicates different habituation from wild type when they were given repeated stimulation. Compared with wild type, *eat-4* mutant was habituated more rapid and recovered more slowly. For wild type, the habituated response recovery in 5 min, but the mutant did not even over 10 min. The result suggested that glutamate release from sensory neuron has an important role for short-term habituation. As well as short-term habituation, EAT-4 is also required for long-term habituation (Rose et al., 2002). The importance of glutamate transmission for long-term habituation was also supported by other researches. The mutation of *glr-1*, which encoded a receptor subunit of a non-NMDA excitatory ionotropic glutamate receptor subtype, caused deficits of the formation of long-term memory (Rose, Kaun, Chen, & Rankin, 2003). In addition to the glutamate transmission, dopamine signal also plays a crucial function for retention of memory. *dop-1* gene, which encode D1-like dopamine receptor is expressed in the mechanosensory neurons, ALM and PLM (Sanyal et al., 2004). The *dop-1* mutant shows the frequency of reversal response to tap stimulus decrease more rapidly than the wild type, but the distance of reversal did not change. Further investigation revealed the dopamine pathway including DOP-1 (Kindt et al., 2007). Dopamine modulated a Gq/PLC-b signaling pathway in ALM, which controlled habituation. Interestingly, the experiment showed that the pathway affected habituation only when food was presented.

TERMOSENSATION

For C. *elegans*, temperature is a one of the important factors for behavior, development, physiology and reproduction. In 1975, Hedgecock and Russell first reported thermotaxis of *C. elegans*. When *C. elegans* were grown at from 16℃ to 25℃, animals moved to their cultivation temperature and showed isothermal movement (Hedgecock & Russell, 1975). In the assay plate, a stable linear temperature gradient (approximately 0.5 ℃ /cm) was established. The preference of temperature was expressed from counting the number of worms in the specific area. Originally, another thermotaxis assay was also used, which is a radial non-linear gradient assay. The gradient was established on the agar in 9 cm plastic Petri dish. Mainly, uniform temperature gradient assay was utilized for population of worms; on the other hand, radial non-gradient assay was for tracking of single worm (Kimata, Sasakura, Ohnishi, Nishio, & Mori, 2012; Mori, Sasakura, & Kuhara, 2007). In the assay, the steepness of the temperature gradient is a crucial key to dissect thermotaxis behavior (Jurado, Kodama, Tanizawa, & Mori, 2010). At thermal gradient of < 1℃/cm, worms show similar behavior in the experiment of Hegecock and Russell (1975). However, at the thermal gradient of more 1 ℃/cm, worms grown at low temperature moved to their cultivation temperature but worms grown at high temperature indicated almost athermotactic behavior.

The original thermotaxis assay introduced by Hegecock and Russell (1975) was revealed memory depending on experience of *C. elegans*. When worms grown at 16℃ were transferred to 25℃, they rewrote their memory of cultivation temperature as 25℃ in 2-4 hr and *vice versa*. The memory consolidation was strongly influenced by growth environment condition. After worms were under starved and overcrowded condition in 2-3 hr, they dispersed from their cultivation temperature. The result indicated that food was a crucial factor for thermal memory consolidation. The stable radial temperature gradient assay also supported the result of Hegecock and Russell (1975) and described memory retention of temperature in detail (Mohri et al., 2005). Interestingly, well-fed animals retained their new cultivation temperature memory with food in 2-4 hr as previous researches shown. However, without food, the time course was different from the well-fed condition. Worms pre-cultivated at 17℃ with food memorized their new temperature in 4 hr, but worms pre-cultivated at 25℃ did in 20 min. In addition, starved worms transferred to previous condition with food formed their new cultivation temperature memory in 10 min. The difference between a well-fed and a food-starved state was modulated by serotonin and octopamine. The monoaminses modulated the learning and memory of cultivation temperature with food condition.

Neural circuits of thermosensation were first revealed by killing neurons with micro laser beam (Mori & Ohshima, 1995). Temperature was sensed by AFD sensory neuron mainly. The circuits were included in some interneurons, AIY, AIZ and RIA. AIY, which is projected from AFD, is responsible for the migration to warmer regions (thermophilic movement). On the other hand, AIZ has an opposite feature of AIY, migration to cooler regions (cryphilic movement). RIA interneurons are located the downstream of two interneurons and integrated the information from AIY and AIZ. The features of these neurons in thermosense circuits were revealed by optical imaging techniques. There are some suggestions that temperature memory is stored in the AFD sensory neuron (Clark, Biron, Sengupta, & Samuel, 2006; Kimura, Miyawaki, Matsumoto, & Mori, 2004; Samuel, Silva, & Murthy, 2003). If ambient temperature (T_{amb}) was warmer or cooler than cultivation temperature (T_{cult}), the synaptic vesicles in AFD were released but if T_{amb} was almost equal to T_{cult}, the release was low (Samuel et al., 2003). The result indicated that AFD have some roles in thermal memory. Then, calcium imaging showed that cytosolic calcium ion (Ca^{2+}) in AFD increased when the ambient temperature was increasing but not decreasing (Kimura et al., 2004). The threshold of temperature causing increase in Ca^{2+} was dependent on cultivation temperature. Further investigation showed how precisely AFD sensory neurons sensed temperature and transmitted signals to AIY interneurons (Clark et al., 2006). Interestingly, an olfactory sensory neuron, AWC, is also required for thermosensation (Biron, Wasserman, Thomas, Samuel, & Sengupta, 2008; Kuhara et al., 2008). AWC sensory neuron works as a secondary thermosensing neuron and also remembers cultivation temperature. On the other hand, associative learning with temperature and food was modulated in AIY interneuron (Gomez et al., 2001).

To date, molecular pathways in AFD are described in detail (Kimata et al., 2012; Mori et al., 2007). Putative thermo receptor actives putative G protein, and then G protein activates the guanylate cyclases GCY-8/18/23. The guanylate cyclases produced cGMP, of which concentration regulates cGMP-dependent cation (CNG) channel including in TAX-2/4 subunits. In addition, temperature signals in AFD are regulated by TAX-6 (calcineurin), DGK-1 (DAG kinase), TTX-4 (protein kinase C) and CMK-1 (CaMK I/IV). The signals are also regulated through transcription by CRH-1 (CREB) and NHR-69 (nuclear receptor). Of thermal memory molecules, DGK-3 diacylglycerol kinase has an interesting function, which is the regulation of the rate of cultivation temperature resetting (Biron et al., 2006). Worms grown at 15℃ or 25℃ rewrote their cultivation memory in 2-4 hr when they were transferred to new cultivation condition at 25℃ or 15℃, respectively. However, *dgk-3* mutants reset their memory more slowly than wild type. AIY calcium response in *dgk-3* mutant was rescued by the *dgk-3* expression in AFD and the *dgk-3* deficit did not affected sensitivity to temperature, thus indicating that DGK-3 regulated plasticity by modulating synapse outputs of AFD. On the other hand, the secondary thermo sensory neuron AWC also revealed thermosensory signal pathways (Kimata et al., 2012). SRTX-1 G protein-coupled receptor (GPCR) is regulates rate of temperature response in AWC. G α protein ODR-3 activates guanylate cyclase ODR-1, which produces cGMP. cGMP concentration regulates TAX-2/4 leading Ca^{2+} influx. RGS EAT-16 regulates temperature sensing negatively. Interestingly, thermosensing pathways are overlapping with olfactory sensing pathways (Kuhara et al., 2008).

CHEMOSENSATION

Chemosensatin in *C. elegans* is required for finding food, avoiding noxious condition, performing appropriate development and mating. So far, although chemosensation to various cues have been investigated in *C. elegans*, especially chemosensation to volatile (olfactory) and water-soluble (gustatory) substrate are widely studied (Bargmann, 2006). *C. elegans* is capable of detecting a wide variety of olfactory and gustatory cues with some distinctive sensory neurons. For olfactory sensation, just left/right pairs of AWC and AWA sensory neurons mainly detect odorants (Bargmann et al., 1993). Each neuron exhibits odor specific perception. For example, AWC senses isoamyl alcohol (IAA), benzaldehyde (Bz) and 2-butanone (Bu) while AWA senses diacetyl (DA) and pyrazine (Pz). On the other hand, for gustatory sensation, a single pair of ASE detects water-soluble substances (Bargmann & Horvitz, 1991; Ward, 1973). ASE has asymmetry function of gustatory sensation, which ASEL is mainly sensitive to Na^+ while ASER is mainly sensitive to Cl^- and K^+ (Pierce-Shimomura, Faumont, Gaston, Pearson, & Lockery, 2001). The odorants and gustatory cues are so tractable that a lot of assays and procedures have been developed and utilized for researchers' purpose. Therefore, here, we introduce learning and memory along the behavioral assay and procedures.

After *C. elegans* were exposed to an odorant for a while, response to the odorant diminished for a few hours (Colbert & Bargmann, 1995). Worms were adapted to the odorant on the agar plate with the lid, where the odorant was spotted in advance. After prolonged exposure to the odorant, worms were transferred to the center of a new agar plate, where the odorant was spotted at the end of the plate and the control substance was at the opposite end of the plate. After 60 min, worms at the area of odorant and control spot were counted and Chemotaxis Index (C.I.) was calculated. C.I. varied from -1 to +1, indicating that high C.I. meant that worms were attracted to the odorant. Worms pre-exposed to Bz or IAA diminished responses to the odorant gradually until 60-90 min. The diminished responses recovered in 3 hr after worms were away from odorants. Interestingly, the adaptation to an odorant affected the response to some specific odorants. For example, the adaptation to Bz induced the adaptation to IAA, which was sensed by the same olfactory neuron AWC. Meanwhile, it did not induce the adaptation to Bu, which is also sensed by AWC, and other odorants sensed by AWA sensory neurons. The adaptation was called cross adaptation. Similarly, the adaptation to IAA decreased only the response to Bz, thus indicating that cross adaptation was odorants selective. To date, the gene containing olfactory adaptation was investigated, which indicated that the adaptation was occurred in sensory neurons (Bargmann, 2006; Hart & Chao., 2010).

Naturally, the adaptation is influenced by environmental conditions such as food (Colbert & Bargmann, 1997). Worms starved for an hour before the adaptation procedure to Bz diminished the response to Bz more completely after the adaptation. Therefore, the starvation enhanced the adaptation to the odorant. The adaptation enhanced by 2 hr starvation returned to the same level of well-fed animals when worms cultivated with food for 3 hr after starvation, indicating that the adaptation could recover from the starvation effect. The starvation-induced enhancement was involved in exogenous serotonin, which was known to mimic the presence of food (Avery & Horvitz, 1990). When worms were incubated without food on serotonin containing plates, the effect of starvation on adaptation was inhibited. In addition, at the presence of food, the extent of the adaptation to Bz was suppressed by serotonergic signalings (Nuttley, Atkinson-Leadbeater, & van der Kooy, 2002). As well as food condition, contextual taste cues also affected the olfactory adaptation (Law, Nuttley, & van der Kooy, 2004). When worms were adapted to Bz at the presence of Na^+ or Cl^-, the response to Bz was decreasing on the assay plate containing Na^+ or Cl^- as mentioned above. However, if the plate condition of taste cues changed between training term of adaptation and testing term, the adaptation to Bz was suppressed. These results indicated that the olfactory adaptation of *C. elegans* was closely related to the associative learning. In addition to adaptation, habituation for olfactory cues was also investigated (Bernhard & van der Kooy, 2000). When worms were exposed to low concentration of DA (0.001%) in 15 min, the response to the same concentration of DA decreased. But the decreased response recovered to the basal response level after the dishabituating stimulus (twice 250 g centrifugation for a total of 3 min). Meanwhile, worms also exhibited the originally explained adaptation, indicating *C. elegans* showed two forms of olfactory plasticity: habituation and adaptation. The associative learning for olfactory was also investigated in detail (Amano & Maruyama, 2011). When worms were exposed to an attractive odorant 1-propanol (conditioned stimulus) and hydrochloride (pH 4.0), worms did not attract to 1-propanorl. Spaced training with 10 min intertrial interval (ITI) induced the retention of memory for 24 hr while massed training did only for 3 hr.

C. elegans also showed more various types of plasticity. One of the examples is early adaptation (Hirotsu & Iino, 2005). When worms collected in a microfuge tube were exposed to applied solution containing an odorant for 5 min, worms exhibited the adaptation to the odorant. This short-term adaptation was called early adaptation. Although some futures of early adaptation were similar to the past adaptation assay, the adaptation was different from previous described long-term adaptation. Compared with the long-term adaptation, recovery time in early adaptation was shorter: 30 min is enough to recover from the adaptation. In addition, early adaptation is necessary for AIY interneurons, but not sensory neuron, although sensory neuron senses the odorants. The short-term adaptation was strongly influenced by densities of worms at growth condition (Yamada et al., 2010). The densities of worms were closely related to pheromone signalings, indicating that early adaptation was modulated by pheromone. Another example is imprinting of olfactory sensation (Remy & Hobert, 2005). When *C. elegans* was exposed to an odorant from egg to late larva 1 stage (0-24 hr), the response to the odorant increased, indicating that imprinting was occurred in *C. elegans*. Imprinting was occurred for odorants sensed by AWC sensory neurons but not AWA. For olfactory imprinting, AWC-AIY connectivity is thought to be necessary.

For gustatory adaptation, *C. elegans* changes its behavior to adapt its environment. The early experiments of chemotaxis revealed that worms exhibited rapid adaptation to water-soluble resulting in its behavioral changes (Dusenbery, 1980; Ward, 1973). Further features of the gustatory adaptation were described by checking chemotaxis on the agar plates as well as checking olfactory adaptation (Jansen, Weinkove, & Plasterk, 2002). When worms were exposed to specific concentration of sodium acetate (NaAc) in 15 min, chemotaxis to 25 mM NaCl was significantly reduced. The adaptation enhanced to some extent with increasing pre-exposure time and the concentration of NaAc. The adaptation occurred by the optimized condition (100 mM NaAc pre-exposure in 15 min) recovered in 5 min. The condition with higher concentration and longer pre-exposure time caused worms avoidance to gustatory cues rather than adaptation (Jansen et al., 2002; Tomioka et al., 2006). Therefore, the adaptation to gustatory cues may be not tractable experimental procedure.

On the other hand, associative learning and memory of gustatory cues has been well studied. The gustatory learning was also influenced by food condition (Wen et al., 1997). The associative learning was disrupted by a cold shock in 15-30 min after training, which meant that memory retrieval and memory retention in the specific time of post-training was important for the learning (Morrison & van der Kooy, 1997). Then, the features of the gustatory associative learning were well characterized (Saeki, Yamamoto, & Iino, 2001). If *C. elegans* was cultured with NaCl and food, preferences to NaCl increased; without food, one to NaCl decreased. The associative learning to NaCl affected chemotaxis to other gustatory cues such as cAMP, biotin and lysine sensed by ASE sensory neurons. The preference was strongly dependent on the cultivation condition of *C. elegans* (Kunitomo et al., 2013). Because worms could memorize concentration at cultivation environment with food precisely, they exhibited the strong behavioral preference to the specific concentration on a salt gradient. The memory of salt concentration at growth environment was rewritten in 4 hr.

The associative learning with starvation and salt concentration is required for the insulin signaling pathway (Tomioka et al., 2006). In ASER, an insulin receptor DAF-2 received insulin INS-1 and regulates phosphoinositide 3-kinase AGE-1, modulating the amount of PIP3. PIP3 controls homologs of phosphoinositide-dependent protein kinase PDK-1 and AKT-1. INS-1 is synthesized and secreted from AIA interneurons, thus suggesting the salt associative learning is included in the role of AIA interneurons. Meanwhile, attractive salt learning is necessary for Gq/DAG/PKC signaling pathway, which antagonizes the insulin signaling pathway (Adachi et al., 2010). Recent study revealed the Ras-MAPK pathway responding to food signals regulates the localization of an insulin receptor isoform DAF-2c, which induces the salt learning (Ohno et al., 2014).

OTHERS

Some types of experiences in *C. elegans* are inherited to its offspring, suggesting the existence of inherited memory. The chromatin modifiers, the histone H3 lysine 4 trimethylation (H3K4me3) complex, are known to regulate *C. elegans* life span (Greer et al., 2010). The life extension in H3K4me3 defective mutant was inherited to the third generation although mutated gene did not inherited to following generation by mating (Greer et al., 2011). The inheritance was caused by epigenetic change, indicating an epigenetic memory of longevity in *C. elegans*. In addition to epigenetics, small-RNA also plays important roles of the inherited memory (Rechavi et al., 2014; Rechavi, Minevich, & Hobert, 2011). The virus-deprived, small-interfering RNAs (viRNAs) induced by the expression of the Flock House virus are inherited to multigeneration through the germline (Rechavi et al., 2011). The offspring inheriting vsRNAi is capable of the viral silencing even though the animals could not produce their own viRNAs. The starvation environment induced the production small RNAs, which played the role in nutrition and were inherited through at least three generations (Rechavi et al., 2014). Meanwhile, although mechanisms are not yet revealed, imprinting for olfactory sensation is also inherited through some generations (Remy, 2010).

COMPACT NEURAL CIRCUITS UNDERLYING VARIOUS KINDS OF TIME COUSES IN LEARNING AND MEMORY

C. elegans showed a variety of time courses and sensory modalities in learning and memory. We have described worms store memory for each sensory modality with compact neural circuits, suggesting that the storage of memory is mainly occurred in sensory neurons, which modulated by interneurons. For starvation learning with salt, insulin from AIA regulates memory in ASE (Tomioka et al., 2006). Meanwhile, interestingly, an insulin signal from AIA was also regulates chemotaxis behavior and adaptation to odorants (Chalasani et al., 2010). These results show that roles of each neuron are extremely myriad although the number of neurons related with learning and memory.

Overlapping such as insulin signal from AIA is also observed in AIY interneurons. For early adaptation to odorants, Ras-MAPK pathway including Ca^{2+} signaling in AIY are required (Hirotsu & Iino, 2005). On the other hand, for cultivation temperature learning with food condition, function of Ca^{2+} in AIY is also necessary to control synapse strength between AFD and AIY, regulating memory storage of AFD (Gomez et al., 2001). So, the same molecular in single neuron works distinctively to obtain and retain memory. In addition, sensation of transmitter in AIY has a remarkable feature. For odor sensation signal pathways, AIY receives glutamate released from AWC through the glutamate-gated chloride channel GLC-3 (Chalasani et al., 2007). The same inhibitory glutamate signal pathway is thought to be utilized in thermo sensation (Kuhara, Ohnishi, Shimowada, & Mori, 2011). These examples of overlapping are mainly observed in ciliated sensory neurons and head interneurons.

Though the number of neurons involved in learning and memory was small, time courses in learning and memory is also varied (Figure 1). Short-term memory is observed in minutes while long-term memory is over 24 hr. In *C. elegans*, even memory transmitted over generation exists. The learning and memory have different features in each sensory modality. The olfactory adaptation is observed from minutes to hours depending on time of pre-exposure odorant solution. Compared with olfactory adaptation, gustatory adaptation retains for only 5 min though worms are exposed to gustatory cues for 15 min. These results show that worms are more easily adapted to odorants than gustatory cues. The difference between to sensory modalities may indicate an important role for worms' life. On the other hand, the similar points are also monitored in learning and memory of *C. elegans*. At least, food condition is closely related to obtaining and retaining memory of mechanosensation, thermosensation, chemosensation, and even epigenetics. Especially, in the associative learning of temperature or gustatory cues with food, the time courses were similar such that memory obtained in about 4 hr, indicating that similar mechanisms and signal pathways may be facilitated.

As mentioned above, multiple modulations in single neuron and combination of some neurons underlie the variety types of learning and memory in *C. elegans*. So, how are the modulations depending on sensory modalities carried out? Recent researches showed that the compartmentalized functions exist in single neuron of *C. elegans* (Hendricks, Ha, Maffey, & Zhang, 2012; Shidara et al., 2013). These results support that single neuron is regulated and modulated in localized area and engages in different memory. Memory consolidation is accompanied by plasticity of synapses. Therefore, to understand complex function of neurons and neural circuits more, we might have to investigate neurons with smaller compartment.

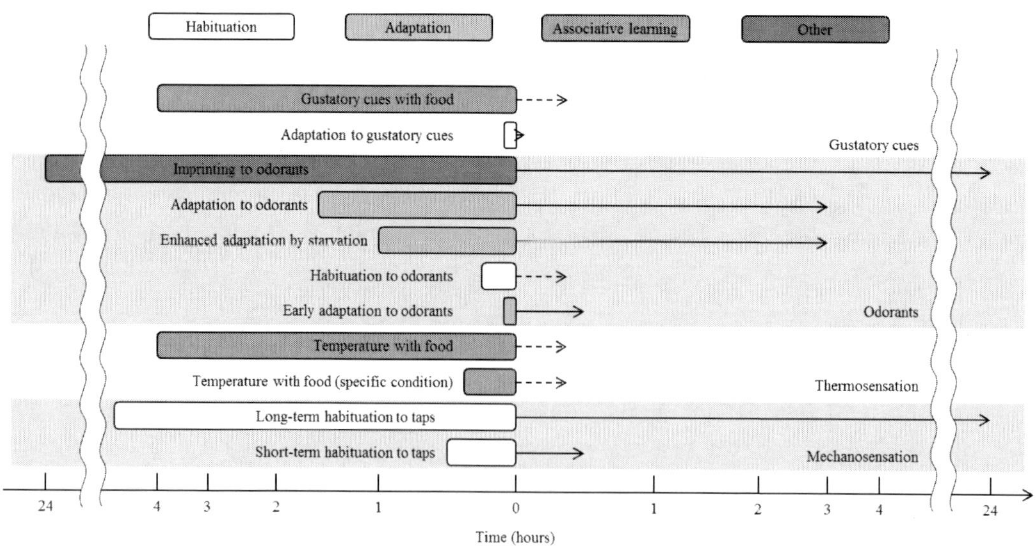

Figure 1. The time courses of learning and memory in *C. elegans*. Box shows training time and arrow shows retention time of memory. Dotted lines mean that retention time was not clarified explicitly in original papers.

FUTURE CONTRIBUTION TO LEARNING AND MEMROY RESEARCHES IN *C. ELEGANS*

C. elegans has just 302 neurons, which connectivity has already been known, and is easy to introduce gene manipulations. In addition, worms exhibit at most four types of behavior: run (forward movement), reverse (backward movement), turn (directional change) and omega turn (reversal + turn), so it is easy to evaluate their behavior. For these merits, *C. elegans* have been widely used in neuroscience researches to reveal the relationship between neural function and behavior. In this last session, we introduce the recent methodology in *C. elegans* and mention how *C. elegans* will contribute to learning and memory researches.

The relationship between behavior and adaptation is mentioned through the ages (Dusenbery, 1980). Worms change their behavior to search food depending on their circumstances (Gray, Hill, & Bargmann, 2005). When worms are removed from food, they gradually suppressed reversal and turn, and move widely after 30 min. Worms are known to use two different strategies on gradient of gustatory or olfactory cues: klinokinesis and klinotaxis (Iino & Yoshida, 2009; Pierce-Shimomura, Morse, & Lockery, 1999). In klinokinesis, when worms move on gradient vertically, they change the frequency of turn by sensing the rate of change of concentration (dC/dt) (Pierce-Shimomura et al., 1999). In klinotaxis, when worms move on gradient parallel, they gradually curve to high concentration (Iino & Yoshida, 2009). To evaluate these behaviors precisely, the devices to stimulate worms and monitor worms' behavior are needed. One of the examples is microfluidic devices (McCormick, Gaertner, Sottile, Phillips, & Lockery, 2011). In the devices, head swings of a semi-fixed worm responding to stimulation are observed. The previous research mentioned that under high concentration of odor some worms got adapted to the odorants and gradually change their behavior (Nuttley, Harbinder, & van der Kooy, 2001). So, with microfluidic devices, behavioral changes corresponding to obtaining memory are observed along time courses.

On the other hand, neural functions are conventionally revealed by killing neurons with leaser ablation. However, with this conventional method we cannot ablate neurons in fixed time because the leaser ablation method should be carried out to larva worms. To overcome the problem, optical ablation methods were introduced (Kobayashi et al., 2013; Qi, Garren, Shu, Tsien, & Jin, 2012). When the phototoxic protein such as KillerRed and miniSOG expressed in target neurons is exposed to light, it damages the neurons. Therefore, researchers decide timing of killing cell. Recent research indicated that phototoxic proteins effectively kill not only neurons but also specific protein. Worms expressing miniSOG-VAMP2 after light irradiation could not move over night because synapse release decreased (Lin et al., 2013).

In addition to ablation methods, optical methods to measure and control neural activities have been developed. Conventionally, Ca^{2+} imaging is mainly used to evaluate neural activities because Ca^{2+} is thought as main carrier in *C. elegans*, whose neurons do not have Na^{+} channel (Goodman, Hall, Avery, & Lockery, 1998; Kimura et al., 2004). Recent researches showed usability of other genetic indicators to measure membrane potential (Shidara et al., 2013), Ras activation (Uozumi et al., 2012) and glutamate release (Marvin et al., 2013). Meanwhile, light-gated channels such as channelrhodopsin-2 (ChR2) and halorhodopsin (NpHR) are used to understand relationship between two neurons or between neurons and behavior (Kocabas, Shen, Guo, & Ramanathan, 2012; Piggott et al., 2011). Because the body of *C. elegans* is transparent, optical methods are useful to understand neural functions. In addition, the connectivity of *C. elegans* neurons has already been known, so combination of these methods may help us elucidation of neural plasticity.

Further to reveal relationship between behavior and neural circuits, techniques to record and control neural activity of unrestricted worms have been developed (Faumont et al., 2011). The tool to record two neurons automatically of freely moving worm is also exploited (Zheng, Cao, Yang, Xu, & Feng, 2012). As mentioned above, to reveal mechanisms of plasticity in *C. elegans*, understanding neural circuits is important, so these tools become more powerful in researches of learning and memory. Although the neural circuits and behavior of *C. elegans* are very simple, a variety of learning and memory are observed. Therefore, contributions of this tractable animal also will play a great role from now on.

REFERENCES

Adachi, T., Kunitomo, H., Tomioka, M., Ohno, H., Okochi, Y., Mori, I., & Iino, Y. (2010). Reversal of salt preference is directed by the insulin/PI3K and Gq/PKC signaling in Caenorhabditis elegans. *Genetics*, *186*(4), 1309–19. doi:10.1534/genetics.110.119768.

Amano, H., & Maruyama, I. N. (2011). Aversive olfactory learning and associative long-term memory in Caenorhabditis elegans. *Learning & Memory (Cold Spring Harbor, N.Y.)*, *18*(10), 654–65. doi:10.1101/ lm.2224411.

Avery, L., & Horvitz, H. R. (1990). Effects of starvation and neuroactive drugs on feeding in Caenorhabditis elegans. *The Journal of Experimental Zoology*, *253*(3), 263–70. doi:10.1002/jez.1402530305.

Bargmann, C. I. (2006). Chemosensation in C. elegans. *WormBook : The Online Review of C. Elegans Biology*, 1–29. doi:10.1895/ wormbook.1.123.1.

Bargmann, C. I., Hartwieg, E., & Horvitz, H. R. (1993). Odorant-selective genes and neurons mediate olfaction in C. elegans. *Cell*, *74*(3), 515–27.

Bargmann, C. I., & Horvitz, H. R. (1991). Chemosensory neurons with overlapping functions direct chemotaxis to multiple chemicals in C. elegans. *Neuron*, *7*(5), 729–42.

Beck, C. D. O., & Rankin, C. H. (1997). Long-term habituation is produced by distributed training at long ISIs and not by massed training or short ISIs inCaenorhabditis elegans. *Animal Learning & Behavior*, *25*(4), 446–457. doi:10.3758/BF03209851.

Beck, C. D., & Rankin, C. H. (1995). Heat shock disrupts long-term memory consolidation in Caenorhabditis elegans. *Learning & Memory (Cold Spring Harbor, N.Y.)*, *2*(3-4), 161–77.

Bernhard, N., & van der Kooy, D. (2000). A behavioral and genetic dissection of two forms of olfactory plasticity in Caenorhabditis elegans: adaptation and habituation. *Learning & Memory (Cold Spring Harbor, N.Y.)*, *7*(4), 199–212.

Biron, D., Shibuya, M., Gabel, C., Wasserman, S. M., Clark, D. A., Brown, A., Sengupta, P., &Samuel, A. D. T. (2006). A diacylglycerol kinase modulates long-term thermotactic behavioral plasticity in C. elegans. *Nature Neuroscience*, *9*(12), 1499–505. doi:10.1038/nn1796.

Biron, D., Wasserman, S., Thomas, J. H., Samuel, A. D. T., & Sengupta, P. (2008). An olfactory neuron responds stochastically to temperature and modulates Caenorhabditis elegans thermotactic behavior. *Proceedings of the National Academy of Sciences of the United States of America*, *105*(31), 11002–7. doi:10.1073/pnas.0805004105.

Bounoutas, A., & Chalfie, M. (2007). Touch sensitivity in Caenorhabditis elegans. *Pflügers Archiv : European Journal of Physiology*, *454*(5), 691–702. doi:10.1007/s00424-006-0187-x.

Chalasani, S. H., Chronis, N., Tsunozaki, M., Gray, J. M., Ramot, D., Goodman, M. B., & Bargmann, C. I. (2007). Dissecting a circuit for olfactory behaviour in Caenorhabditis elegans. *Nature*, *450*(7166), 63–70. doi:10.1038/nature06292.

Chalasani, S. H., Kato, S., Albrecht, D. R., Nakagawa, T., Abbott, L. F., & Bargmann, C. I. (2010). Neuropeptide feedback modifies odor-evoked dynamics in Caenorhabditis elegans olfactory neurons. *Nature Neuroscience*, *13*(5), 615–21. doi:10.1038/nn.2526.

Chalfie, M., & Sulston, J. (1981). Developmental genetics of the mechanosensory neurons of Caenorhabditis elegans. *Developmental Biology*, *82*(2), 358–70.

Chalfie, M., Sulston, J. E., White, J. G., Southgate, E., Thomson, J. N., & Brenner, S. (1985). The neural circuit for touch sensitivity in Caenorhabditis elegans. *The Journal of Neuroscience*, *5*(4), 956–64.

Clark, D. A., Biron, D., Sengupta, P., & Samuel, A. D. T. (2006). The AFD sensory neurons encode multiple functions underlying thermotactic behavior in Caenorhabditis elegans. *The Journal of Neuroscience*, *26*(28), 7444–51. doi:10.1523/JNEUROSCI.1137-06.2006.

Colbert, H. A., & Bargmann, C. I. (1995). Odorant-specific adaptation pathways generate olfactory plasticity in C. elegans. *Neuron*, *14*(4), 803–12.

Colbert, H. A., & Bargmann, C. I. (1997). Environmental signals modulate olfactory acuity, discrimination, and memory in Caenorhabditis elegans. *Learning & Memory (Cold Spring Harbor, N.Y.)*, *4*(2), 179–91.

Dusenbery, D. B. (1980). Responses of the nematodeCaenorhabditis elegans to controlled chemical stimulation. *Journal of Comparative Physiology · A*, *136*(4), 327–331. doi:10.1007/BF00657352.

Faumont, S., Rondeau, G., Thiele, T. R., Lawton, K. J., McCormick, K. E., Sottile, M., Griesbeck, O., Heckscher, E. S., Roberts, W. M., Doe, C. Q., & Lockery, S. R. (2011). An image-free opto-mechanical system for creating virtual environments and imaging neuronal activity in freely moving Caenorhabditis elegans. *PloS One*, *6*(9), e24666. doi:10.1371/journal.pone.0024666.

Funabashi, Y., Kawamura, K., Oshio, K., Morita, S., Osana, Y., Akiyama, E., & Oka, K. (2001). Native Response of C. elegans Encoded in Its Neuron Network. *Journal of the Physical Society of Japan*, *70*(4), 1154–1161. doi:10.1143/JPSJ.70.1154.

Gomez, M., De Castro, E., Guarin, E., Sasakura, H., Kuhara, A., Mori, I., Bartfai, T., Bargmann, C. I., & Nef, P. (2001). Ca^{2+} signaling via the neuronal calcium sensor-1 regulates associative learning and memory in C. elegans. *Neuron*, *30*(1), 241–8.

Goodman, M. B., Hall, D. H., Avery, L., & Lockery, S. R. (1998). Active currents regulate sensitivity and dynamic range in C. elegans neurons. *Neuron*, *20*(4), 763–72.

Gray, J. M., Hill, J. J., & Bargmann, C. I. (2005). A circuit for navigation in Caenorhabditis elegans. *Proceedings of the National Academy of Sciences of the United States of America*, *102*(9), 3184–91. doi:10.1073/ pnas.0409009101.

Greer, E. L., Maures, T. J., Hauswirth, A. G., Green, E. M., Leeman, D. S., Maro, G. S., Han, S., Banko, M. R., Gozani, O., & Brunet, A. (2010). Members of the H3K4 trimethylation complex regulate lifespan in a germline-dependent manner in C. elegans. *Nature*, *466*(7304), 383–7. doi:10.1038/nature09195.

Greer, E. L., Maures, T. J., Ucar, D., Hauswirth, A. G., Mancini, E., Lim, J. P., Benayoun, B. A., Shi, Y., & Brunet, A. (2011). Transgenerational epigenetic inheritance of longevity in Caenorhabditis elegans. *Nature*, *479*(7373), 365–71. doi:10.1038/nature10572.

Hart, A. C., & Chao., M. Y. (2010). From Odors to Behaviors in Caenorhabditis elegans. In A. Menini (Ed.), *The Neurobiology of Olfaction.* (Chapter 1). CRC Press.

Hedgecock, E. M., & Russell, R. L. (1975). Normal and mutant thermotaxis in the nematode Caenorhabditis elegans. *Proceedings of the National Academy of Sciences of the United States of America*, *72*(10), 4061–5.

Hendricks, M., Ha, H., Maffey, N., & Zhang, Y. (2012). Compartmentalized calcium dynamics in a C. elegans interneuron encode head movement. *Nature*, *487*(7405), 99–103. doi:10.1038/nature11081.

Hirotsu, T., & Iino, Y. (2005). Neural circuit-dependent odor adaptation in C. elegans is regulated by the Ras-MAPK pathway. *Genes to Cells : Devoted to Molecular & Cellular Mechanisms*, *10*(6), 517–30. doi:10.1111/j.1365-2443.2005.00856.x.

Iino, Y., & Yoshida, K. (2009). Parallel use of two behavioral mechanisms for chemotaxis in Caenorhabditis elegans. *The Journal of Neuroscience*, *29*(17), 5370–80. doi:10.1523/JNEUROSCI.3633-08.2009.

Jansen, G., Weinkove, D., & Plasterk, R. H. A. (2002). The G-protein gamma subunit gpc-1 of the nematode C.elegans is involved in taste adaptation. *The EMBO Journal*, *21*(5), 986–94. doi:10.1093/emboj/21.5.986.

Jurado, P., Kodama, E., Tanizawa, Y., & Mori, I. (2010). Distinct thermal migration behaviors in response to different thermal gradients in Caenorhabditis elegans. *Genes, Brain, and Behavior*, *9*(1), 120–7. doi:10.1111/j.1601-183X.2009.00549.x.

Kawamura, K., Funabashi, Y., Morita, S., Omata, K., Oshio, K., Osana, Y., & Oka, K. (2001). Does a Randomly Organized Electrical Circuit Function as a Neuronal System? *Japanese Journal of Applied Physics*, *40*(Part 1, No. 3B), 2095–2099. doi:10.1143/JJAP.40.2095.

Kimata, T., Sasakura, H., Ohnishi, N., Nishio, N., & Mori, I. (2012). Thermotaxis of C. elegans as a model for temperature perception, neural information processing and neural plasticity. *Worm*, *1*(1), 31–41. doi:10.4161/worm.19504.

Kimura, K. D., Miyawaki, A., Matsumoto, K., & Mori, I. (2004). The C. elegans thermosensory neuron AFD responds to warming. *Current Biology : CB*, *14*(14), 1291–5. doi:10.1016/j.cub.2004.06.060.

Kindt, K. S., Quast, K. B., Giles, A. C., De, S., Hendrey, D., Nicastro, I., Rankin, C. H., &Schafer, W. R. (2007). Dopamine mediates context-dependent modulation of sensory plasticity in C. elegans. *Neuron*, *55*(4), 662–76. doi:10.1016/j.neuron.2007.07.023.

Kobayashi, J., Shidara, H., Morisawa, Y., Kawakami, M., Tanahashi, Y., Hotta, K., & Oka, K. (2013). A method for selective ablation of neurons in C. elegans using the phototoxic fluorescent protein, KillerRed. *Neuroscience Letters*, *548*, 261–4. doi:10.1016/j.neulet.2013.05.053.

Kocabas, A., Shen, C.-H., Guo, Z. V, & Ramanathan, S. (2012). Controlling interneuron activity in Caenorhabditis elegans to evoke chemotactic behaviour. *Nature*, *490*(7419), 273–7. doi:10.1038/nature11431.

Kuhara, A., Ohnishi, N., Shimowada, T., & Mori, I. (2011). Neural coding in a single sensory neuron controlling opposite seeking behaviours in Caenorhabditis elegans. *Nature Communications*, *2*, 355. doi:10.1038/ ncomms1352.

Kuhara, A., Okumura, M., Kimata, T., Tanizawa, Y., Takano, R., Kimura, K. D., Inada, H., Matsumoto, K., & Mori, I. (2008). Temperature sensing by an olfactory neuron in a circuit controlling behavior of C. elegans. *Science (New York, N.Y.)*, *320*(5877), 803–7. doi:10.1126/science.1148922.

Kunitomo, H., Sato, H., Iwata, R., Satoh, Y., Ohno, H., Yamada, K., & Iino, Y. (2013). Concentration memory-dependent synaptic plasticity of a taste circuit regulates salt concentration chemotaxis in Caenorhabditis elegans. *Nature Communications*, *4*, 2210. doi:10.1038/ncomms3210.

Law, E., Nuttley, W. M., & van der Kooy, D. (2004). Contextual taste cues modulate olfactory learning in C. elegans by an occasion-setting mechanism. *Current Biology : CB*, *14*(14), 1303–8. doi:10.1016/j.cub. 2004.06.066.

Lee, R. Y., Sawin, E. R., Chalfie, M., Horvitz, H. R., & Avery, L. (1999). EAT-4, a homolog of a mammalian sodium-dependent inorganic phosphate cotransporter, is necessary for glutamatergic neurotransmission in caenorhabditis elegans. *The Journal of Neuroscience*, *19*(1), 159–67.

Lin, J. Y., Sann, S. B., Zhou, K., Nabavi, S., Proulx, C. D., Malinow, R., Jin, Y., & Tsien, R. Y. (2013). Optogenetic inhibition of synaptic release with chromophore-assisted light inactivation (CALI). *Neuron*, *79*(2), 241–53. doi:10.1016/j.neuron.2013.05.022.

Marvin, J. S., Borghuis, B. G., Tian, L., Cichon, J., Harnett, M. T., Akerboom, J., Gordus, A., Renninger, S. L., Chen, T-W., Bargmann, C. I., Orger, M. B., Schreiter, E. R., Demb, J. B., Gan, W-B., Hires, S. A., & Looger, L. L. (2013). An optimized fluorescent probe for visualizing glutamate neurotransmission. *Nature Methods*, *10*(2), 162–70. doi:10.1038/nmeth.2333.

McCormick, K. E., Gaertner, B. E., Sottile, M., Phillips, P. C., & Lockery, S. R. (2011). Microfluidic devices for analysis of spatial orientation behaviors in semi-restrained Caenorhabditis elegans. *PloS One*, *6*(10), e25710. doi:10.1371/journal.pone.0025710.

Mohri, A., Kodama, E., Kimura, K. D., Koike, M., Mizuno, T., & Mori, I. (2005). Genetic control of temperature preference in the nematode Caenorhabditis elegans. *Genetics*, *169*(3), 1437–50. doi:10.1534/ genetics.104.036111.

Mori, I., & Ohshima, Y. (1995). Neural regulation of thermotaxis in Caenorhabditis elegans. *Nature*, *376*(6538), 344–8. doi:10.1038/ 376344a0.

Mori, I., Sasakura, H., & Kuhara, A. (2007). Worm thermotaxis: a model system for analyzing thermosensation and neural plasticity. *Current Opinion in Neurobiology*, *17*(6), 712–9. doi:10.1016/j.conb.2007.11.010.

Morita, S., Oshio, K., Osana, Y., Funabashi, Y., Oka, K., & Kawamura, K. (2001). Geometrical structure of the neuronal network of Caenorhabditis elegans. *Physica A: Statistical Mechanics and Its Applications*, *298*(3-4), 553–561. doi:10.1016/S0378-4371(01)00266-7.

Morrison, G. E., & van der Kooy, D. (1997). Cold shock before associative conditioning blocks memory retrieval, but cold shock after conditioning blocks memory retention in Caenorhabditis elegans. *Behavioral Neuroscience*, *111*(3), 564–78.

Nuttley, W. M., Atkinson-Leadbeater, K. P., & Van Der Kooy, D. (2002). Serotonin mediates food-odor associative learning in the nematode Caenorhabditiselegans. *Proceedings of the National Academy of Sciences of the United States of America*, *99*(19), 12449–54. doi:10.1073/ pnas.192101699.

Nuttley, W. M., Harbinder, S., & van der Kooy, D. (2001). Regulation of distinct attractive and aversive mechanisms mediating benzaldehyde chemotaxis in Caenorhabditis elegans. *Learning & Memory (Cold Spring Harbor, N.Y.)*, *8*(3), 170–81. doi:10.1101/lm.36501.

Ohno, H., Kato, S., Naito, Y., Kunitomo, H., Tomioka, M., & Iino, Y. (2014). Role of synaptic phosphatidylinositol 3-kinase in a behavioral learning response in C. elegans. *Science (New York, N.Y.)*, *345*(6194), 313–7. doi:10.1126/science.1250709.

Pierce-Shimomura, J. T., Faumont, S., Gaston, M. R., Pearson, B. J., & Lockery, S. R. (2001). The homeobox gene lim-6 is required for distinct chemosensory representations in C. elegans. *Nature*, *410*(6829), 694–8. doi:10.1038/35070575.

Pierce-Shimomura, J. T., Morse, T. M., & Lockery, S. R. (1999). The fundamental role of pirouettes in Caenorhabditis elegans chemotaxis. *The Journal of Neuroscience*, *19*(21), 9557–69.

Piggott, B. J., Liu, J., Feng, Z., Wescott, S. A., & Xu, X. Z. S. (2011). The neural circuits and synaptic mechanisms underlying motor initiation in C. elegans. *Cell*, *147*(4), 922–33. doi:10.1016/j.cell.2011.08.053.

Qi, Y. B., Garren, E. J., Shu, X., Tsien, R. Y., & Jin, Y. (2012). Photo-inducible cell ablation in Caenorhabditis elegans using the genetically encoded singlet oxygen generating protein miniSOG. *Proceedings of the National Academy of Sciences of the United States of America*, *109*(19), 7499–504. doi:10.1073/pnas.1204096109.

Rankin, C. H., Beck, C. D., & Chiba, C. M. (1990). Caenorhabditis elegans: a new model system for the study of learning and memory. *Behavioural Brain Research*, *37*(1), 89–92.

Rankin, C. H., & Broster, B. S. (1992). Factors affecting habituation and recovery from habituation in the nematode Caenorhabditis elegans. *Behavioral Neuroscience*, *106*(2), 239–49.

Rankin, C. H., & Wicks, S. R. (2000). Mutations of the caenorhabditis elegans brain-specific inorganic phosphate transporter eat-4 affect habituation of the tap-withdrawal response without affecting the response itself. *The Journal of Neuroscience*, *20*(11), 4337–44.

Rechavi, O., Houri-Ze'evi, L., Anava, S., Goh, W. S. S., Kerk, S. Y., Hannon, G. J., & Hobert, O. (2014). Starvation-Induced Transgenerational Inheritance of Small RNAs in C. elegans. *Cell*. doi:10.1016/j. cell.2014.06.020.

Rechavi, O., Minevich, G., & Hobert, O. (2011). Transgenerational inheritance of an acquired small RNA-based antiviral response in C. elegans. *Cell*, *147*(6), 1248–56. doi:10.1016/j.cell.2011.10.042.

Remy, J.-J. (2010). Stable inheritance of an acquired behavior in Caenorhabditis elegans. *Current Biology : CB*, *20*(20), R877–8. doi:10.1016/j.cub.2010.08.013.

Remy, J.-J., & Hobert, O. (2005). An interneuronal chemoreceptor required for olfactory imprinting in C. elegans. *Science (New York, N.Y.)*, *309*(5735), 787–90. doi:10.1126/science.1114209.

Rose, J. K., Kaun, K. R., Chen, S. H., & Rankin, C. H. (2003). GLR-1, a non-NMDA glutamate receptor homolog, is critical for long-term memory in Caenorhabditis elegans. *The Journal of Neuroscience*, *23*(29), 9595–9.

Rose, J. K., Kaun, K. R., & Rankin, C. H. (2002). A new group-training procedure for habituation demonstrates that presynaptic glutamate release contributes to long-term memory in Caenorhabditis elegans. *Learning & Memory (Cold Spring Harbor, N.Y.)*, *9*(3), 130–7. doi:10.1101/lm.46802

Saeki, S., Yamamoto, M., & Iino, Y. (2001). Plasticity of chemotaxis revealed by paired presentation of a chemoattractant and starvation in the nematode Caenorhabditis elegans. *The Journal of Experimental Biology*, *204*(Pt 10), 1757–64.

Samuel, A. D. T., Silva, R. A., & Murthy, V. N. (2003). Synaptic activity of the AFD neuron in Caenorhabditis elegans correlates with thermotactic memory. *The Journal of Neuroscience*, *23*(2), 373–6.

Sanyal, S., Wintle, R. F., Kindt, K. S., Nuttley, W. M., Arvan, R., Fitzmaurice, P., … Van Tol, H. H. M. (2004). Dopamine modulates the plasticity of mechanosensory responses in Caenorhabditis elegans. *The EMBO Journal*, *23*(2), 473–82. doi:10.1038/sj.emboj.7600057.

Shidara, H., Kobayashi, J., Tanamoto, R., Hotta, K., & Oka, K. (2013). Odorant-induced membrane potential depolarization of AIY interneuron in Caenorhabditis elegans. *Neuroscience Letters*, *541*, 199–203. doi:10.1016/j.neulet.2013.02.016.

Tomioka, M., Adachi, T., Suzuki, H., Kunitomo, H., Schafer, W. R., & Iino, Y. (2006). The insulin/PI 3-kinase pathway regulates salt chemotaxis learning in Caenorhabditis elegans. *Neuron*, *51*(5), 613–25. doi:10.1016/j.neuron.2006.07.024.

Uozumi, T., Hirotsu, T., Yoshida, K., Yamada, R., Suzuki, A., Taniguchi, G., Iino, Y., & Ishihara, T. (2012). Temporally-regulated quick activation and inactivation of Ras is important for olfactory behaviour. *Scientific Reports*, *2*, 500. doi:10.1038/srep00500.

Varshney, L. R., Chen, B. L., Paniagua, E., Hall, D. H., & Chklovskii, D. B. (2011). Structural properties of the Caenorhabditis elegans neuronal network. *PLoS Computational Biology*, *7*(2), e1001066. doi:10.1371/journal.pcbi.1001066.

Ward, S. (1973). Chemotaxis by the nematode Caenorhabditis elegans: identification of attractants and analysis of the response by use of mutants. *Proceedings of the National Academy of Sciences of the United States of America*, *70*(3), 817–21.

Watts, D. J., & Strogatz, S. H. (1998). Collective dynamics of "small-world" networks. *Nature*, *393*(6684), 440–2. doi:10.1038/30918.

Wen, J. Y., Kumar, N., Morrison, G., Rambaldini, G., Runciman, S., Rousseau, J., & van der Kooy, D. (1997). Mutations that prevent associative learning in C. elegans. *Behavioral Neuroscience*, *111*(2), 354–68.

White, J. G., Southgate, E., Thomson, J. N., & Brenner, S. (1986). The structure of the nervous system of the nematode Caenorhabditis elegans. *Philosophical Transactions of the Royal Society of London. Series B, Biological Sciences*, *314*(1165), 1–340.

Wicks, S. R., & Rankin, C. H. (1995). Integration of mechanosensory stimuli in Caenorhabditis elegans. *The Journal of Neuroscience*, *15*(3 Pt 2), 2434–44.

Yamada, K., Hirotsu, T., Matsuki, M., Butcher, R. A., Tomioka, M., Ishihara, T., Clardy, J., Kunitomo, H., & Iino, Y. (2010). Olfactory plasticity is regulated by pheromonal signaling in Caenorhabditis elegans. *Science (New York, N.Y.)*, *329*(5999), 1647–50. doi:10.1126/science.1192020.

Zheng, M., Cao, P., Yang, J., Xu, X. Z. S., & Feng, Z. (2012). Calcium imaging of multiple neurons in freely behaving C. elegans. *Journal of Neuroscience Methods*, *206*(1), 78–82. doi:10.1016/ j.jneumeth. 2012. 01.002.

In: Memory Consolidation
Editors: Manabu Sakakibara and Etsuro Ito
ISBN: 978-1-63482-596-2

Chapter 2

ASSOCIATIVE PLASTICITY OF MUSHROOM BODY-EXTRINSIC NEURONS IN THE HONEYBEE OLFACTORY LEARNING

Ryuichi Okada*
School of Human Science and Environment, University of Hyogo, Japan

ABSTRACT

The honeybee has been extensively used for the investigation of the brain mechanism on learning and memory as a suitable model animal. So far, the mushroom body, a higher center of an insect brain, has been behaviorally and physiologically shown a crucial neuropil for the establishment of learning and formation of memory for the insect olfactory learning. Compared to behavioral work, there have been little success to provide physiological evidence, especially electrophysiologically, for the involvement of the mushroom body in olfactory learning, Several years ago, a long-lasting extracellular recording method was developed, in which neural activity was recorded while the honeybee is receiving a learning paradigm. Using the new method, the PE1 neuron, an identified mushroom body extrinsic neuron of the honeybee, was found to reduce its response to an odor (CS+) paired with rewarding sucrose solution and did not change its response to unpaired odor after learning. The response reduction of PE1 neuron correlated with the probability of behavioral response to CS+. In this chapter, the learning-related plasticity of PE1 neuron will be introduced and a current hypothetical neural network model for honeybee olfactory learning will be proposed. Finally, the source of the associative reduction of odor response will be discussed.

INTRODUCTION

Learning and memory formation lead to behavioral changes. So far, numerous behavioral and physiological research have been trying to show the neural mechanism. Ideally, the

* Correspndence to: Dr. Ryuichi Okada, address: Shinzaike-Honcho, Himeji 670-0092, Hyogo, Japan. Phone: +81 79-292-1523; E-mail: okd@shse.u-hyogo.ac.jp.

search for the neural basis of learning and memory is performed in an animal that can directly report its behavioral change. Neural substrates of learning and memory are expected to be related to plasticity of neural functions that should thus be measured at the single-neuron level under conditions when the animal learns. For such measurements, extracellular recordings have been successfully used to monitor neural processes during learning and memory retrieval at the single-neuron level (e.g., place cells in the rodent hippocampus [1], prefrontal neurons and working memory in primates [2], orbito-frontal neurons in olfactory learning[3]; dopamine neurons in the ventral tegmentum of the monkey [4]).

The honeybees are known to have a high capability to learn odors [5] and have been used as a suitable model animal for the investigation of the brain mechanism underlying learning and memory [6-9]. Once an odor stimulus (conditioned stimulus, CS) is presented followed by sucrose solution (unconditioned stimulus, US), the honeybees learn the relationship between the odor and sucrose solution reward. This procedure is called proboscis extension response (PER) conditioning. Afterward, the conditioned animal will reliably extend the proboscis if the CS is applied. Only one conditioning trial is required for bees to learn an odor and several trials lead to long-term memory which is kept for several days at a high level of retention. Using this learning paradigm, learning ability and characteristics of olfactory memory in honeybees have been studied extensively (behaviorally [11-15] and physiologically [16-21]). The honeybee olfactory memory consists of multiple phases, early short-term, late short-term, middle-term, early long-term, and late long-term memory [22-24]. Spaced training leads to longer memory than massed training [14]. The memory will extinct by repeated presentations of CS without US and recover next day [25-27]. Those primary characteristics are rather similar to the corresponding associative phenomena of memory formation and retrieval in vertebrates. In addition, there are also many common molecular mechanisms involved in memory formation in the honeybee and laboratory mammals (e.g., cAMP, PKA, NOS, CREB, NMDA receptors) [19, 28-31].

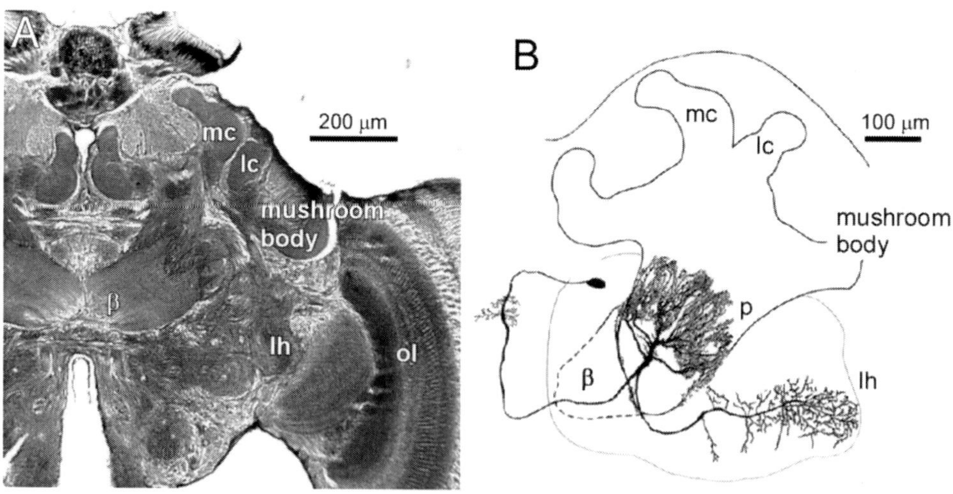

Figure 1. Bee brain and the PE1 neuron A: a frontal section of the bee brain. The brain was stained with Bodian impregnation method. The dorsal side is at the top. B: Morphology of the PE1 neuron. A scheme of the left hemisphere of the bee brain with an intracellular staining (after [41]). mushroom body (lc: lateral calyx, mc: medial calyx, p: pedunculus), lh: lateral horn, ol: optic lobe, ß: ß lobe.

Therefore, the honeybee provides an excellent model for the investigation of the mechanism underlying learning and memory on various levels of complexity ranging from molecular, cellular, network and behavior level [9, 32].

The mushroom body (Figure 1A), one of the higher centers for the insect brain [33], is the crucial neuropil for olfactory learning and memory formation in both honeybee memory [34] and *Drosophila* [35]. In the bee, the transition from short-term memory to long-term memory after appetitive olfactory learning requires normal functions of the mushroom boy [36]. Two mushroom body neurons, the PE1 neuron and the VUMmx1 neuron, were found to relate to the honeybee olfactory learning [37]. The VUMmx1 neuron (ventral unpaired median neurons number 1 of the maxillary neuromere) conveys the reinforcing function of sucrose reward [38]. The VUMmx1 neuron has axonal terminals that converge with the olfactory pathway at the mushroom body calyces and two other locations, the antennal lobe and the lateral horn [38, 39]. The PE1 neuron (pedunculus-extrinsic neuron 1) was characterized both anatomically and physiologically [40-42]. The PE1 neuron is a mushroom body output neuron and has rich arborized dendritic trees in the pedunculus of the mushroom body [41, 43]. Thus the PE1 neuron reads out from the results after the mushroom body processed neural information. Additionally, this neuron changes its odor responses by olfactory conditioning [40]. It is, however, usually rather difficult to assign neural activity to a particular type of neuron - or even to an identified single neuron. Insects provide conditions that allow us to overcome these limitations and to trace associative plasticity to single identified neurons when an animal learns a stimulus and subsequently responds to it. As described above, in only two cases, the VUMmx1 neuron and the PE1 neuron in the bee brain has so far been possible to return to the same individual neuron in subsequent recordings from different animals exposed to associative learning schedules, but in both cases the animal's behavior was considerably restricted and the recordings lasted only a few minutes. Recently, a long-lasting extracellular recordings from mushroom body-extrinsic neurons during olfactory learning and retention was developed and has given important insights [44-46]. In this chapter, the new method for extracellular recording will be introduced, which allows us to measure neural activities over several hours, and training the animals to an odor. Using this method, learning-related plasticity of a mushroom body neuron will be described, especially focusing on the PE1 neuron, particularly changes in responses to CS by the olfactory learning. In conclusion, either enhanced inhibition of the input to the PE1 neuron or associative long-term depression (LTD) within the PE1 neuron is responsible for the associative reduction.

METHODS

Animals

Forager honeybees were collected at the entrance to the hive one day before the experiment. The bees were anesthetized on ice, and then harnessed in a plastic tube. The heads were immobilized with wax. After the all bees were finished to be treated, the bees were fed with 4 or 5 drops of 1.25 M sucrose solution and kept at room temperature until the next day

Electrophysiology

Electrodes were manufactured as described previously [45, 47, 48]. A polyurethane-coated copper wire, 14 μm in diameter (Electrisola, Escholzmatt, Switzerland), was used as an electrode. Two wires were glued together with wax. The copper wires were connected to a differential amplifier (npi, Tamm, Germany) after removing the insulation at the other end. As a ground electrode, a silver wire with a diameter of 50 μm (Nilaco, Tokyo, Japan) was inserted into the right compound eye. A small window was opened between the compound eyes, head glands and only the trachea sacks above one α-lobe were removed. No muscles were injured or stretched, and thus the bee could open its mandibles, extend its proboscis, drink sugar water, and freely move the flagellum of its antennae. The brain surface at the ventral border of the α-lobe was carefully scratched to facilitate inserting the electrode. After insertion of the electrode, the cuticle removed from the head capsule was divided into two pieces and placed back onto the head capsule. All gaps between the cuticle pieces were sealed with silicon (wpi, Sarasota, FL) in order to prevent the brain from drying out. The animal was left for 15-20 minutes to allow the silicon to harden, the recording to stabilize, and the animal to recover.

The electric signals were amplified by a differential amplifier (npi, Tamm, Germany). Hum noise (50 Hz) was eliminated by a filter (HumBug, Digitimer, Hertfordshire, UK). Neural activity was stored at 20 kHz through an AD-converter (1401 micro MKII, CED, Cambridge, UK) using Spike2 software (Cambridge Electronic Design, Cambridge, UK).

The tip of recording electrode was marked (Figure 3). After recordings ended, copper ions were released from the tip of the recording electrode by passing a positive current (typically 20-40 nA for 20 sec) through the electrode and were precipitated by Rubeanic acid. The bee's head was isolated, and the brain was dissected under saline, and then fixed in 4% paraformaldehyde in honeybee saline or Carnoy's fixative for 45-50 min. Dissected brains were hydrated and intensified with silver [49]. Subsequently, the specimens were embedded in 6-7% agarose (Sigma, St Louis, MO), sectioned at 100 μm with a vibratome, and observed under a Nomarski interference or a conventional light microscope.

Odor Conditioning

Three odors, peppermint, limonene, and 2-octanol (Sigma), were used as olfactory stimuli. 4 μl of each odor was applied to a 1 cm x 1 cm filter paper. An air pulse was applied to the antennae through a plastic syringe containing the odor-soaked filter paper for odor stimuli or a filter paper alone as control (blank). Stimulation was controlled by a computer. The odor was exhausted through a tube positioned behind the animal.

The experiment has three phases: pre-test phase, conditioning phase (acquisition phase), and test phase (extinction phase). The pre-test consisted of five 3 sec long repetitions of each of the three odors presented in a pseudo-random sequence. The inter-stimulus interval was 1 min. 15-20 min after the pre-test, the conditioning phase was started. One odor was selected as the conditioned stimulus (CS+) and was applied for 3 sec followed by an application of 1.25 M sucrose solution (unconditioned stimulus, US, for 3 sec overlapping with the CS+ by 1 sec) first to the antennae and then to the proboscis. Another odor was delivered for 3 sec without sucrose as CS-. The interval between these two odor or odor/sucrose stimulations was

2.5 min. This form of differential conditioning was repeated 5 times. The third odor, a control odor (Ctr), was not applied during the conditioning phase but only in the pre-test and test phase. In the test phase, the animal was exposed to each of three odors once without sucrose 5 min, 30 min, and 60 min after the last conditioning trial. Inter-stimulus interval was 2.5 min. The odor sequence was fixed during the conditioning phase and test-phase (CS+, CS-, and Ctr). Since repeated exposures to the CS+ led to extinction, only a single stimulation of each of the three odors was applied at each of the three test intervals.

For control animals the same sequence of odors as experimental animals was applied, but no sucrose stimulation was provided. The first odor applied to control animals is, for convenience, called CS+, the subsequent one CS-. Since the same procedure as in the experimental animals was applied during the test phase the first odor tested ("CS+") was the same odor as in the conditioning, keeping the sequence of odor stimulation the same in experimental and test animals.

For all animals, CS+, CS-, and Ctr were balanced although not fully equally balanced. Since bees could freely extend their probosces, and since the CS+ pulses during the conditioning phase preceded the US, the response to the CS+ could be visually monitored and expressed as acquisition functions.

Analysis

Electrophysiology: In general, extracellular recordings show more than one unit simultaneously. Therefore, recorded units were separated from each other by the Spike 2 program. The results of the method were inspected and visually confirmed afterwards.

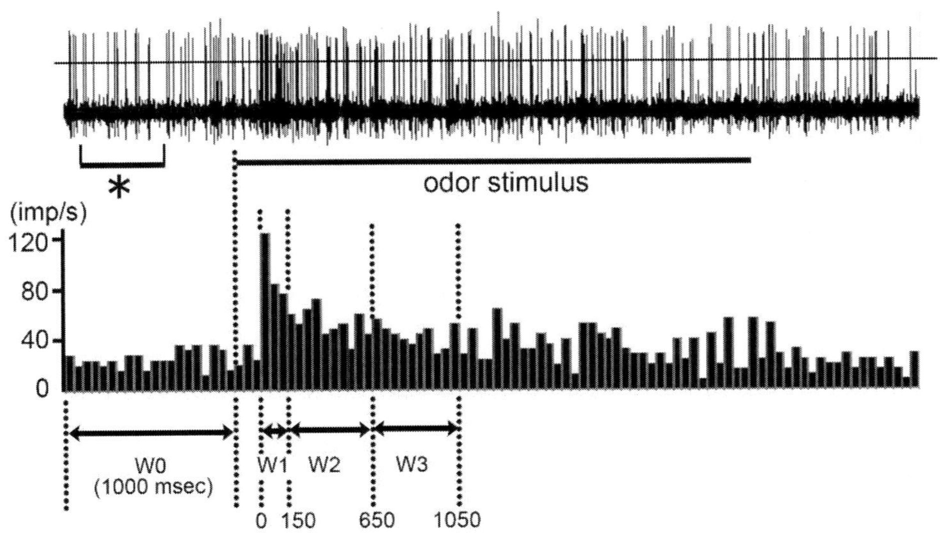

Figure 2. Representative example of a recording of a unit assigned as PE1 neuron and the corresponding peristimulus time histogram (after [45]). The time windows for the data analysis are indicated by: W0 (1 sec before the onset of stimulation), W1 (phasic response, 0-150 msec of the odor response as defined by the response initiation time, W2 (tonic period, 150-650 msec of the odor response), and W3 (late tonic response, 650-1050 msec of the odor response). A horizontal line shows a threshold for spike sorting.

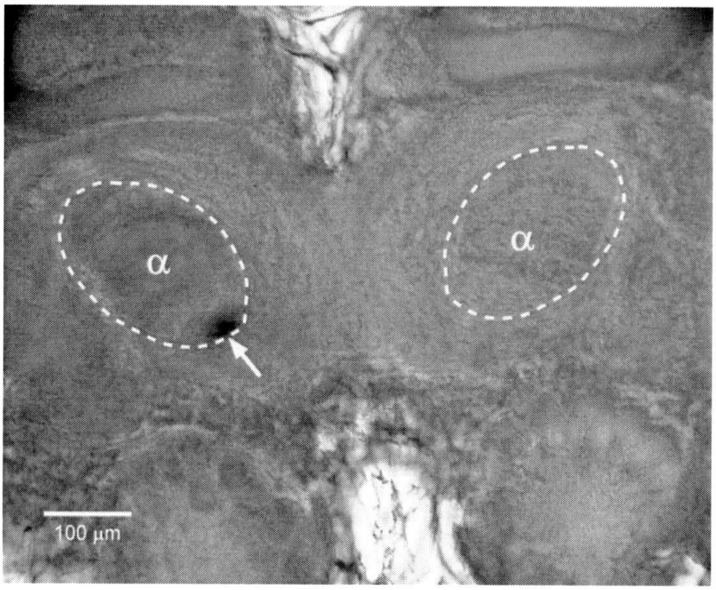

Figure 3. An electrode mark. The vicinity of the tip of the electrode (arrow) was marked by copper-deposit amplified with silver. The dorsal side is at the top. α: α-lobe.

Only reliably isolated units with large spikes were used for further analysis. All responses were converted to Peri-Stimulus Time Histograms (PSTHs) between 1 sec prior to an onset of and 1 sec after the termination of odor stimulation (bin width: 50 msec). Since the spontaneous activity fluctuated, activities during the 1 sec prior to stimulation onset for each odor were calculated and subtracted from the odor response (designated as "rate difference"). The first bin after the onset of stimulation was defined as the response initiation time; spike frequency at this time was always at least twice spontaneous activity. We calculated separately the mean of rate differences in four different time windows in each of the 3 test phases for each odor: window 0 (W0) refers to 1 sec before the onset of stimulation, window 1 (W1, phasic period) to 0-150 msec of the odor response as defined by the response initiation time, window 2 (W2, tonic period) 150-650 msec of the odor response, and window 3 (W3, late tonic response) 650-1050 msec of the odor response (Figure 2).

Statistics

Behavior: animals which extended their probosces during the 3-sec period of odor stimulation prior to sucrose stimulation were counted as showing the PER. The Wilcoxon matched-pair test was applied to examine the responses of the respective groups for homogeneity.

Electrophysiology: The change as the consequence of conditioning was statistically evaluated by two procedures applied both for within group comparison (responses to CS+, CS- and Ctr of the experimental animals) and between group comparison (experimental animals vs control animals): (1) Rate difference was calculated for each of the three time windows and the respective values of these normalized responses to CS+, CS-, and Ctr were compared using one-way ANOVA and post-hoc Tukey-Kramer test using STATISTICA

(Statsoft, Tulsa, UK). (2) A combined measure of the conditioning effect on spike activity was calculated by subtracting the spike rate differences during the pretest and test phase.

This value is called Δ. Δ is calculated for each time window of the odor response and expressed as e.g., Δ_{W1} for time window W1. Again Δ values were analyzed in all possible combinations by repeated measures ANOVA and post-hoc Tukey-Kramer test.

Behavioral Performance of the Bees with Electrode in the Brain

Totally 13 animals were conditioned. The three odors, peppermint, limonene, and 2-octanol, were balanced with respect to these three functions, CS+, CS-, and Ctr. Because no odor-specific effect was found in this conditioning experiment, the odors were pooled with respect to their functions.

Figure 4 shows the learning scores. Extinction tests show significantly higher CS+ responses as compared to the pre-test (5 min: $p = 0.008$, at 30 min: $p = 0.012$, and 60 min: $p = 0.012$), and no significant decrease during the 3 extinction tests ($p > 0.5$). A slight but not significant ($p = 0.109$) increase was found for responses to CS- during the 3 extinction tests. There is a significant increase in the response to Ctr (generalization) between the pretest and the 5 min extinction test ($p = 0.028$). Generalization shows a tendency to drop during the three extinction tests but the effect is not significant ($p = 0.22$).

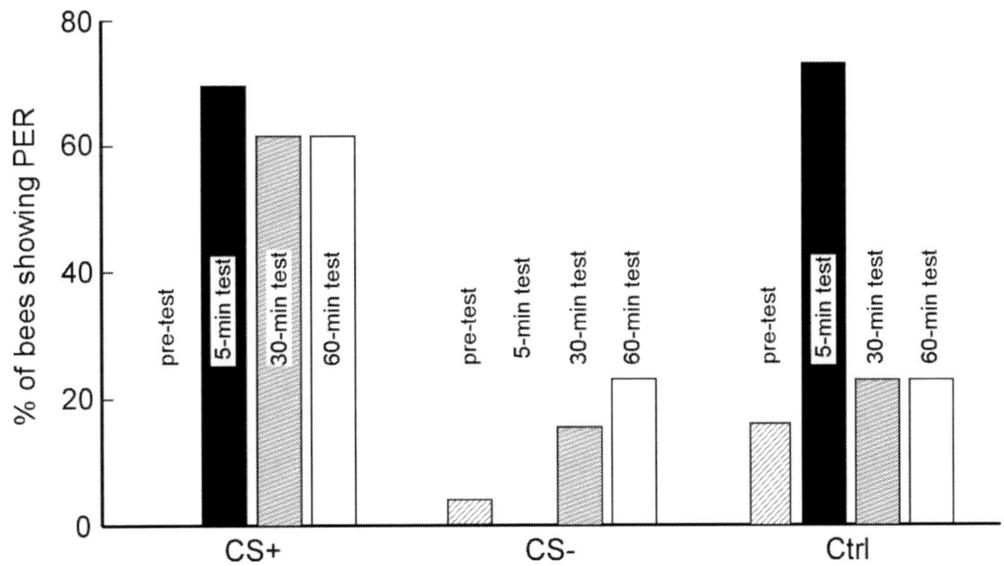

Figure 4. Behavioral response probability of all experimental animals (n = 13) during an initial pre-test phase (wider-hatched bar) and an extinction phase with tests at 5 (black bar), 30 (narrow-hatched bar), and 60 (white bar) mins after the end of acquisition. CS+: odor paired with sucrose stimulation, CS-: odor without sucrose solution, and Ctr: control odor presented only during the pre-test and the extinction phase.

The behavioral responses to the three odors were analyzed when each odor was used as CS+ or Ctr separately. No bias for either of the three odors as CS+, CS- or Ctr was found. Therefore, we pooled the data for the different odors according to their functions, and proceeded in the same way with the spike responses.

General Physiological Properties of the PE1

Extracellular recordings were performed from the ventral region of the α-lobe (Figure 3), at a location where the PE1 neuron exits. All recorded spikes were isolated and analyzed. Since the PE1 neuron shows a double or triplet spike pattern [40], characteristic for only this neuron in the region of the α exit [50], PE1 neurons could be identified in extracellular recordings with additional criterions [45].

The spike rate difference between odor responses in three time windows and the spontaneous activity (spike rate difference,) were calculated. As shown in Figure 2, the phasic-tonic odor response was divided up into 3 response windows (W1, W2, W3), and response differences were calculated separately. Long-lasting recordings (> 3 hours) indicated that spontaneous activity fluctuated irregularly. There was a tendency that it increased after sensory stimulation, particularly when the spontaneous activity was rather low before stimulation, indicating a kind of arousal.

The PE1 neuron responded to all kinds of odors tested. The odors used, peppermint, limonene, and 2-octanol, were equally effective in both the PE1 neuron and non-PE1 neurons. The average response rate of PE1 neuron in the two windows W1 and W2 were, respectively: peppermint: W1: 86.7 Hz, W2: 49.2 Hz; limonene: W1: 94.5 Hz, W2: 55.9 Hz; 2-octanol: W1: 76.8 Hz, W2: 48.9 Hz. There is a tendency that spike rate difference declines in the course of multiple sequential odor stimulations, but the effect is not significant. Sucrose stimulation was applied in our experiments only during conditioning, following CS+ stimulation and partially overlapping with it. The mean firing rate was 56.0 Hz.

Changes in the Responses of the PE1 Neuron: On Average

All animals were included in the within group comparison. Out of 13 conditioned animals, the PE1 neuron was recorded from 5 animals. The response changes of the PE1 neurons were statistically analyzed. As a first step, the spontaneous frequency was subtracted from the rate response during odor stimulation (rate difference for the three time windows W1, W2, W3). The first procedure compared directly the rate differences of CS+, CS- and Ctr responses before and after conditioning separately. There were no significant differences in the respective values of responses to CS+, CS- and Ctr before conditioning (all p values > 0.95). After conditioning the rate differences were significantly reduced for CS+ as compared to CS- in W1 at all three test intervals (5-min test: $p = 0.0059$, 30-min test: $p = 0.0077$, 60-min test: $p = 0.0392$). The respective p values for the comparison between CS+ and Ctr are 0.0394 for 5-min test, 0. 0365 for 30-min test, and 0.2773 for 60-min test. This result indicates CS+ was generalized to Ctr the same as in the behavioral responses. In time windows W2 and W3 there are only significant differences for the 30-min test for CS+ vs. CS- (W2: CS+ vs CS-: $p = 0.0036$; W3: CS+ vs CS-: $p = 0.0294$).

Next, another statistical analysis was performed (Figure 5) based on Δ which refers to the normalized spike rate differences during the pretest and test phase and which was calculated for CS+, CS- and Ctr and each time windows W1, W2 and W3 (Δ_{w1}, Δ_{w2}, Δ_{w3} respectively). The responses in PE1 neurons to CS+ declined significantly (during time window W1: $p = 0.0059$ in pre-test vs 5-min test, $p = 0.0154$ in pre-test vs 30-min test, $p = 0.0528$ in pre-test vs 60-min test; during time window W2 the respective p-values are: 0.0068, 0.0011, 0.0379)

(Figure 5), and did not change for CS- and Ctr. (CS-: during time window W1: $p = 0.9300$ in pre-test vs 5-min test, $p = 0.4379$ in pre-test vs 30-min test, $p = 0.9617$ in pre-test vs 60-min test; during time window W2 the respective p-values are: 0.3138, 0.4803, 0.4708; for Ctr: during the time window W1: $p = 0.8108$ in pre-test vs 5-min test, $p = 1.0000$ in pre-test vs 30-min test, $p = 0.8724$ in pre-test vs 60-min test, during time window W2 the respective p-values are: 0.6611, 0.7917, 0.9977). No significant differences in the odor responses were found during the time window W3 neither for CS+ nor for CS- and Ctr. The significant correlation ($p = 0.025$) was found between the behavioral CS+ response probability (%PER) and the reduction in responses of PE1 neuron to CS+ [45].

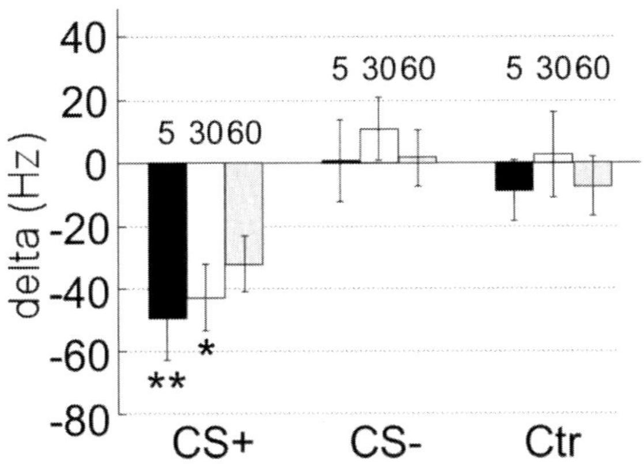

Figure 5. Response changes Δ_{w1} of neurons identified as PE1 neurons in the three extinction phases (after [45]). Average Δ_{w1} for responses to CS+, CS-, and Ctr for the three extinction tests at 5 mins after the end of acquisition (black bar), and at 30 mins after the end of acquisition (white bar), and at 60 mins after the end of acquisition (hatched bar). Error bars are SEM. (**: $p < 0.01$, *: $p < 0.05$).

A comparison between the intracellular recordings [40] and extracellular recordings [45] shows that the PE1 neuron reduces its responses to CS+ not only for a few minutes but for at least 1 hour. Surprisingly, the associative response reduction is remained even after two preceding extinction trials. Extinction might also have affected the PE1 neuron response. Thus, PE1 neuron response reduction might have been even stronger if the animal had not been given the repeated extinction tests. In addition, possibly isolated heads such as Mauelshagen used for the intracellular recording [40] lack a modulatory input to the brain necessary for the formation of longer lasting memory.

Changes in the Responses of the PE1 Neuron: On Individual Basis

An analysis of the units was performed on an individual basis by comparing the average response rates before conditioning and during the three extinction trials. Since all the significant changes found were seen during the time windows W1 and W2 the following results apply for both W1 and W2. No PE1 neuron significantly enhanced the response to CS+ at any of the three test trial sessions. Four out of 5 PE1 neurons significantly reduced their CS+ responses. The PE1 neuron that did not reduce its responses to CS+ enhanced its responses to CS-. Another PE1 neuron reduced its responses in time window W1 to both CS+ and CS- but only in the 5 min test.

Since all animals behaviorally discriminated CS+ from CS- in the 5 min-test, animals were grouped according to their conditioned PERs in the second and third test. Based on behavioral responses to the respective odors (% PER, proboscis extension response), bees could be classified as discriminators or non-discriminators. A discriminator was defined as an animal which responded to CS+, but not CS-. The non-discriminators responded either to both CS+ and CS-, or to neither CS+ nor CS-. In the second test (30 min after the last acquisition trial) the distribution was 5 discriminators, 0 non-discriminators, and in the third test (60 min after the last acquisition trial) 2 animals were discriminators and from the non-discriminators, 1 responded neither to CS+ nor CS- (non-responders) and 2 to both CS+ and CS- (responders). The average Δ values of PE1 neuron responses to CS+ decreased and for CS- increased in the 2 discriminators of the third extinction test but not in the 3 non-discriminators. Thus the PE1 neuron response changes to CS- is opposite to that of CS+, a property which is not seen in non-PE1 neurons (see below). Control animals did not show any significant response changes in either of the three time windows in the course of multiple odor stimulation.

Changes in the Responses of the Non-PE1 Neuron

In non-PE1 neurons, the same statistical analysis was applied to the response of non-PE1 neurons. No significant changes, on average, in their responses for any combinations of odor functions CS+, CS-, and Ctr, and time windows, W1 and W2. However, there were a few cases in which the individual neurons exhibited significant changes for CS+. Unfortunately, such neurons could not be identified. Because characteristics of such neurons cannot be confirmed by repeated investigation among animals, the possibility that the response change was coincidental cannot be excluded.

The non-PE1 neurons cannot be related to any anatomical or functional type yet. It is still noteworthy that both decreased and increased responses to CS+ have been found alongside an absence of associative plasticity. These associative effects were only seen in discriminators, animals that had learned to respond to CS+ and CS-. Indeed, some non-mushroom body neurons show learning plasticity [18, 51]. It will be important in future work to trace the connectivity of the neurons with antagonising associative plasticity.

The Learning-Related Network

The current best interpretation for the learning-related network model is summarized in Figure 6. The model assumes that both the PE1 neuron and the GABA-ir PCT (GABA-immunoreactive protocerebral calycal tract) neurons receive input both from Kenyon cells coding the odors to be learned (KCo), and from Kenyon cells transmitting sucrose-related excitation (KCs). Odor-coding Kenyon cells were anatomically and physiologically reported [16, 52, 53]. Gustatory input come to the mushroom body from the mouth parts [54]. Odor learning causes PE1 neuron to be less excited by the learned odor. Because PE1 axonal branches are closely attached to GABAir profiles these local inhibitory neurons in the lateral horn may transmit less inhibition to descending fibers [55].

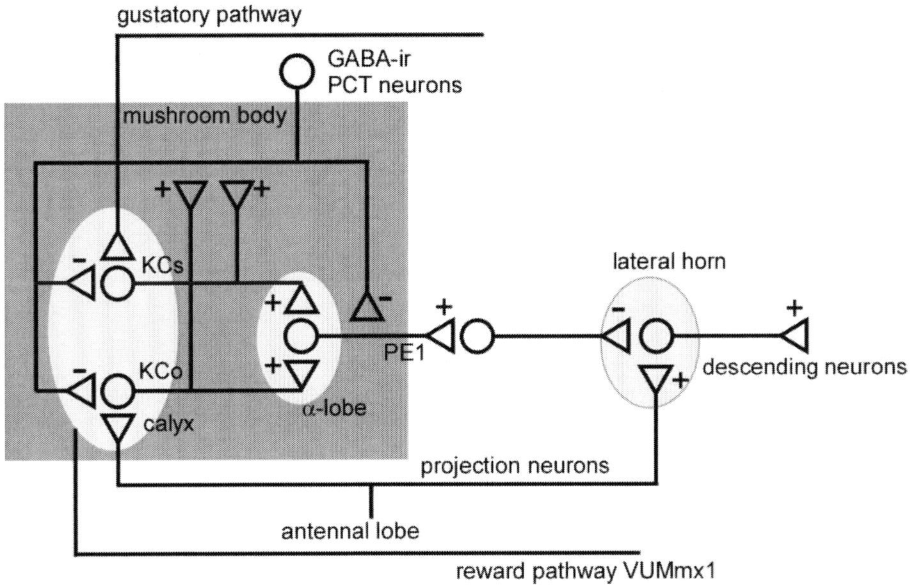

Figure 6. A model of the neural network as it relates to the PE1 neuron. The olfactory information is transmitted via the projection neurons from the antennal lobe to the mushroom body calyx where they terminate on Kenyon cells (KCo). These projection neurons also read out the associative changes within the antennal lobe. The sugar reinforcing signal is transmitted by the VUMmx1 neuron that converges with the projection neurons and the Kenyon cells in the calyx (left bright oval area). In addition, sucrose-related excitation is transmitted to the calyx via an ascending gustatory pathway terminating at a subpopulation of Kenyon cells (KCs). Both KCo and KCs feed the PE1 neuron and PCT neurons. The PE1 neuron also receives input from PCT neurons and projects to the lateral horn via inhibitory interneuron, e.g those projection neurons that are GABA immunoreactive. Descending neurons control the animal's motor responses. + indicate excitatory, and - inhibitory synapses.

The Source of Associative Plasticity in the PE1

Where does the associative reduction emerge from? PE1 neuron receives excitatory inputs from Kenyon cells of the mushroom body and indicates the multimodal sensitivity [40, 41, 56]. Only EPSPs and action potentials were elicited in PE1 neuron when Kenyon cells were electrically stimulated, and long-term potentiation (LTP) in the PE1 neuron was led by the tetanic stimulation to Kenyon cells together with the depolarization of the PE1 neuron [50]. Therefore, it is not likely that the Kenyon cells provided the associative reduction. Possibly associative plasticity originates upstream of inhibitory neurons, i.e., within the upstream inhibitory neurons (GABA-ir PCT neurons) or reflects an intrinsic property of PE1 neuron itself.

Immunohistochemistry exhibited localizing GABA-ir profiles in the immediate vicinity of marked PE1 neurons [45, 57]. The PCT neurons type A3 were GABA-ir neurons, exiting the α-lobe around its lateral midline and project to the input region of the mushroom body, the calyces [58, 59]. These neurons are, therefore, both local and recurrent inhibitory neurons. If the PCT neurons are responsible for the associative learning effect (either by developing associative plasticity themselves or by receiving it from upstream networks), PCT neurons were expected to enhance their responses to a learned odor. Indeed some PCT neurons were found to change their response to CS+ after conditioning [55, 60]. However, it has not yet be concluded that they are the source of associative plasticity as seen in PE1 neuron. Although we cannot exclude the possibility that other than PCT neurons may constitute the GABA-ir contacts with PE1 neuron, it is likely that they are the major source of these close appositions, which is strongly supported by the morphology of GABA-ir profiles in the peduncle.

Associative reduction of CS+ response in PE1 neuron could also reflect an intrinsic property of this neuron. The PE1 neuron develops associative long-term potentiation (LTP) [50]. Associative LTP and its counterpart long-term depression (LTD) are closely related and were found that they converted from one to the other form of lasting synaptic plasticity [61-63]. It was found that fear conditioning involves both LTP and LTD at parallel fibre-Purkinje cell synapses [64].

The Target of the Reduced Responses of PE1

The PE1 neuron projects to the lateral horn and is likely to make a synapse directly or via interneurons onto descending neurons. Since the mushroom body is a crucial for both the development and possibly storage of long-term olfactory memory [23] and PE1 neuron is one of the read-out neurons of memory, the descending neurons receive less inputs via PE1 neuron for learned odors. However, it is not yet known whether the PE1 neuron is presynaptic to these neurons, and whether the descending neurons are postsynaptic to them. It might be concluded that one learning effect of PE1 neuron may reduce inhibition, a property which would be executed if the PE1 neuron drives inhibitory neurons. Thus neurons like the PE1 neuron might not transmit specific information about the learned stimuli, but rather the information “these particular stimulus combinations have been learned”.

Conclusion

The search for neural correlates of learning and memory requires recording from neurons during the behavioral process of acquisition and retention. Ideally, the neurons should be identifiable on the level of the individual neuron, and experiments performed with different animals should allow us to repeatedly record from the same neurons. One of such a favorable cases is the VUMmx1 neuron in the honeybee brain. VUMmx1 neuron was found to serve the function of reward-related reinforcement in appetitive odor learning in honeybees [38]. Another neuron in the bee brain is the PE1 neuron, a single identified mushroom body-extrinsic neuron. The PE1 neuron changed its responses to odors during an associative olfactory learning paradigm [40, 45]. Eventually, multiple individually identifiable neurons need to be recorded simultaneously, with the aim of understanding the network properties in establishing and storing memories. Extracellular recording techniques were applied because only such approaches will allow us to perform multi-neuron recordings with the necessary time resolution [65]. The PE1 neuron can be identified in extracellular recording due to its spiking properties [45] therefore, the PE1 neuron can be reliably isolated from non-PE1 neurons. Consequently, their differences after the differential conditioning in response to CS+ was found, i.e., PE1 neurons reduce their responses to CS+, and non-PE1 neurons may not change, increase or decrease their responses to CS+ (please note that non-PE1 neurons are not indentifiable!).

The accumulative knowledge on learning related plasticity in the mushroom body of the bee suggests that associative plasticity is distributed between several types of neurons, Kenyon cells, PCT neurons, α-lobe extrinsic neurons (PE1 neuron and none-PE1 neuron). It is possible that any of these three classes of neurons is composed of a heterogenic collection of neurons with different forms of associative plasticity. It is also possible that brain state conditions as defined by the action of modulatory systems change the forms of associative plasticity in any of these neurons types. Recently, an aversive olfactory learning in the honeybee was reported [66-68]. It must be interesting to compare the neural mechanisms on between appetitive and aversive olfactory learning, including whether PE1 neuron as well as the other classes of neurons is involved in both or only appetitive olfactory learning. In any cases, the unique situation with a single identified neuron, the PE1 neuron, allows us to detect these ranges of associative plasticity, and provides us with the opportunity to trace the potential sources for these forms of plasticity. To investigate the brain mechanism with a behaving animal, the extracellular recording must be still a strong tool for neuroscience [65], never an antique method.

Acknowledgments

The study was partially supported by the Humboldt Foundation and KAKENHI (Nos. 22570078 and 25440175). The author thanks Drs Menzel, Rybak, and Ms. Manz for fruitful discussion.

References

[1] Sutherland GR, McNaughton B (2000) Memory trace reactivation in hippocampal and neocortical neuronal ensembles. *Curr. Opin. Neurobiol.* 10:180-186.

[2] Goldman-Rakic PS (1995) Cellular basis of working memory. *Neuron* 14:477-485.

[3] Rolls ET, Critchley HD, Mason R, Wakeman EA (1996) Orbitofrontal cortex neurons: Role in olfactory and visual association learning. *J. Neurophysiol.* 75:1970-1981.

[4] Schultz W (1998) Predictive reward signal of dopamine Neurons. *J. Neurophysiol.* 80:1-27.

[5] Takeda K (1961) Classical conditioned response in the honey bee. *J. Ins Physiol.* 6: 168-179.

[6] Giurfa M (2013) Cognition with few neurons: higher-order learning in insects. *Trend Neurosci.* 36: 285-294.

[7] Giurfa M, Sandoz JC (2012) Invertebrate learning and memory: fifty years of olfactory conditioning of the proboscis extension response in honeybees. *Learn Mem.* 19: 54-66.

[8] Himmelreich S, Grünewald B (2012) Cellular physiology of olfactory learning in the honeybee brain. *Apidologie* 43: 308-321.

[9] Menzel R (2012) The honeybee as a model for understanding the basis of cognition. *Nat. Rev. Neurosci.* 13: 758-768.

[10] Bitterman ME, Menzel R, Fietz A, Schäfer S (1983) Classical conditioning of proboscis extension in honeybees (*Apis mellifera*). *Journal of Comparative Psychology* 97:107-119.
[11] Hadar R, Menzel M (2010) Memory formation in reversal learning of the honeybee. *Front Behav. Neurosci.* 4:186. doi: 10.3389/fnbeh.2010.00186
[12] Hussaini SA, Komischke B, Menzel R, Lachnit H (2007) Forward and backward second-order Pavlovian conditioning in honeybees. *Learn Mem* 14: 678-683.
[13] Matsumoto Y, Menzel R, Sandoz JC, Giurfa M (2012) Revisiting olfactory classical conditioning of the proboscis extension response in honey bees: a step toward standardized procedures. *J. Neurosci. Meth* 211:159-167.
[14] Menzel R, Manz G, Menzel R, Greggers U (2001) Massed and spaced learning in honeybees: the role of CS, US, the intertribal interval, and the test interval. *Learn Mem.* 8: 198-208.
[15] Stollhoff N, Menzel R, Eisenhardt D (2008) One retrieval trial induces reconsolidation in an appetitive learning paradigm in honeybees (*Apis mellifera*). *Neurobiol Learn Mem.* 89: 419-425.
[16] Abel R, Rybak J, Menzel R (2001) Structure and response patterns of olfactory interneurons in the honeybee, *Apis mellifera*. *J. Comp. Neurol.* 437:363-383.
[17] Denker M, finke R, Schaupp F, Grün S, Menzel R (2010) Neural correlates of odor learning in the honeybee antennal lobe. *Eur. J. Neurosci.* 31: 119-133.
[18] Haehnel M, Menzel R (2012) Long-term memory and response generalization in mushroom body extrinsic neurons in the honeybee *Apis mellifera*. *J. Exp. Biol.* 215: 559-565.
[19] Menzel R, Mueller U (1996) Learning and memory in honeybees: from behavior to neural substrates. *Annu. Rev. Neurosci.* 19: 379-404.
[20] Szyszka P, Galin A, Menzel R (2008) Associative and non-associative plasticity in Kenyon cells of the honeybee mushroom body. *Front Syst. Neurosci.* 2:3-
[21] Szyszka P, Demmler C, Oemisch M, sommer L, Blergans S, Birnbach B, Silbering AF, Balizia G (2011) Mind the gap: olfactory trace conditioning in honeybees. *J. Neurosci.* 31: 7229-7239.
[22] Menzel R (1999) Memory dynamics in the honeybee. *J. Comp. Physiol.* A 185: 323-340.
[23] Menzel R (2001) Searching for the memory trace in a mini-brain, the honeybee. *Learn Mem.* 8:53-62.
[24] Menzel R, Giurfa M (2001) Cognitive architecture of a mini-brain: the honeybee. *Trends Cogn* Sci 5:62-71.
[25] Eisenhardt D, Menzel R (2007) Extinction learning, reconsolidation and the internal reinforcement hypothesis. *Neurobiol. Learn Mem.* 87: 167-173.
[26] Stollhoff N, Eisenhardt D (2009) consolidation of an extinction memory depends on the unconditioned stimulus magnitude previously experienced during training. *J. Neurosci.* 29: 9644-9650.
[27] Stollhoff N, Menzel R, Eisenhardt D (2005) Spontaneous recovery from extinction depends on the reconsolidation of the acquisition memory in an appetitive learning paradigm in the honeybee (*Apis mellifera*). *J. Neurosci.* 25: 4485-4492.

[28] Eisenhardt D (2014) Molecular mechanisms underlying formation of long-term reward memories and extinction memories in the honeybee (*Apis mellifera*). *Learn Mem.* 21: 534-542.

[29] Friedrich A, Thomas U, Müller U (2004) learning at different satiation levels reveals parallel functions for the cAMP-protein kinase A cascade in formation of long-term memory. *J. Neurosci.* 24: 4460-4468.

[30] Müller U (2012) The molecular signaling processes underlying olfactory learning and memory formation in honeybees. *Apidologie* 43: 322-333.

[31] Müßig L, Richltzkl A, Rößler R, Eisenhardt D, Menzel R, Leboulle G (2010) Acute disruption of the NMDA receptor subunit NR1 in the honeybee brain selectively impairs memory formation. *J. Neurosci.* 30: 7817-7825.

[32] Menzel R, Brembs B, Giurfa M (2006) Cognition in Invertebrates. In: Evolution of Nervous Systems, Vol. II: Evolution of Nervous Systems in Invertebrates (Kaas JH, ed), pp 403-422. Oxford: Academic Press.

[33] Okada R, Sakura M, Mizunami M (2003) Distribution of dendrites of descending neurons and its implications for the basic organization of the cockroach brain. *J. Comp. Neurol.* 458: 158-174.

[34] Erber J, Masuhr T, Menzel R (1980) Localization of short-term memory in the brain of the bee, *Apis mellifera. Physiol. Entomol.* 5:343-358.

[35] Heisenberg M (2003) Mushroom body memoir: from maps to models. *Nature Reviews Neuroscience* 4:266-275.

[36] Menzel R, Erber J, Masuhr T (1974) Learning and memory in the honeybee. In: Experimental analysis of insect behaviour (Barton-Browne L, ed), pp 195-217. Berlin: Springer.

[37] Hammer M (1997) The neural basis of associative reward learning in honeybees. *Trends Neurosci.* 20: 245-252.

[38] Hammer M (1993) An identified neuron mediates the unconditioned stimulus in associative olfactory learning in honeybees. *Nature* 366:59-63.

[39] Schröter U, Malun D, Menzel R (2007) Innervation pattern of suboesophageal ventral unpaired median neurons in the honeybee brain. *Cell Tissue Res.* 327: 647-667.

[40] Mauelshagen J (1993) Neural correlates of olfactory learning in an identified neuron in the honey bee brain. *J. Neurophysiol.* 69:609-625.

[41] Rybak J, Menzel R (1998) Integrative properties of the Pe1-neuron, a unique mushroom body output neuron. *Learn Mem.* 5:133-145.

[42] Brandt R, Rohlfing T, Rybak J, Krofczik S, Maye A, Westerhoff M, Hege H-C, Menzel R (2005) A three-dimensional average-shape atlas of the honeybee brain and its applications. *J. Comp. Neurol.* 492:1-19.

[43] Strausfeld NJ (2002) Organization of the honey bee mushroom body: representation of the calyx within the vertical and gamma lobes. *J. Comp. Neurol.* 450:4-33.

[44] Hussaini SA, Menzel R (2013) Mushroom body extrinsic neurons in the honeybee brain encode cues and contexts differently. *J. Neurosci.* 33: 7154-7164.

[45] Okada R, Rybak J, Manz G, Menzel R (2007) Learning-related plasticity in PE1 and other mushroom body-extrinsic neurons in the honeybee brain. *J. Neurosci.* 27: 11736-11747.

[46] Strube-Bloss MF, Nawrot MP, Menzel R (2011) Mushroom body output neurons encode odor-reward associations. *J. Neurosci.* 31: 3129-3140.

[47] Mizunami M, Okada R, Li YS, Strausfeld NJ (1998) Mushroom bodies of the cockroach: Activity and identities of neurons recorded in freely moving animals. *J. Comp. Neurol.* 402: 501-519.

[48] Okada R, Ikeda J, Mizunami M (1999) Sensory responses and movement-related activities in extrinsic neurons of the cockroach mushroom bodies. *J. Comp. Physiol. A* 185:115-129.

[49] Bacon JP, Altman JS (1977) A silver intensification method for cobalt-filled neurones in wholemount preparations. *Brain Res.* 138:359-363.

[50] Menzel R, Manz G (2005) Neural plasticity of mushroom body-extrinsic neurons in the honeybee brain. *J. Exp. Biol.* 208:4317-4332.

[51] Strube-Bloss MF, Herrera-Valdez MA, Smithi BH (2012) Ensemble response in mushroom body output neurons of the honey bee outpaces spatiotemporal odor processing two synapses earlier in the antennal lobe. *PLoS ONE* 7: e50322. doi: 10.1371/journal.pone.0050322.

[52] Kirschner S, Kleineidam CJ, Zube C, Rybak J, Grünewald B, Rössler W (2006) Dual olfactory pathway in the honeybee *Apis mellifera*. *J. Comp. Neurol.* 499:933-952.

[53] Szyszka P, Ditzen M, Galkin A, Galizia CG, Menzel R (2005) Sparsening and temporal sharpening of olfactory representations in the honeybee mushroom bodies. *J. Neurophysiol.* 94:3303-3313.

[54] Schröter U, Menzel R (2003) A new ascending sensory tract to the calyces of the honeybee mushroom body, the subesophageal-calycal tract. *J. Comp. Neurol.* 465:168-178.

[55] Haehnel M, Menzel R (2010) Sensory representation and learning-related plasticity in mushroom body extrinsic feedback neurons of the protocerebral tract. *Front Syst. Neurosci.* 4: 161 doi: 10.3389/fnsys.2010.00161.

[56] Iwama A, Shibuya T (1998) Physiology and morphology of olfactory neurons associating with the protocerebral lobe of the honeybee brain. *J. Insect Physiol.* 44:1191-1204.

[57] Rybak J, Kuß A, Lamecker H, Zachow S, Hege HC, Lienhard M, Singer J, Neubert K, Menzel R (2010) The digital bee brain: integrating and managing neurons in a common 3D reference system. *Front Syst. Neurosci.* 4: 30. doi: 103389/fnsys.2010.00030.

[58] Grünewald B (1999) Morphology of feedback neurons in the mushroom body of the honeybee, *Apis mellifera*. *J. Comp. Neurol.* 404:114-126.

[59] Rybak J, Menzel R (1993) Anatomy of the mushroom bodies in the honey bee brain: the neuronal connections of the alpha-lobe. *J. Comp. Neurol.* 334:444-465.

[60] Grünewald B (1999) Physiological properties and response modulations of mushroom body feedback neurons during olfactory learning in the honeybee *Apis mellifera*. *J. Comp. Physiol.* A 185:565-576.

[61] Bear MF, Cooper LN, Ebner FF (1987) A physiological basis for a theory of synapse Modification. *Science* 237:42-48.

[62] Bear MF, Malenka RC (1994) Synaptic plasticity: LTP and LTD. *Curr. Opin. Neurobiol.* 4:389-399.

[63] Vetter P, Roth A, Häusser M (2001) Propagation of action potentials in dendrites depends on dendritic morphology. *J. Neurophysiol.* 85:926-937.

[64] Zhu L, Scelfo B, Hartell NA, Strata P, Sacchetti B (2007) The effects of fear conditioning on cerebellar LTP and LTD. *Eur. J. Neurosci.* 26: 219-227.

[65] Menzel R (2014) The insect mushroom body, an experience-dependent recording device. *J. Physiol.* doi: 10.10.16/j.jphysparis.2014.07.004

[66] Giurfa M, Fabre E, Flanven-Pouchon J, Groll H, overwallner B, Vergoz V, Roussel E, Sandoz JC (2009) Olfactory conditioning of the sting extension reflex in honeybees: memory dependence on trial number, interstimulus interval, intertribal interval, and protein synthesis. *Learn Mem.* 16: 761-765.

[67] Tedjakumala SR, Giurfa M (2013) Rules and mechanisms of punishment learning in honey bees: the aversive conditioning of the sting extension response. *J. Exp. Biol.* 216: 2985-2997.

[68] Vergoz V, Roussel E, Sandoz JC, Giurfa M (2007) Aversive learning in honeybees revealed by the olfactory conditioning of the sting extension reflex. *PLoS ONE* 2:e288. doi: 10.1371/ournal.pone.0000288.

In: Memory Consolidation
Editors: Manabu Sakakibara and Etsuro Ito

ISBN: 978-1-63482-596-2

Chapter 3

MEMORY MEDIATED BY INTERNAL STATE: MEMORY OF LOST SUPRESSES MOTIVATION OF FIGHT IN THE CRICKET *GRYLLUS BIMACULATUS*

Hitoshi Aonuma,[1,2,*] Midori Sakura[3] and Daisuke Kurabayashi[4]

[1]Research Institute for Electronic Science, Hokkaido University, Sapporo, Japan
[2]CREST, Japan Science and Technology Agency, Honcho Kawaguchi, Saitama, Japan
[3]Department of Biology, Graduate School of Science, Kobe University, Rokkodai, Nada-ku, Kobe, Japan
[4]Department of Control and Mechanical Engineering, Tokyo Institute of Technology, Ookayama, Meguro-ku, Tokyo, Japan

ABSTRACT

Aggression with a physical attack is a common behavior in animals. Male crickets exhibit intensively aggressive behavior towards conspecific males, often culminating in fighting. The result of the fight is the establishement of dominant-subordinate hierarchy between male crickets. Chemical cues are a dominant stimulus for them to initiate aggressive behavior. Crickets use their antenna to detect these cuticular substances on the body of conspecific males, which initiates the aggressive behavior. After a fight, the losing male no longer exhibits aggressiveness in subsequent bouts with the same male or in separate encounters with another male; rather the defeated male exhibits avoidance behavior. This phenomenon is known as the "loser's effect". The loser's effect is a kind of memory of previously losing a fight. Therefore, aggressive behavior in crickets provides a good model system to understand physiological mechanisms underlying how animals modify internal state to exhibit sophisticated behavior.

[*] Corresponding author address. Research Institute for Electronic Science, Hokkaido University, Sapporo 060-0812, Japan; E-mail: aon@es.hokudai.ac.jp.

Keywords: loser's effect, aggression, previous experience, internal state, neuromodulator, biogenic amine, brain

INTRODUCTION

Sophisticated behaviors in animals are related to the memory of previous experiences. Animals do not always respond in the same ways to the same stimulus. Instead, they alter their behavior in response to the specific demands of their circumstances. The memory of previous experiences is thus one of the important factors to decide on the appropriate behavioral response. In this chapter, we focus on aggressive agonistic behavior. The loser's effect is based on a memory of the previous experience of lost agonistic encounter (e.g., a fight with a conspecific). After the loss of a fight, losers' depress motivation of aggressive behavior for some time. The motivation of aggressive behavior is modified by the internal state that is regulated by complex interactions among physiological, motivational and behavioral systems. Changing social and physical environments are major factors that regulate the motivation of aggressive behavior, which makes us difficult to fully understand aggressive behavior. The cricket provides us a good model systems to investigate how animals modify their behavioral motivation. In particular, the subordinate cricket (loser of a fight) helps us to understand how aggressive motivation changes depending on social status.

It is common in animals that a dominant-subordinate hierarchy is established after an agonistic aggressive encounter between individuals. In some cases, they fight utilize violent attacks in order to acquire resources such as food, territories, mating partners and so on. The aggressive behavior of male crickets towards conspecific males is impressive and has been extensively studied (Alexander, 1961; Phillips and Konishi, 1973; Adamo and Hoy, 1995; Hack, 1997; Hoffmann and Schildberger, 2001; Stevenson et al., 2005). Male crickets fight in order to acquire both terrotires and mating partners (Simmons, 1986). Aggressive behavior often occurs when two males encounter each other for the first time. If neither of the males retreats after the initial encouther, they consequently engage in a battle. A fight consists of a sequentially escalating series of behaviors beginning with antennal fencing, which leads to spreading and later the engaging of the mandibles and grappeling. The fight continues until one of the males surrenders. At that point, a behavioral hierarchy is established between the crickets, with the winner initiating the singing of an aggressive song and chasing the loser who flees away from the winner. The loser in subsequent encounters with the winner exhibits avoidance behavior and thus avoids additional fights (Adamo and Hoy, 1995). This depressive period in the losing cricket continues for several hours or more (Hofmann and Stevenson, 2000). This phenomenon is so-called loser's effect. Interestingly, some factors modify the motivation of aggressive behavior of male crickets. These include recent experience such as: copulation, flight, the opponent's size and behavior, and population density (Alexander, 1961; Hofmann and Stevenson, 2000; Rillich et al., 2007; Ashikaga et al., 2009). Thus, whether males decide to engage in aggressive behaviors with each other may depend on their ability to compare their own motivation to fight with their perception of a potential opponent's fighting performance (Hack, 1997; Rillich et al., 2007).

While many of the ethological aspects of aggressive behavior and fighting are known, the causal neural mechanisms underlying aggressive behavior, including the identity of sensory systems involved, remain largely unknown. Recently, progress has been made in delineating which sensory information is involved in the mediation of fighting behavior. However, all these previous studies focused on the various sensory cues for the 'assessment' process during the fight but did not attempt to identify the key stimuli that elicit aggressive behavior. Furthermore, it is also unclear how losing individuals accumulate the momory of a previous fight to depress aggressive motivation. We have built a hypothesis that neuromodulators in the brain mediate internal state as a memory of a fight of the crickets, which regulate the motivation of aggressive behavior. In this chapter, we demonstrate how neuromodulators, nitric oxide and the biogenic amine octopamine, modulate the motivation of aggressive behavior and how the loser's memory of a fight regulates group size dependent aggressive behavior in the crickets.

Aggressive Behavior in the Cricket

Adult male cricket (*Gryllus bimaculatus*) exhibit intense aggressive behavior when it encounters a conspecific male whereas it exhibit courtship behavior to a conspecific female (Alexander, 1961). Sex discrimination by male crickets is based on cuticular chemical substances on the body surface. The components of cuticular hydrocarbons are different between males and females (Tregenza and Wedell, 1997), and the cuticular hydrocarbons are critical to expression of aggressive and courtship behavior in male cricket (Nagamoto et al., 2005 ; Ryan and Sakaluk, 2009). Cuticular substances play a role of pheromones that initiate aggressive or courtship behavior in male cricket. Behaviors induced by releaser pheromones have been thought to be hard-wired: behaviors that could be turned on and off without plasticity. However, aggressive behavior in crickets is modified by the outcome of their previous agonistic encounters. For example, while cuticular pheromones induce winners (dominant crickets) to be aggressive, losers (subordinate crickets) respond to cuticular pheromones with avoidance instead of aggression (Sakura and Aonuma, 2013).

An agonistic encouter between male crickets starts out slowly and escalates into a fierce struggle (Figure 1). Aggressive behavior can be separated into two types, offensive and defensive aggressive behavior. Chemical and tactile information from antennae are important for males to express offensive aggressive behavior, identifying conspecific males and initiating an attack. However, antennal information is not always necessary to elicit defensive aggression (Sakura and Aonuma, 2013). Antennal sensory information that consists of chemical and tactile sensory information is important for crickets when they seek foods and mating partners, as well as when they detect threats. Males discriminate a conspecific's sex using sensory information from antennae. Both chemical and tactile cues are integrated in the brain to decide whether to release aggressive or courtship behavior. The crickets without antennae can release defensive aggressive behavior but not offensive behavior. Antennectomized males, or males who have had their antenna removed, express less aggression toward antennectomized opponents whereas they continue to exhibit typical fighting to an intact opponent. In addition, antennectomized losers exhibit significantly higher aggressiveness towards the opponent than intact losers do in a second bout.

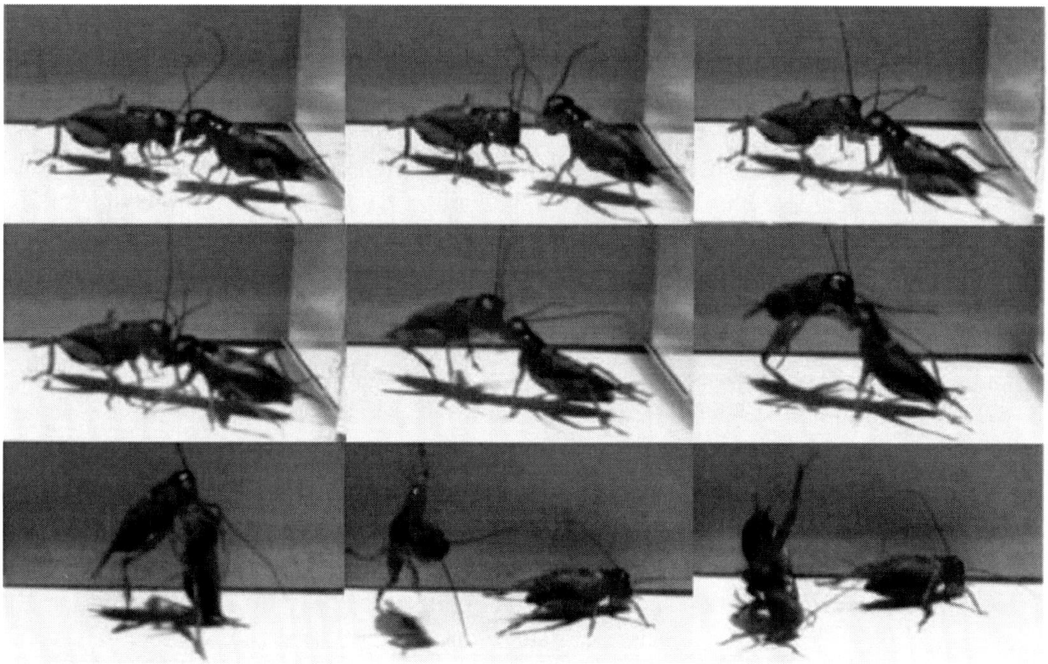

Figure 1. Cricket fight between males. Fighting escalates until an opponent gives up attacking and escapes by moving away. Male crickets start antennal fencing when they perceive cuticular substances through antennal contact. Mandible spreading follows.

These suggest to us that defensive aggression can be elicited without antennal information. Antennectomized crickets need not use visual or pulpal sensory input to elicit defensive aggression. In contrast, intact males show aspects of aggressive behavior in response to male cuticular substances before and after winning a fight, however if these males lose a fight, then they show avoidance behavior. Visual inputs play an important role, since blinded males exhibit more intense fighting than sighted males. Visual information concerning body size, weight, and mandible display behavior of the opponent can suppress aggressiveness (Rillich et al., 2007). Differences in the frequency of antennal fencing between the eventual winners and losers of a fight suggest that male crickets may 'decide' on the opponent's willingness to fight based on tactile information delivered to the CNS via the antennae in cricket (Hoffmann and Schildberger, 2001). Thus, it appears that males make decisions whether to fight against or to flee from a potential opponent based on different sensory modalities.

Population density also strongly affects aggressiveness in the cricket (Funato et al., 2011) (Figure 2). Pairs of male crickets fight locally to establish a dominant–subordinate relationship in a group. Once this relationship is established, victorious males (dominant) retain their aggressiveness. On the other hand, defeated males (subordinate) exhibit reduced aggressiveness. If dominant males encounter other dominant males, they fight again to establish dominant–subordinate relationships.

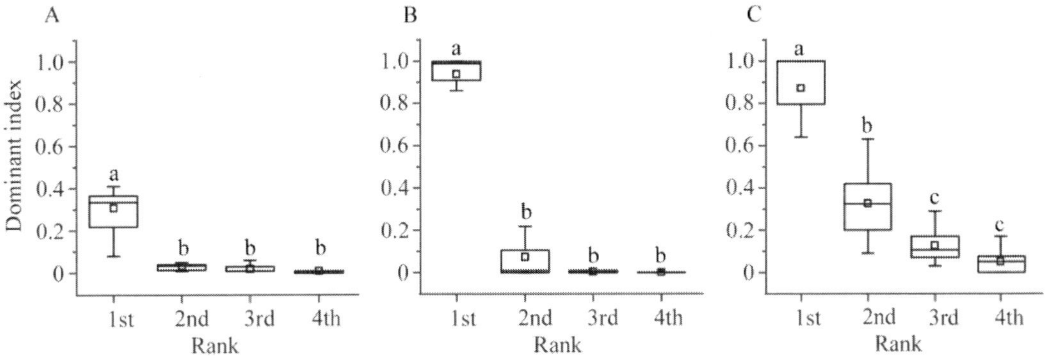

Figure 2. Change in aggressiveness dependent on the population density of cickets. The aggressiveness is evaluated by using a dominance index. The dominance index, *Xi*, is defined as

$$Xi = \frac{Di}{Di + Si + Ni} \quad (i = 1, 2, 3, ..., Num),$$

where *i* is the social rank of the cricket. The total number of crickets is *Num*, and *Di* indicates the total number of dominance behaviours performed by cricket *i*. *Si* indicates the total number of subordinate behaviours performed by cricket *i*, and *Ni* indicates the number of non-responsive behaviours when cricket *i* encounters another male. A: Four male crickets are placed in a 75 cm^2 arena (N = 12). B: Four male crickets are placed in a 300 cm^2 arena (N = 12). C: Four male crickets are placed in a 1200 cm^2 arena. (Different letters denote significant differences between treatments, $^*P < 0.05$, two-tailed Kruskal-Wallis ANOVA with Bonferroni-type multiple nonparametric comparison). Modified from Funato et al., 2011.

However, when subordinate males encounter a dominant male within a short period, they exhibit avoidance behavior to retreat from the opponent. Subordinate males also exhibit avoidance if they encounter victorious males in other fights, naive males, or even males defeated in other fights. After repeated fights, a repeatedly subordinate male decreases aggressiveness. In due course, one or more males are established as the dominant status in the group. Depending on the number of individuals in a group, the loss of individual aggressiveness limits the number of fighting crickets at any given time. Thus, overall aggressive behavior in a group is changed as a consequence of group size, but is ultimately determined by individual fights.

Higher population density increases the probability that individuals might encounter conspecifics and in turn engage in aggressive behavior. Subordinates mainly exhibit avoidance behavior when they encounter other males in a high population group. On the other hand, the highest-ranked males could maintain aggressiveness more than other males, but they suppress aggressiveness in a condition of high population density. Therefore, all crickets in a condition of high population density behave with lower aggression. This phenomenon is called the crowd effect. Alternatively, a decrease in local population density decreases the chance to encounter another crickets. The dominant cricket always exhibits aggressive behavior if it encounters other crickets in a lower population density group. In contrast, all subordinate crickets exhibit avoidance behavior when they encounter other males. Dominant males establish a despotic hierarchy within a group. This relationship likely depends on local population density. As the local population density decreases, an increasing number of dominant crickets that exhibit aggression will increase in the group.

NEUROMODULATORS REGULATING AGGRESSIVENESS

Cricket aggressiveness changes according to the result of a fight, copulation, crowd effect and so on. Aggressiveness is depressed for some time in a loser cricket. This suggests that neuromodulation within a brain must play important role in makeing the decision of fight or flee. Nitric oxide (NO) and biogenic amine octopamine (OA) are major neuromodulators that regulate aggressiveness in the crickets.

Endogenous nitric oxide (NO) is a free radical signaling molecule that diffuses across cell membranes in the nervous system of a wide variety of animals, both vertebrate and invertebrate. It is generated from L-arginine by NO synthase (NOS) (Moncada et al., 1991) and activates the heme-containing enzyme soluble guanylate cyclase (SGC) to generate the second messenger cyclic guanosine monophosphate (cGMP) in target cells (Bredt and Synder, 1989). This results in activation of cGMP-activated protein kinase (PKG) that phosphorylates downstream target proteins and evokes a cellular response (Bicker, 2001). Cyclic-nucleotide-gated channels (Kaupp and Seifert, 2002) and cyclic-nucleotide phosphodiesterase (Bender and Beavo, 2006) are also major targets of cGMP. Furthermore, evidence for cGMP-independent NO signaling has also been found. NO acts directly, without producing cGMP, on ion channels through S-nitrosylation, which regulates the electrical activity of the target neurons (Ahern et al., 2002; Wilson et al., 2007). S-nitrosylation is thought to require a higher concentration of NO than does activation of soluble guanylate cyclase, and tends to proceed with slower kinetics than cGMP-mediated actions.

NO is continuously generated at a basal level in the cricket nervous system, and cholinergic activation accelerates NO generation (Aonuma et al., 2008). Therefore, relative changes in the concentration of NO likely mediate physiological properties of local circuits in nervous systems. Experimental manipulations of NO level in the nervous systems and histochemical experiments indicate that NO acts as a multi-functional mediator. In mollusks, NO modulates synaptic efficacy of cholinergic synapses (Mothet et al., 1996), and mediates oscillatory neuronal activities underlying chemical information processing in the central nervous system (Gelperin, 1994). In crayfish, components of NO-cGMP signaling pathway contribute to neuronal network functions that underlie escape behavior (Aonuma et al., 2000), mechanosensory processing (Aonuma and Newland, 2001; 2002; Aonuma et al., 2008), and olfactory processing (Johansson and Mellon, 1998; Fujie et al., 2002). In insects, NO is believed to function as a crucial component in motor control (Qazi and Trimmer, 1999), pheromone processing (Seki et al., 2005), olfactory processing (Wilson et al., 2007) and olfaction-related learning behavior (Müller, 1997; Matsumoto et al., 2006).

The presence of NO-cGMP signaling in the brain is reported in cricket (Aonuma and Niwa, 2004). Pharmacological and behavioral experiments demonstrate that NO signaling mediates the motivation of aggressive behavior in the male cricket (Aonuma et al., 2004; Iwasaki et al., 2007). Inhibition of NO/cGMP signaling pathways using NOS inhibitor L-NAME or SGC inhibitor ODQ restores aggressiveness of subordinates. The aggressiveness of a cricket that is injected with L-NAME or with ODQ exhibits normal aggression at the first engagement. However, the loser cricket that is injected with L-NAME or with ODQ exhibits significantly increases aggressiveness in subsequent encounters. This suggests that NO/cGMP signaling play important role in memory of the previous fight and decision-making of fight or fleeing. Interestingly, the behavior of subordinate whose antennae are removed is similar to the behavior of subordinate whose NO/cGMP signaling is inhibited (Sakura and Aonuma, 2013). This suggests to us that NO/cGMP signaling in the antennal sensory-information processing pathway could participate in the neuronal mechanism underlying aggressive behavior. Indeed, there are NO donor and target neurons in antennal lobes of the cricket (Aonuma et al., 2004).

Octopamine (OA) is another modulator that affects aggressive behavior in the cricket. It has been demonstrated that the biogenic amine level in hemolymph mediates cricket aggression (Adamo et al., 1995). OA titer in the brain of a defeated cricket decreases after fighting (Aonuma et al., 2009). The OAergic system is generally involved in the motivation of aggressive behavior in insects (Rillich et al., 2007; Aonuma and Watanabe, 2012). In cricket, head injection of the OA antagonist mianserin decreases the aggressiveness of male cricket (Aonuma et al., 2009). In a sequential encounter, the mianserin-injected cricket exhibits significantly lower aggressiveness in the first encounter as compared to control. Mianserin is widely used as OA antagonist in insect. In a second encounter, after a few hours, the mianserin-injected cricket still still display reduced aggressive behavior. Therefore, OA antagonist slows recovery of aggressiveness in the male cricket.

The NOergic and OAergic system regulate experience-dependent aggressive behavior of male crickets. The NOergic and OAergic systems have mutually opposite effects on initiating aggressiveness in the cricket (Figure 3). The depression of the subordinates' aggressiveness is alleviated by NOS inhibitor, whereas an OA antagonist prolongs the reduction of the subordinates' aggressiveness. We thus hypothesize that NO/cGMP system mediates the OAergic system in the brain, which in turn regulates loser's effect in the cricket.

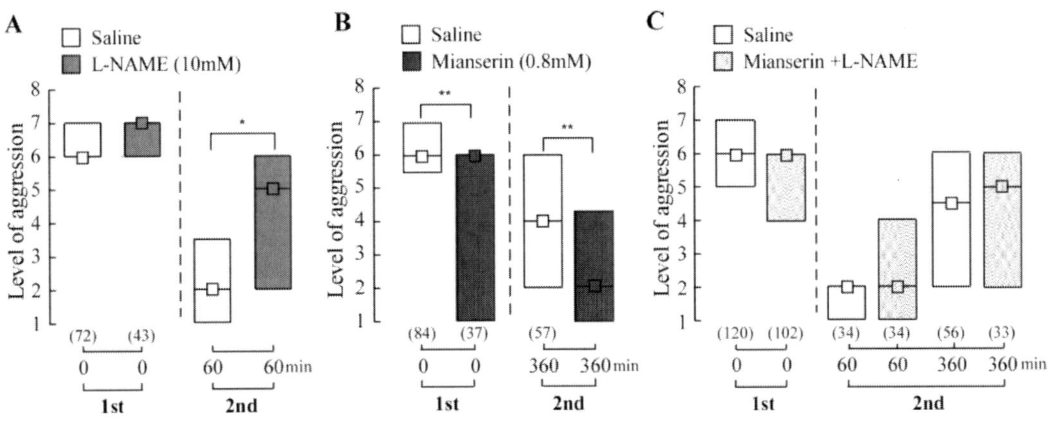

Figure 3. Effect of NOS inhibitor and OA antagonist on cricket aggression. A. NOS inhibitor L-NAME restored loser's aggressiveness in the 2nd bout. B. OA antagonist mianserin prolonged loser's effect. C. L-NAME rescues the effect of mianserin on loser's effect. Aggresion level, 1: Avoidance without antennal contacts with the opponent; 2: Avoidance after antennal contacts with the opponent; 3: Antennal lashing, 4: Antennal fencing with threat posture, 5: Mandible flare, 6: Mandible engagement (single); 7: Mandible engagement (multiple); 8: Tactile combat.

Modeling and Computor Simulation for Understanding Change in Agressiveness

Reproducing animal behavior artificially helps us to understand the mechanisms underlying sophisticated behaviors (Funato et al., 2008; Funato et al., 2011). By using computer simulation and embedding the argorism into a simple robot, we can examine how the function focued work in animals, and can compare the differences between artificial systems and biological systems. One of the most important things is to build as simple of a model as even if it does not fully represent every intracy of the organism's behavior. Simple models suggests to us what is missing to reproduce animal function or what neuronal functionions are underlying focused behavior.

Modeling aggressive behavior in crickets has been performed based on observation and neurophysiology experiments (Ashikaga et al., 2009; Funato et al., 2011; Kawabata et al., 2012; Mizuno et al., 2012). In order to examine how crickets behave in a group, a finite automaton model was built based on the observation (Ashikaga et al., 2009). The finite automaton model interestingly reproduces group size dependent aggressiveness in real crickets suggesting the importance of the internal state of individual (Figure 4). In this model, a fight between male crickets is considered as a two-body problem. The artificial crickets are designed as follows: each one can be identified but agents cannot recongnize each other, it can detect obstacles and avoid them using antennal input information. Additionally they can detect other agents to select fight or flee depending on their own internal state. A fight in an agent is designed as follows. The result of a fight is fed back to the agent and the fight does not affect the third agent. It is assumed that the agent requires an internal state to select select proceeding behaviors because the result of a fight affects the decision-making to fight or to flee in real crickets. The parameter of the internal state "α" plays role of a memory of the previous fight in the model. The agents make decision of fight or flee dependent on the value of internal state "α". Then, they are given three basic behaviors: wandering, fighting, and obstacle avoidance. In this model, an agent detects another agent and exhibits aggressive behavior if the opponent comes within its personal field. Aggressive behavior will decrease if the opponent flees from the agent. The probability P of losing (fleeing from the opponent) is defined as $P=\alpha$. Therefore, an agent loses with probability P and continues attacking at probability $(1-P)$. Since real crickets modify aggressiveness depending on the result of previous fight and loser depresses aggressive motivation for some time, the internal state α is updated.

The simulation of this finite automaton model suggests that the internal state of individual agent, in particular losers' regulates total aggression in a group, and that the internal state of each agent must be maintained by individual interaction, which must be a key of social adaptability.

We then examined how the memory of a fight in a loser affects total aggression in a group by using oscillator robots. Two levels of approaches are performed to elucidate the neuronal mechanism underlying adaptive behavior, *i.e.*, modeling neuronal mechanism underlying decision-making of a behavior and identification of a functional component in the brain that responds to changing circumstance at the individual behavioral level. To correlate them, a robot controlled by brain functional model is built. The movement of the robot is generated by an interaction between the individual interaction and the brain function model, allowing both levels to be studied simultaneously.

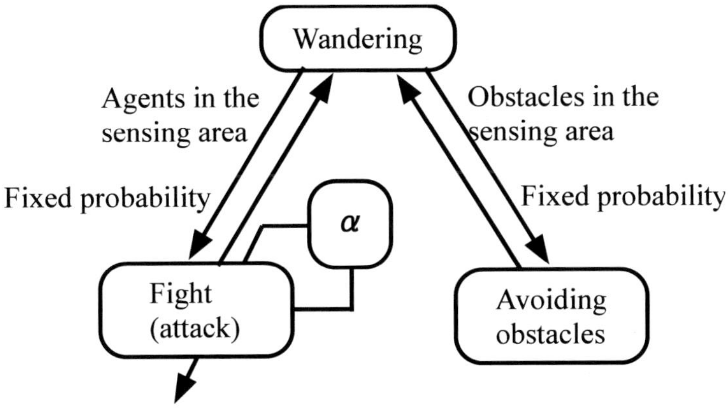

Figure 4. Finite automaton model of a fight between crickets. Behavior is determined depending on a finite-state machine model where 3 motion primitives are as follow, wandering (agent can walk randomly while going straight and can stop and turn ±π/2 rad periodically), fighting (it can choose fighting if it detect another one in its personal field) and avoiding a obstacle. The parameter of the internal state α in each agent is determined; it stores the memory of won or lost as its internal state α. The internal state α expresses the level of passivity and has a value of 0 through 1. Agent selects behavior dependent on the internal state. The artificial crickets update the internal state α when they interact with others as follow.

$$\alpha_{t+1} = (1-\omega)\,\alpha_t + \varepsilon_{lose}\eta_{lose} + \varepsilon_{win}\eta_{win}$$

Here, α_t signifies the values of the internal states α at time t. The coefficient of memory of loss is ω. The coefficient effects of fighting with other crickets are ε_{lose} and ε_{win}. These coefficients are set by trial and error, $\omega=10^{-2.5}$, $\varepsilon_{lose}=0.5$, and $\varepsilon_{win}=-0.5$. Then, η_{lose} and η_{win} are the functions as follw. $\eta_{lose}=1$ (if agent loses) and $\eta_{lose}=0$ (other). $\eta_{win}=1$ (if agent wins) and $\eta_{win}=0$ (other).

Neuronal oscillation in the brain is an important function for processing information, and thus this phenomenon is focused upon, when building a brain function model. Oscillation in the brain is evoked by sensory inputs in both vertebrate and invertebrate (Gelperin, 1994; Laurent and Naraghi, 1994; Fujie et al., 2002; Schadow et al., 2007; Tanaka et al., 2009). The function of the neuronal oscillation is still disputed, and insect olfactory system has provided a good experimental model system for elucidating the function. In the insect brain, neuronal oscillations evoked by chemical stimulation are observed as local field potential (Laurent and Naraghi, 1994; Tanaka et al., 2009). The synchrony between coupled oscillators is influenced by the network structure (Barahona and Pecora, 2002). This phenomenon can be interpreted as a temporal transformation of a neural network. Inspired from this, oscillators network, where output changes depending on the network transformation, is employed as a platform to examine the role of memory of a previous result of a fight in decide to attack or flee.

Using neuronal oscillation, cricket aggressive behavior is modeled. As described above, when a battle between male crickets is settled, dominant crickets maintiain their aggression while subordinate crickets avoid the victorious opponents for some time. Futhermore, the proportion of aggressive individuals in a group changes according to the size of the group (Funato et al., 2011). The effect of repeated encounters is to alter group aggression. The speed of the emergence of group level behavior depends on both density of a group and local result of a fight. Group size-dependent aggression is also known in many animals, including crayfish (Bovbjerg, 1953; Patullo et al., 2009), and domestic fowl (Pagel and Dawkins, 1997; Estevez et al., 2003). However, few studies have focused on the neuronal mechanisms underlying group size-dependent behavior. Using an oscillator network model, the basis of a neuronal factor elicited during group size-dependent aggression in crickets is investigated. The question is whether the behavior of an individual cricket in a group is controlled solely by available environmental information, or whether crickets need to maintain a memory of the previous loss in a fight. To better address this question, robots possessed either an embedded memory of previous fight or not.

Chemical cues are the major information source used in the cricket decision-making (Nagamoto et al., 2005). On the other hand, the diffusive neuromodulator NO modulates neuronal oscilation in the chemical information-processing center (Gelperin, 1994; Gelperin et al., 1996; Fujie et al., 2002). Neuronal oscillation might be involved in neuronal mechanisms underlying a decision-making whether to attack or to flee. We built a hypothesis that the following three main characteristics are crucial components of the agonistic aggression process: (1) aggression is linked to diffusive neuromodulator NO in the brain; (2) NO changes synaptic connection; and (3) the information governing aggressive behavioral response is coded as an oscillation. Suppose the oscillators comprise an oscillator network, and then the information can be controlled by change in synaptic connection. The strategy for controlling the synchrony of oscillations by the manipulation on the network connection is proposed (Funato and Kurabayashi, 2008). This method allows the oscillator network to encode alternative states, such as aggression and avoidance, by considering synchrony as behavioral states.

The proposed model is shown in Figure 5A where circles are oscillators and the control signal of motor unit is determined by the synchrony of oscillations. New links between oscillators, determined by the level of 'capacitor' on the circuit, play a role in the switching of behavior, which can change after a certain period of a fight. Here, new links mimic neuromodulation that mediates behavior change. Then robot is constructed and behavior is validated before testing the conditions that induce interaction-dependent behavior (Figure 5B). The purpose of the simple model of brain function and robot is to elucidate the major factors controlling adaptive behavior, in particular identifying correlations between brain mechanism and behavior. It is examined if oscillator network robot can induce variations in the number of victorious crickets according to the group size.

A computer simulation of a group of agents whose action rules keep consistencies with the mechanisms of the constructed robot is performed. Four agents are placed in a two-dimensional square field matching the area used in a behaviour experiments in real crickets. The simulated behavior of agents is observed and the number of victorious agents is measured (Figure 6). Each agent behaves by using the rules described for the oscillator network robot, *i.e.*, during agent encounters, each agent either fighting or avoiding, and a change in behavior occurred if the charge level of a 'capacitor' exceeds a threshould. Then, the loser's effect as a key element for group aggression is focused. To simulate the memory of lost at a fight, a mechanism to accelerate the charge speed and reduce the discharge speed after agent selected avoidance behavior are provided. The recovery time is postponed by this operation. The behavior of agent groups equipped with a memory of losses is compared with those whose actions are not affected by memories of past fights. This allows us to examine the effect of the memory of losing on group aggression.

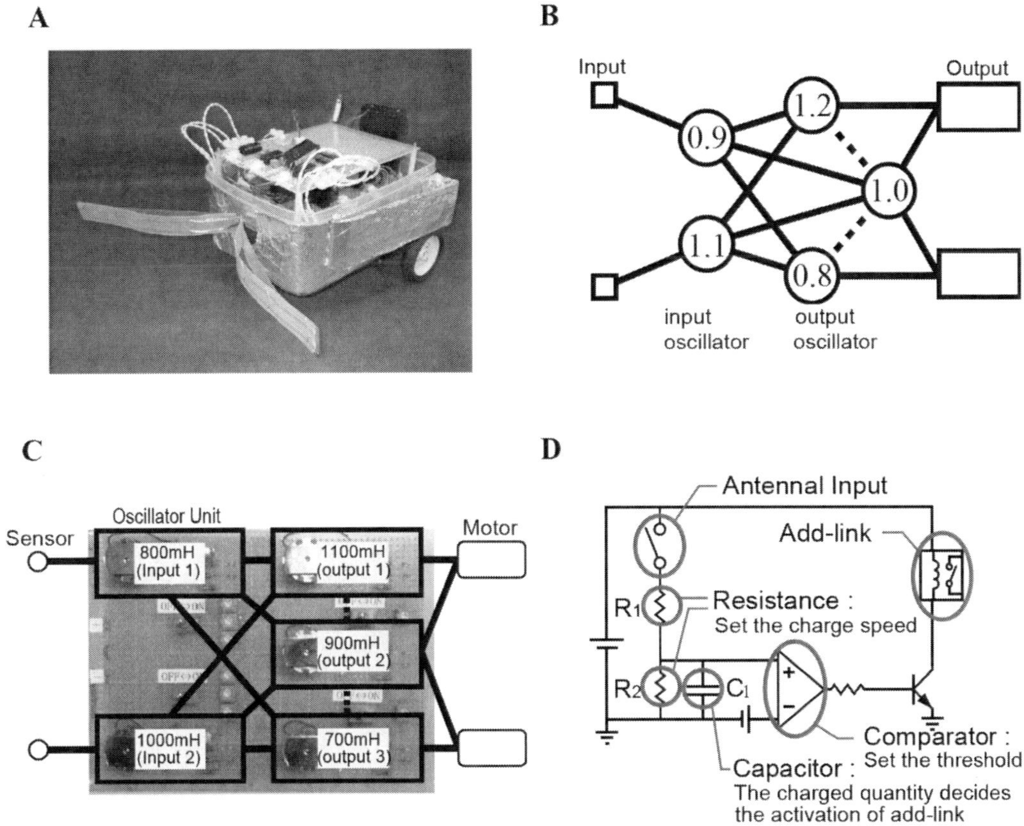

Figure 5. Cricket robots. A: The robot model is composed of an oscillator network that transfers the network structure to robot behavior. The circles represent oscillators and the lines represent connections between oscillators. The robot has two states depending on the existence of cross-links (dotted lines). The activation of motors connected to the oscillators is changed by the synchrony of the oscillators, whilst the synchrony of the oscillators is modified by cross-links. Thus, cross-link connections change robot behavior. The connections indicated as cross-links change several minutes after robot starts fighting, and change the robot behavior. B: Cricket robots. C, D: Robot control circuit. The robot has two motors with wheels, two sensor inputs and a network based on the model of a chemical processing center. The outputs of the oscillators are connected to motors such that they activate only when they are synchronised, making the robot move in the same manner.

Robot model of behavior selection using an oscillator network is then examined. If there are no cross-links, the robot changes its direction and headed towards the obstacle. This behavior corresponds to the cricket behavior of heading towards the opponent, which is regarded as aggressive behavior. If cross-links are present (the neural network is transformed after a fight), the robot heads away from the obstacle. Thus, the robot behaves in a submissive manner as a consequence of the structural change in the network.

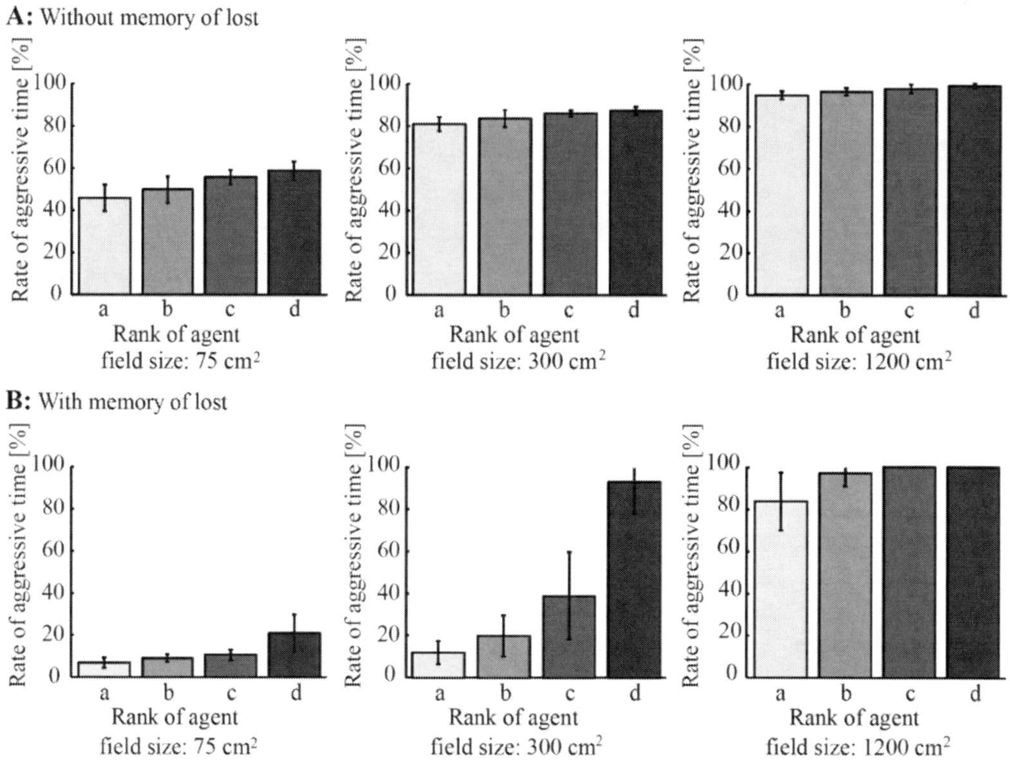

Figure 6. Duration of aggressive behavior (as a percentage of total simulation time) for each agent. The duration of aggressive behavior includes the time that agents behave in an agonistic manner when they encounter other agents; this includes walking and fighting. The simulation is repeated five times with different initial conditions and the bars are shown in order of the rank (1–4) for each trial. Values are means ± s.d. for the five trials. Results for agents without using the memory of their previous fights to decide their behavior, where aggression decreases at an almost equal rate according to the size of field (**A**). Results for agents equipped with memory, where the number of aggressive agents change depending on the field size (**B**). Only one agent is continually aggressive with the medium field ($300cm^2$), every agent is aggressive with the large field ($1200cm^2$), and no agents are aggressive with the small field ($75cm^2$).

This indicates that a change in robot behavior as a consequence of network structure is successfully achieved. The delay circuit alternates the connection after a certain period of a fight, so a robot initially exhibits aggressive behavior by heading towards the obstacle, but then avoid it. Thus, the robot model simulates the individual interaction of crickets when the first contact with another cricket induces fighting, followed by avoidance behavior with network transformation after a certain period. Then the behavioral transitions in simulated agents with variable group density are then examined. Tests of group behavior using four agents on a computer that possess the proposed mechanism implemented in the robots investigate a group behavior of crickets, i.e., the number of aggressive crickets change according to the field size. Performance of the two agents group can be tested using simulation to compare group behavior, i.e., the recovery time of one group is not affected by the result of a fight (without memory) (Figure 6A), whereas the recovery time of the other group is postponed in the defeated agent (with memory) (Figure 6B). Without the memory of lost in a previous fight, no specific agent appears dominant in the simulation and the aggressiveness of every agent is uniformly suppressed by reducing the field size. On the other hand, with the memory of losing in a previous fight, the simulation results demonstrate that only the highest-ranked agent fight for most of the simulation period, whereas the other agents continuously avoid the highest ranked agent. With the large field, every agent fights continuously for most of the simulation time, whereas with the small field they stop fighting. This differentiation among agents that possess a memory of previous lost in a fight is similar to that observed in real cricket groups. This robot experiment and modelling suggest to us that loser's effect plays important role to determine group size dependent aggression in the grouped crickets. In other words, the mechanism of suppression in a loser animal must be a key of social adaptation. Futhermore, local interaction must be a key mechanism for the organization of group behavior in animals. The results of the experiments highlight two main functions in local interactions: the automatic transition of behavior and the memory of the previous experience.

CONCLUTION

Aggressive agonistic behavior in the crickets is described in this chapter. Male cricket exhibits intensive aggression when they encounter conspecific males. A loser cricket in a fight (subordinate) depresses aggression for some time. This phenomenon is the so-called loser's effect. Loser's effect can be a kind of memory. This type of memory is regulated by the internal state of an animal. The internal state is modulated by aminergic control in a brain. In cricket fights, octopaminergic system in the brain regulates internal state to decide whether to continue attacking or to flee. Decreases in the titer of brain OA of the loser cricket leads avoinding conspecific males. An interesteing point is that the internal state is mediated by individual interaction that influences homeostasis of aminergic system in the brain. The loser's effect is a key to understand social adaptability in animals. Group size dependent aggression is reproducted by oscillator network model, where memory of a previous fight and local interaction between individuals are important factors. Thus, function of an internal state in animals must be one of the important factors for understanding memory consolidation. Understanding the neuronal mechanisms underlying aggressive behavior will help us better understand the role of neuronal mechanisms in learing and memory.

REFERENCES

Adamo S, Hoy R. 1995. Agonistic behaviour in male and female field crickets, Gryllus bimaculatus, and how behavioural context influences its expression. *Anim. Behav. Vol.* 49. p 1491-1501.

Adamo SA, Linn CE, Hoy RR. 1995. The role of neurohormonal octopamine during 'fight or flight' behaviour in the field cricket Gryllus bimaculatus. *J. Exp. Biol.* 198(Pt 8):1691-1700.

Ahern GP, Klyachko VA, Jackson MB. 2002. cGMP and S-nitrosylation: two routes for modulation of neuronal excitability by NO. *Trends Neurosci.* 25(10):510-517.

Alexander DR. 1961. Aggressiveness, territoriality, and sexual behavior in field crickets (Orthoptera: Gryllidae). *Behaviour* 17:130-223.

Aonuma H, Iwasaki M, Niwa K. 2004. Role of NO signaling in switching mechanisms in the nervous system of insect. *Proc. SICE Ann. Conf. CD-ROM*:2477-2482 (ISBN 2474-907764-907722-907767).

Aonuma H, Kitamura Y, Niwa K, Ogawa H, Oka K. 2008. Nitric oxide-cyclic guanosine monophosphate signaling in the local circuit of the cricket abdominal nervous system. *Neuroscience.*

Aonuma H, Nagayama T, Takahata M. 2000. Modulatory effects of nitric oxide on synaptic depression in the crayfish neuromuscular system. *J. Exp. Biol.* 203(Pt 23):3595-3602.

Aonuma H, Newland P. 2001. Opposing actions of nitric oxide on synaptic inputs of identified interneurones in the central nervous system of the crayfish. *Journal of Experimental Biology* 204(7):1319-1332.

Aonuma H, Newland P. 2002. Synaptic inputs onto spiking local interneurons in crayfish are depressed by nitric oxide. *Journal of neurobiology* 52(2):144-155.

Aonuma H, Niwa K. 2004. Nitric oxide regulates the levels of cGMP accumulation in the cricket brain. *Acta Biol. Hung.* 55(1-4):65-70.

Aonuma H, Sakura M, Ota J, Asama H. 2009. Social adaptive functions in animals -learning from ansect social behaviors-". Proc of the 2009 *IEEE/RSJ Int. Conf. on Intelligent. Robots and Systems* (Workshops/ Tutorials, CD):10-15.

Aonuma H, Watanabe T. 2012. Octopaminergic system in the brain controls aggressive motivation in the ant, Formica japonica. *Acta Biol. Hung.* 63 Suppl. 2:63-68.

Ashikaga M, Sakura M, Kikuchi M, Hiraguchi T, Chiba R, Aonuma H, Ota J. 2009. Establishment of Social Status without Individual Discrimination in the Cricket. *Advanced Robotics* 23(5):563-578.

Barahona M, Pecora LM. 2002. Synchronization in small-world systems. *Physical review letters* 89(5):054101.

Bender AT, Beavo JA. 2006. Cyclic nucleotide phosphodiesterases: molecular regulation to clinical use. *Pharmacological reviews* 58:488-520.

Bicker G. 2001. Sources and targets of nitric oxide signalling in insect nervous systems. *Cell and tissue research* 303(2):137-146.

Bovbjerg RV. 1953. Dominance order in the crayfish *Orconectes virilis* (Hagen). *Physiol. Zool.* 26:173-178.

Bredt DS, Synder SH. 1989. Nitric oxide mediates glutamate-linked enhancement of cGMP levels in the cerebellum. *Proc. Natl. Acad. Sci. USA* 86:9030-9033.

Estevez I, Keeling LJ, Newberry RC. 2003. Decreasing aggression with increasing group size in young domestic fowl. *Appl. Anim. Behav. Sci.* 84:213-218.

Fujie S, Aonuma H, Ito I, Gelperin A, Ito E. 2002. The nitric oxide/cyclic GMP pathway in the olfactory processing system of the terrestrial slug *Limax marginatus*. *Zoological science* 19(1):15-26.

Funato T, Kurabayashi D. 2008. Network structure for control of coupled multiple nonlinear oscillators. *IEEE Trans Syst. Man. Cybern. B Cybern.* 38(3):675-681.

Funato T, Kurabayashi D, Nara M, Aonuma H. 2008. Switching mechanism of sensor-motor coordination through an oscillator network model. *IEEE Trans Syst. Man Cybern. B Cybern.* 38(3):764-770.

Funato T, Nara M, Kurabayashi D, Ashikaga M, Aonuma H. 2011. A model for group-size-dependent behaviour decisions in insects using an oscillator network. *J. Exp. Biol.* 214(Pt 14):2426-2434.

Gelperin A. 1994. Nitric oxide mediates network oscillations of olfactory interneurons in a terrestrial mollusc. *Nature* 369(6475):61-63.

Gelperin A, Kleinfeld D, Denk W, Cooke I. 1996. Oscillations and gaseous oxides in invertebrate olfaction. *Journal of neurobiology* 30(1):110-122.

Hack MA. 1997. Assessment strategies in the contests of male crickets, Acheta domesticus (L.). *Anim. Behav.* 53:733-747.

Hoffmann HA, Schildberger K. 2001. Assessment of strength and willingness to fight during aggressive encounters in crickets. Anim Behav 62:337-348.

Hofmann HA, Stevenson PA. 2000. Flight restores fight in crickets. *Nature* 403(6770):613.

Iwasaki M, Nishino H, Delago A, Aonuma H. 2007. Effects of NO/cGMP signaling on behavioral changes in subordinate male crickets, Gryllus bimaculatus. *Zoological science* 24(9):860-868.

Johansson KU, Mellon D, Jr. 1998. Nitric oxide as a putative messenger molecule in the crayfish olfactory midbrain. *Brain research* 807(1-2):237-242.

Kaupp UB, Seifert R. 2002. Cyclic nucleotide-gated ion channels. *Physiol. Rev.* 82:769-824.

Kawabata K, Fujii T, Aonuma H, Suzuki T, Ashikaga M, Ota J, Asama H. 2012. A neuromodulation model of behavior selection in the fighting behavior of male crickets. *Robotics and Autonomous Systems* 60(5):707-713.

Laurent G, Naraghi M. 1994. Odorant-induced oscillations in the mushroom bodies of the locust. *J. Neurosci.* 14(5 Pt 2):2993-3004.

Matsumoto Y, Unoki S, Aonuma H, Mizunami M. 2006. Critical role of nitric oxide-cGMP cascade in the formation of cAMP-dependent long-term memory. *Learn Mem.* 13(1): 35-44.

Mizuno T, Sakura M, Ashikaga M, Aonuma H, Chiba R, Ota J. 2012. Model of a sensory-behavioral relation mechanism for aggressive behavior in crickets. *Robotics and Autonomous Systems* 60(5):700-706.

Moncada S, Palmer RMJ, Higgs EA. 1991. Nitric oxide: physiology, phathophysiology, and pharmacology. *Pharmacol. Rev.* 43(2):109-142.

Mothet JP, Fossier P, Tauc L, Baux G. 1996. NO decreases evoked quantal ACh release at a synapse of Aplysia by a mechanism independent of Ca2+ influx and protein kinase G. *Journal of Physiology* - London 493(3):769-784.

Müller U. 1997. The nitric oxide system in insects. *Prog. Neurobiol.* 51(3):363-381.

Nagamoto J, Aonuma H, Hisada M. 2005. Discrimination of conspecific individuals via cuticular pheromones by males of the cricket *Gryllus bimaculatus*. *Zoological science* 22(10):1079-1088.

Pagel M, Dawkins MS. 1997. Peck orders and group size in laying hens: 'futures contracts' for non-aggression. *Behav. Processes* 40:13-15.

Patullo BW, Baird HP, Macmillan DL. 2009. Altered aggression in different sized groups of crayfish supports a dynamic social behaviour model. *Appl. Anim. Behav. Sci.* 120: 231-237.

Phillips LH, Konishi M. 1973. Control of aggression by singing in crickets. *Nature* 241: 64-65.

Qazi S, Trimmer BA. 1999. The role of nitric oxide in motoneuron spike activity and muscarinic-evoked changes in cGMP in the CNS of larval Manduca sexta. *J. Comp. Physiol.* [A] 185(6):539-550.

Rillich J, Schildberger K, Stevenson PA. 2007. Assessment strategy of fighting crickets revealed by manipulating information exchange. *Animal Behaviour* 74(4):823-836.

Ryan KM, Sakaluk SK. 2009. Dulling the senses: the role of the antennae in mate recognition, copulation and mate guarding in decorated crickets. *Anim. Behav.* 77:1345-1350.

Sakura M, Aonuma H. 2013. Aggressive behavior in the antennectomized male cricket Gryllus bimaculatus. *J. Exp. Biol.* 216(Pt 12):2221-2228.

Schadow J, Lenz D, Thaerig S, Busch NA, Frund I, Herrmann CS. 2007. Stimulus intensity affects early sensory processing: sound intensity modulates auditory evoked gamma-band activity in human EEG. International journal of psychophysiology : *official journal of the International Organization of Psychophysiolo*gy 65(2):152-161.

Seki Y, Aonuma H, Kanzaki R. 2005. Pheromone processing center in the protocerebrum of Bombyx mori revealed by nitric oxide-induced anti-cGMP immunocytochemistry. *The Journal of comparative neurology* 481(4):340-351.

Simmons LW. 1986. Inter-male competition and mating success in the field cricket, Gryllus bimaculatus (De Geer). *Anim. Behav.* 34:567-579.

Stevenson PA, Dyakonova V, Rillich J, Schildberger K. 2005. Octopamine and experience-dependent modulation of aggression in crickets. *J. Neurosci.* 25(6):1431-1441.

Tanaka NK, Ito K, Stopfer M. 2009. Odor-evoked neural oscillations in Drosophila are mediated by widely branching interneurons. *J. Neurosci.* 29(26):8595-8603.

Tregenza T, Wedell N. 1997. Definitive evidence for cuticular pheromones in a cricket. *Anim. Behav.* 54(4):979-984.

Wilson CH, Christensen TA, Nighorn AJ. 2007. Inhibition of nitric oxide and soluble guanylyl cyclase signaling affects olfactory neuron activity in the moth, Manduca sexta. *J. Comp. Physiol. A Neuroethol. Sens. Neural Behav. Physiol.* 193(7):715-728.

In: Memory Consolidation
Editors: Manabu Sakakibara and Etsuro Ito
ISBN: 978-1-63482-596-2

Chapter 4

CONSOLIDATION AND RECONSOLIDATION OF OLFACTORY MEMORY IN THE TERRESTRIAL PULMONATES

Ryota Matsuo*
International College of Arts and Sciences,
Department of Environmental Sciences,
Fukuoka Women's University, Japan

ABSTRACT

Pulmonate gastropods living terrestrial lives (land slugs and snails) have a remarkable ability to acquire a long-term olfactory memory by a single conditioning session. Their brains are equipped with a pair of specialized structures called procerebrum (PC). The PC contains numerous small interneurons that are involved in the higher olfactory functions such as odor discrimination and olfactory associative learning. Time-dependent change of the memory state has long been studied at the behavioral level using the terrestrial slug *Limax* since many years ago. *Limax* will serve as an ideal model system for the study of memory consolidation and reconsolidation because of (1) the simplicity of the brain, (2) the easiness of the behavioral study, (3) the availability of inexhaustible number of slugs as experimental animal, and (4) the feasibility of *in vitro* conditioning system.

Keywords: olfactory memory, *Limax*, procerebrum, cooling-induced amnesia, protein synthesis

INTRODUCTION

Neurobiologists have used gastropods for the study of memory, neuronal plasticity, neural regeneration, and phototaxis/chemotaxis behaviors because of the simplicity and

* E-mail: matsuor@fwu.ac.jp.

robustness of their nervous system and behaviors. A multiple series of studies using *Aplysia* have revealed cellular and molecular bases of non-associative and associative memory during several decades. The pond snail *Lymnaea* has served as a model animal for the study of various forms of operant conditioning. The simplicity of its nervous system allowed the researchers to examine the mechanism of memory at the level of a restricted number of synapses.

On the other hand, the terrestrial pulmonates exhibit a very sophisticated type of memory, which is unique to the terrestrial animals because of the involvement of olfaction.

The terrestrial slugs (*Limax*, Figure 1) and snails (*Helix*) have evolved a peculiar neuronal structure called procerebrum (PC) within its brain, as well as the tentacles that function as a versatile multimodal sensory organ (olfaction, vision, somatosensation). The PC densely receives olfactory inputs, and has been considered to play important roles in the olfactory information processing and memory formation. In this chapter, I firstly overview the past studies on olfactory learning and memory in *Limax* and *Helix*, and discuss the role of the PC. Later, I will review the current knowledge about the memory consolidation and reconsolidation in these animals, with a special focus on the cooling-induced amnesia in *Limax*.

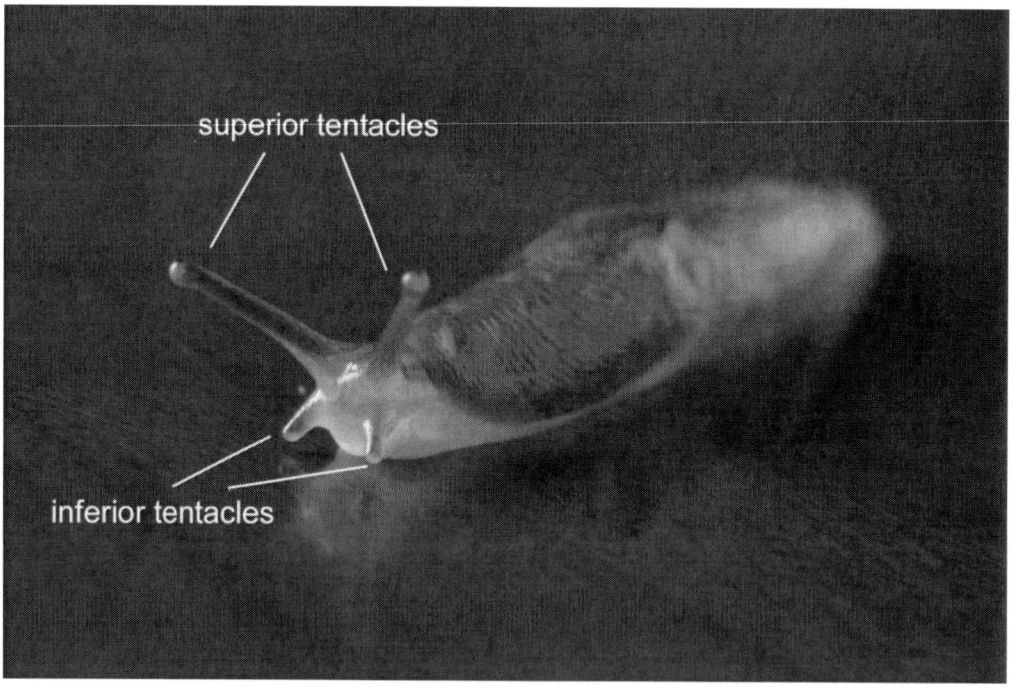

Figure 1. Tentacles function as a multimodal sensory organ.

OLFACTORY CENTER AND OLFACTORY MEMORY IN PULMONATES

The brain of terrestrial pulmonates (*Limax*, *Helix*, and *Achatina*) is equipped with a specialized structure called PC (Figure 2), which is absent from the brain of aquatic pulmonates, such as *Lymnaea* or *Melampus* (Chase, 2000). This structure contributes to olfactory function, and has dense connections with superior and inferior tentacles which have olfactory epithelium on the tip (Chase and Tolloczko, 1993; Kawahara et al., 1997; Matsuo et al., 2011). Almost half the number of the total neurons are located in the PC (Gelperin and Tank, 1990; Chase, 2000). They are interneurons devoted to higher olfactory functions. Involvement of so many neurons in the olfactory information processing would be reasonable because olfaction is the most important sensory modality for the slug's daily life.

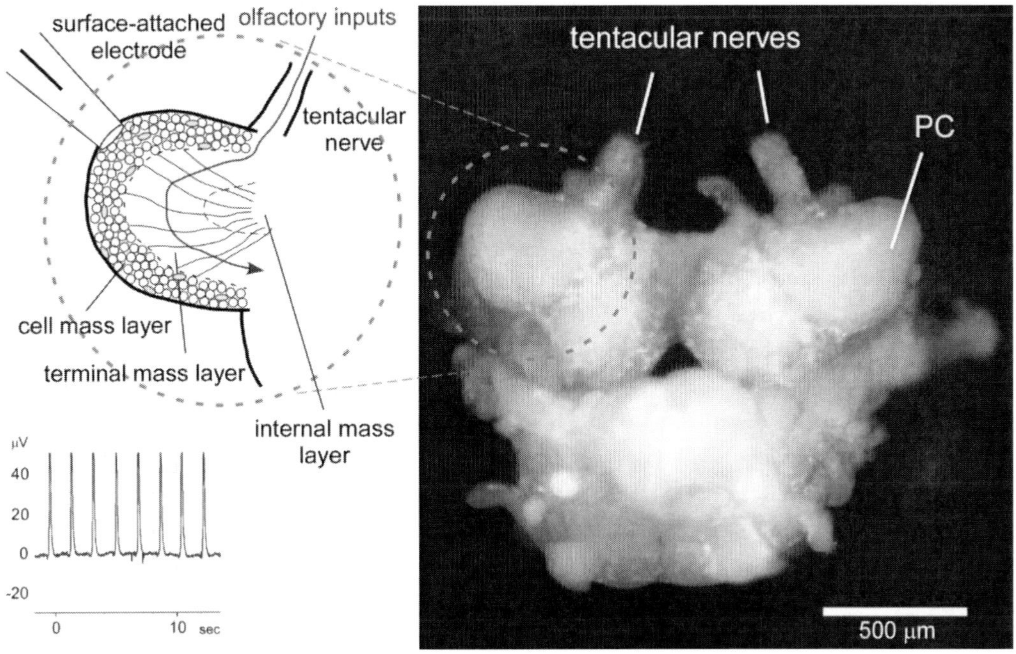

Figure 2. Brain of terrestrial pulmonates equipped with structure called PC.

It has been demonstrated that the terrestrial slugs *Limax* turn to avoid the odor of some food once the food odor is paired with an aversive stimulus such as a bitter taste of quinidine solution. This associative learning can be acquired by a single pairing session (Gelperin, 1975; Sahley et al., 1981a). Moreover, they can learn more complex logical relationships between the stimuli, such as second-order conditioning and blocking (Sahley et al., 1981b).

The reports on such exquisite capabilities of learning were followed by the discovery of the rhythmical wave activity of the local field potential (LFP) recorded on the surface of the PC (Gelperin and Tank, 1990). The LFP oscillation is produced by the synchronous activities of the constituent interneurons (~10^5 neurons in the PC). A minority of neurons (bursting neurons) have local inhibitory projections to the majority neurons (nonbursting neurons), and the LFP reflects the locally synchronized inhibitory currents in the nonbursting neurons evoked by the input from the bursting neurons (Kleinfeld et al., 1994; Watanabe et al., 2008). The LFP oscillation is diminished by pharmacologically suppressing the production of nitric oxide (NO) within the PC, although its mechanism of action has not been fully understood yet. The synchronized activity of the PC neurons is not necessary for odor recognition, but is necessary for the discrimination of similar odors, as revealed by the use of nitric oxide synthase inhibitors (Teyke and Gelperin, 1999; Sakura et al., 2004).

The wave of the LFP oscillation travels from the apical end of the PC to the basal end along the longitudinal axis, and the phase difference between the apical and the basal regions during the wave propagation can be modulated by an odor application to the olfactory epithelium on the tentacle (Delaney et al., 1994; Gervais et al., 1996). Particularly, the changes in the oscillatory frequency is an important indicator of the learning-related activity of the PC (Gervais et al., 1996; Kimura et al., 1998a; 1998c; Toda et al., 2000; Samarova and Balaban, 2009). It has also been reported in *Helix* that application of aversively-conditioned odor produces a difference in the amplitude of LFP waves between the apical and the basal regions of the PC (Nikitin and Balaban, 2000).

It has not been fully understood how the oscillatory activity of the PC neurons contributes to odor discrimination and olfactory learning. Several theoretical models, however, have been proposed (Gelperin, 1999; Ermentrout et al., 2001; Sekiguch et al., 2010). Unlike an insect olfactory center (antennal lobe) or a vertebrate olfactory bulb, there seems to be no built-in structural compartment (such as glomeruli) in the PC. The wave propagation, therefore, may serve to assign the space within the seemingly uniform mass of the small interneurons in the PC for the representation of specific odors in a timing-dependent manner (Gelperin, 1999). In fact, the input efficacy of the olfactory afferents to the PC neurons depends on the phase of the LFP oscillation (Inoue et al., 2000).

Involvement of the PC in olfactory learning has also been suggested from the study that used a fluorescent molecule Lucifer Yellow as a neuronal activity marker (Wilcox and Franceschini, 1984). When this marker molecule was injected into the body cavity of *Limax* after the conditioning session, it specifically labeled a cluster of neurons in the PC aligned in the shape of a band. But when a time interval was given between the conditioned stimulus (CS, e.g., food odor) and the unconditioned stimulus (US, e.g., quinidine solution), the associative memory was not formed because of the lack of temporal contiguity, and at the same time, no labeling occurred (Kimura et al., 1998b; Ermentrout et al., 2001; Sekiguchi et al., 2010). According to the mathematical model by Ermentrout (2001), the formation of such a "memory band" is enabled by the transition of the LFP oscillation from apical-to-basal propagation to synchronization throughout the PC (Gelperin, 1999; Ermentrout et al., 2001). In fact, the odor-evoked phase collapse (i.e., synchronization) of LFP oscillation has been demonstrated in *Limax* (Delaney et al., 1994; Gervais et al., 1996).

Another more direct approach also proved the importance of the PC for olfactory learning. The retention and/or retrieval of olfactory aversive memory was abolished by the surgical ablation of the PC (Kasai et al. 2006; Matsuo et al., 2010a). The PC ablation, however, did not completely abolish the ability to recognize odors (Kasai et al., 2006). Further experiments that exploited the recoverability of the PC from surgical destruction suggested that the PC is the locus of memory storage (Matsuo et al. 2010b). The PC seems, in theory, to serve as a coincidence detector because it receives the inputs of aversive stimulus possibly transmitted through serotonergic nerves (Osborne and Cottrell, 1971; Inoue et al., 2004; Elekes et al., 2013) as well as olfactory inputs via tentacular nerves (Kawahara et al., 1997; Kimura et al., 1998b). The changes of synaptic strength between the PC neurons or between the olfactory inputs and the PC neurons might be the cellular basis for olfactory memory. Further studies are awaited to prove this idea.

Thus far the knowledge on the output route from the PC is relatively scarce. In the snail *Achatina fulica*, long dendrites of the neurons in the pedal ganglion have direct interactions with the PC neurons, suggesting the possibility that the activity of the PC affects the locomotion of the snail (Chase and Tolloczko, 1989). In *Limax*, a neuron in the metacerebrum was identified that is likely to receive inputs from the PC interneurons (Shimozono et al., 2001). The posterior visceral neuron (*p*-VN) was reported to exhibit correlative firing activity with the change in the oscillatory frequency of the LFP. The *p*-VN commands mantle contraction, which is one of the characteristic aversive behaviors observed in the conditioned slugs (Inoue et al., 2004; 2006). In *Helix*, a cerebral motor neuron MtC3 that elicits tentacle contraction was reported to receive phasic inputs synchronous to the LFP oscillation of the PC, suggesting the direct involvement of this neuron in the olfaction-driven motor control of the tentacles in pulmonates (Nikitin et al., 2005).

Memory Consolidation in Pulmonates

Once established, olfactory aversive memory persists for several weeks in *Limax* (Matsuo et al., 2002; 2010b). Memory persistence for over 2 days requires protein synthesis during and/or just following the conditioning (Matsuo et al., 2002). As such, the long-term memory is defined as the memory of over 2 days, and the short-term memory is defined as that of less than 2 days in *Limax*. In other animals, such as rat, *Aplysia* or *Drosophila*, the protein synthesis-independent short-term memory is much shorter (Flood et al., 1972; Montarolo et al., 1986; Tully et al., 1994): the amnesiac effect of protein synthesis inhibition appears earlier than 2 days post conditioning. In *Limax*, moreover, the persistence of short-term memory is dependent on the extent of protein synthesis during the memory formation. If the protein synthesis inhibition is insufficient, the length of the short term memory is prolonged accordingly (Yasui et al., 2004).

In mammals, hippocampus-dependent memory is stabilized through two qualitatively distinct processes. On the order of hours, memory undergoes "synaptic consolidation" gradually becoming resistant to external disturbances. On the order of months (or even years), memory undergoes "systems consolidation" where the memory gradually becomes hippocampus-independent (also called "remote memory", Kim and Fanselow, 1992; Frankland and Bontempi, 2005; Nader and Einarsson, 2010). Since it is highly probable that the procerebrum is the sole locus for the storage of olfactory memory in *Limax* (Matsuo et al., 2010b), only the synaptic consolidation is considered thereafter.

Until the memory is consolidated, it is labile and prone to be disrupted by external disturbances. During the early phase of the memory consolidation, odor-aversion memory is disrupted by cooling the slug to 1°C for 5 min in *Limax* (Yamada et al., 1992). Cooling can effectively cause amnesia only if it is given within less than 2 min after the conditioning. A similar property of cooling-induced amnesia has been reported in the operant conditioning of *Lymnaea* (Sangha et al., 2003a; 2003b). Although it seems that the cooling and protein synthesis inhibition are considered equivalent interventions at the molecular level, hypothermia will affect multiple biochemical processes other than protein synthesis. In addition to the attenuation of protein and mRNA synthesis, their degradation also stops (Roobol et al., 2009). Furthermore, several stress-response genes and proteins are induced by cold shock (Odani et al., 2003; Al-Fageeh et al., 2006). The biophysical properties of lipid membranes and the structure of cytoskeleton are also changed (Aloia and Raison, 1989; Roobol et al., 2009). Potentially all the molecular functions are under the influence of temperature, and may be greatly affected by cold stress. Therefore, the relation between the cooling and the protein synthesis inhibition is not so straightforward as expected. It is recently argued that these two operations affect different cellular processes involved in the memory consolidation because the effective time windows were not mutually overlapping (Fulton et al., 2005; 2008). In the study by Yamada et al. (1992), the memory retention tests were performed on the following day of the conditioning/cooling. Since the 24 hr memory is within the range of the short-term memory in *Limax* with no requirement for protein synthesis at the time of conditioning (Matsuo et al., 2002), it is unlikely that the cooling-induced amnesia was caused solely by the protein synthesis inhibition. Probably, the cooling inhibits some critical molecular steps necessary for the consolidation of memory, such as the kinase cascades that connect neuronal activity to the translational up-regulation of LTM-related genes (Fulton et al., 2008). At the same time, these kinase activity may contribute to the formation of short-term memory through the functional change of synaptic proteins by covalent modifications (Kandel, 2012). Therefore, it is reasonable that the earlier, protein synthesis-independent phase of memory (up to 24 hr) was abolished by cooling (Yamada et al., 1992).

It is largely unknown what type of genes and/or proteins is necessary for the memory consolidation in pulmonates. In an aquatic pulmonate *Lymnaea*, expression of several genes are altered in the brain after taste-aversion learning in a CREB-dependent manner (Azami et al., 2006), and the amount of various proteins changes after pneumostome opening operant conditioning (Silverman-Gavrila et al., 2011). In *Limax*, activity-dependent induction of several genes has been reported in the PC (Fukunaga et al., 2006), and the serotonin-induced change in the protein synthesis has also been observed in the PC (Yamane et al., 1989). It is intriguing to know whether the similar set of genes are activated in the PC neurons as in the sensori-motor system of *Aplysia*, which is a well-studied model system in the research of synaptic plasticity (Bailey et al., 2004; Lee et al., 2008).

Memory Reconsolidation in Pulmonates

In 2000, Nader and colleagues reported a phenomenon of memory reconsolidation failure in tone-dependent fear conditioning in rats (Nader et al., 2000). They found that the animals exhibit amnesia 24 hr after the re-exposure to the CS (tone) when it is combined with the injection of anisomycin into the basolateral amygdala where the fear memory is thought to be formed and stored. This effect was evident even at 14 days after the initial conditioning. Similarly, intra-hippocampal injection of anisomycin just following the return to the conditioning chamber caused amnesia in the context-dependent fear conditioning paradigm that is dependent on the hippocampus (Debiec et al., 2002). These findings drew attention of many neuroscientists, and the reports on similar observations followed in other animal species using various learning paradigms in the succeeding years (Pedreira et al., 2002; Child et al., 2003; Sangha et al., 2003; Gainutdinova et al., 2005). The state of memory is now considered to become unstable by the presentation of the CS (or CS + US) even after the initial consolidation process has already completed.

The report of Nader et al. (2000) gave a strong impact in that they used protein synthesis inhibition in a specific brain area (amygdala) as a disturbing operation. However, reactivation-dependent labilization of the consolidated memory had been reported repeatedly long before their work (reviewed in Sara et al. 2000), although different types of disturbance were adopted to induce retrograde amnesia, such as an electroconvulsive shock (Misanin et al., 1968), rapid cooling (Mactutus et al., 1979), or a systemic injection of anisomycin (Judge and Quartermain, 1982). Indeed, in *Limax*, such a reactivation-dependent amnesia has been reported by Yamada et al. (1992). They found a cooling-induced amnesia in *Limax* that was induced by the re-conditioning (CS + US) followed by an immediate cooling. The effectiveness of this procedure is dependent on the interval between the initial conditioning and re-conditioning. Amnesiac effect was evident only if the re-conditioning/cooling was given within 3 days after the initial food-aversion conditioning, and gradually declined with the increase in this interval. Further studies conducted by the same group revealed that the re-exposure to the CS is enough to induce a cooling-induced retrograde amnesia (Sekiguchi et al., 1997). Here again, the amnesiac effect of cooling exhibited similar dependence on the interval between the initial conditioning and the re-exposure/cooling as in the case of re-conditioning/cooling paradigm. We should keep in mind again that it is elusive whether the cooling and protein synthesis inhibition are equivalent interventions in the molecular cascade leading to the memory reconsolidation. Because an odor-aversion memory can be established by a single pairing of the CS and US in *Limax*, the transition of the internal memory state can be analyzed with a precise time resolution. Therefore, the food-aversion learning of *Limax* is a classical and useful model system for the behavioral study of memory reconsolidation.

The reconsolidation failure is caused by unnaturalistic operations, such as an injection of protein synthesis inhibitors or a rapid cooling of the whole body to 1°C, and seems not to occur in the animals' daily lives. The physiological role of memory reconsolidation has been elusive, and long been debated. Recently, reconsolidation is supposed to have a "memory updating" function, where the stored memory undergoes a minor modification when the animal encounters a similar but slightly different situation after the initial learning (Lee, 2009).

Dudai and Eisenberg (2004) also proposed that the memory consolidation is much slower process than ever thought and that reconsolidation serves in making a new link to other neural representations that should be integrated into the already existing memory trace until it is completely stabilized. According to the notion of Dudai and Eisenberg (2004), it is a prerequisite to use a whole animal or a whole brain system to investigate the mechanism of reconsolidation because the functional role of reconsolidation is relating and connecting the recurrent neuronal network activities distributed within the various parts of the brain.

In contrast, recent reports by Lee et al. (2012) and Cai et al. (2012) showed cellular correlates of memory reconsolidation in siphon-withdrawal reflex of *Aplysia* using a simple *in vitro* preparation of sensori-motor system. They could demonstrate that the reconsolidation failure of synaptic facilitation in the identified synapse successfully occurred when protein synthesis inhibitor was applied in combination with a reminder stimulation (a tail shock or application of 5-HT). These findings indicate that it is feasible to study the molecular/cellular mechanisms of memory reconsolidation in a reduced preparation separated from a whole animal.

In the future, the cellular properties of reconsolidation of olfactory memory in pulmonates, which will be more complex than that of sensitization in siphon-withdrawal reflex of *Aplysia*, may become able to be studied using an *in vitro* learning system developed by Inoue et al. (2004; 2006).

Conclusions and Perspectives

The research on memory consolidation and reconsolidation has been one of the hottest areas in neuroscience. Associative olfactory learning in pulmonates will serve as an ideal model system because the olfactory memory can be easily formed by a single CS-US pairing, making it possible to analyze the temporal transition of the internal memory state with a fine time resolution. Moreover, their nervous system is robust and can function normally in *in vitro* preparation, which will be amenable to artificial interventions at the cellular level. Finally, availability of a lot of animals is one important point that should not be overlooked. The use of enough number of animals ensures the reliable behavioral experiments. In my laboratory, for example, as many as more than 1000 slugs (*Limax valentianus*) are maintained as a closed colony all the time, and the maintenance of them is relatively easy and does not cost too much because of the slugs' vigorous reproductive power.

References

Al-Fageeh, M. B., Smales, C. M., (2006). Control and regulation of cellular responses to cold shock: the responses in yeast and mammalian systems. *Biochem. J.,* 397, 247-59.

Aloia, R. C., Raison, J. K., (1989). Membrane function in mammalian hibernation. *Biochim. Biophys. Acta.,* 988, 123-46.

Azami, S., Wagatsuma, A., Sadamoto, H., Hatakeyama, D., Usami, T., Fujie, M., Koyanagi, R., Azumi, K., Fujito, Y., Lukowiak., K., Ito, E., (2006). Altered gene activity correlated with long-term memory formation of conditioned taste aversion in *Lymnaea. J. Neurosci. Res.*, 84, 1610-20.

Bailey, C. H., Kandel, E. R., Si, K., (2004). The persistence of long-term memory: a molecular approach to self-sustaining changes in learning-induced synaptic growth. *Neuron,* 44, 49-57.

Cai, D., Pearce, K., Chen, S., Glanzman, D. L., (2012). Reconsolidation of long-term memory in *Aplysia. Curr. Biol.*, 22, 1783-8.

Chase, R., (2000). Structure and function in the cerebral ganglion. *Microsc. Res. Tech.,* 49, 511-20.

Chase, R., Tolloczko, B., (1989). Interganglionic dendrites constitute an output pathway from the procerebrum of the snail *Achatina fulica. J. Comp. Neurol.,* 283, 143-52.

Chase, R., Tolloczko, B., (1993). Tracing neural pathways in snail olfaction: from the tip of the tentacles to the brain and beyond. *Microsc. Res. Tech.,* 24, 214-30.

Child, F. M., Epstein, H. T., Kuzirian, A. M., Alkon, D. L., (2003). Memory reconsolidation in *Hermissenda. Biol. Bull.,* 205, 218-9.

Debiec, J., LeDoux, J. E., Nader, K., (2002). Cellular and systems reconsolidation in the hippocampus. *Neuron,* 36, 527-38.

Delaney, K. R., Gelperin, A., Fee, M. S., Flores, J. A., Gervais, R., Tank, D. W., Kleinfeld, D., (1994). Waves and stimulus-modulated dynamics in an oscillating olfactory network. *Proc. Natl. Acad. Sci. U.S.A.,* 91, 669-73.

Dudai, Y., Eisenberg, M., (2004). Rites of passage of the engram: reconsolidation and the lingering consolidation hypothesis. *Neuron,* 44, 93-100.

Elekes, K., Battonyai, I., Kobayashi, S., Ito, E. (2013). Organization of the procerebrum in terrestrial pulmonates (*Helix, Limax*) reconsidered: cell mass layer synaptology and its serotonergic input system. *Brain Struct. Funct.,* 218, 477-90.

Ermentrout, B., Wang, J. W., Flores, J., Gelperin, A., (2001). Model for olfactory discrimination and learning in *Limax* procerebrum incorporating oscillatory dynamics and wave propagation. *J. Neurophysiol.,* 85, 1444-52.

Flood, J. F., Bernnett, E. L., Rosenzweig, M. R., Orme, A. E., (1972). Influence of training strength on amnesia induction by pretraining injections of cycloheximide. *Physiol. Behav.,* 9, 589-600.

Frankland, P. W., Bontempi, B., (2005). The organization of recent and remote memories. *Nat. Rev. Neurosci.,* 6, 119-30.

Fukunaga, S., Matsuo, R., Hoshino, S., Kirino, Y., (2006). Novel-krüppel like factor is induced by neuronal activity and by sensory input in the central nervous system of the terrestrial slug *Limax valentianus. J. Neurobiol.,* 66, 169-81.

Fulton, D., Kemenes, I., Andrew, R. J., Benjamin, P. R., (2005). A single time-window for protein synthesis-dependent long-term memory formation after one-trial appetitive conditioning. *Eur. J. Neurosci.,* 21, 1347-58.

Fulton, D., Kemenes, I., Andrew, R. J., Benjamin, P. R., (2008). Time-window for sensitivity to cooling distinguishes the effects of hypothermia and protein synthesis inhibition on the consolidation of long-term memory. *Neurobiol. Learn. Mem.,* 90, 651-4.

Gelperin, A., (1975). Rapid food-aversion learning by a terrestrial mollusk. *Science,* 189, 567-70.

Gelperin, A., (1999). Oscillatory dynamics and information processing in olfactory systems. *J. Exp. Biol.,* 202, 1855-64.

Gelperin, A., Tank D. W., (1990). Odour-modulated collective network oscillations of olfactory interneurons in a terrestrial mollusc. *Nature,* 345, 437-40.

Gervais, R., Kleinfeld, D., Delaney, K. R., Gelperin, A., (1996). Central and reflex neuronal responses elicited by odor in a terrestrial mollusk. *J. Neurophysiol.,* 76, 1327-39.

Gainutdinova, T. H., Tagirova, R. R., Ismailova, A. I., Muranova, L. N., Samarova, E. I., Gainutdinov, K. L., Balaban, P. M., (2005). Reconsolidation of a context long-term memory in the terrestrial snail requires protein synthesis. *Learn. Mem.,* 12, 620-5.

Inoue, T., Watanabe, S., Kawahara, S., Kirino, Y., (2000). Phase-dependent filtering of sensory information in the oscillatory olfactory center of a terrestrial mollusk. *J. Neurophysiol.,* 84, 1112-5.

Inoue, T., Inokuma, Y., Watanabe, S., Kirino, Y., (2004). In vitro study of odor-evoked behavior in a terrestrial mollusk. *J. Neurophysiol.,* 91, 372-81.

Inoue, T., Murakami, M., Watanabe, S., Inokuma, Y., Kirino, Y., (2006). In vitro odor-aversion conditioning in a terrestrial mollusk. *J. Neurophysiol.,* 95, 3898-903.

Judge, M. E., Quartermain, D., (1982). Characteristics of retrograde amnesia following reactivation of memory in mice. *Physiol. Behav.,* 28, 585-90.

Kandel, E. R., (2012). The molecular biology of memory: cAMP, PKA, CRE, CREB-1, CREB-2, and CPEB. *Mol. Brain,* 5, 14.

Kasai, Y., Watanabe, S., Kirino, Y., Matsuo, R., (2006). The procerebrum is necessary for odor-aversion learning in the terrestrial slug Limax valentianus. *Lean. Mem.,* 13, 482-8.

Kawahara, S., Toda, S., Suzuki, Y., Watanabe, S., Kirino, Y., (1997). Comparative study on neural oscillation in the procerebrum of the terrestrial slugs *Incilaria bilineata* and *Limax marginatus*. *J. Exp. Biol.,* 200, 1851-61.

Kimura, T., Toda, S., Sekiguchi, T., Kirino, Y., (1998a). Behavioral modulation induced by food odor aversive conditioning and its influence on the olfactory response of an oscillatory brain network in the slug *Limax marginatus*. *Learn. Mem.,* 4, 365-75.

Kimura, T., Suzuki, H., Kono, E., Sekiguchi, T., (1998b). Mapping of interneurons that contribute to food aversive conditioning in the slug brain. *Learn. Mem.,* 4, 376-88.

Kimura, T., Toda, S., Sekiguchi, T., Kirino, Y., (1998c). Optical recording analysis of olfactory response of the procerebral lobe in the slug brain. *Learn. Mem.,* 4, 389-400.

Kleinfeld, D., Delaney, K. R., Fee, M. S., Flores, J. A., Tank, D. W., Gelperin, A. (1994). Dynamics of propagating waves in the olfactory network of a terrestrial mollusk: an electrical and optical study. *J. Neurophysiol.,* 72, 1402-19.

Lee, Y. S., Bailey, C. H., Kandel, E. R., Kaang, B. K., (2008). Transcriptional regulation of long-term memory in the marine snail *Aplysia*. *Mol. Brain,* 1, 3.

Lee, J. L. C., (2009). Reconsolidation: maintaining memory relevance. *Trends Neurosci.,* 32, 413-20.

Lee, S. H., Kwak, C., Shim, J., Kim, J. E., Choi, S. L., Kim, H. F., Jang, D. J., Lee, J. A., Lee, K., Lee, C. H., Lee, Y. D., Miniaci, M. C., Bailey, C. H., Kandel, E. R., Kaang, B. K., (2012). A cellular model of memory reconsolidation involves reactivation-induced destabilization and restabilization at the sensorimotor synapse in *Aplysia*. *Proc. Natl. Acad. Sci. U.S.A.,* 109, 14200-5.

Mactutus, C. F., Riccio, D. C., Ferek, J. M., (1979). Retrograde amnesia for old (reactivated) memory: some anomalous characteristics. *Science,* 204, 1319-20.

Matsuo, R., Hitomi, T., Watanabe, S., Kirino, Y., (2002). Delayed-onset anmesia caused by protein synthesis inhibition in odor-taste associative memory of the terrestrial slug *Limax valentianus*. *Neurosci. Lett.,* 334, 201-5.

Matsuo, R., Kawaguchi, E., Yamagishi, M., Amano, T., Ito, E., (2010a). Unilateral memory storage in the procerebrum of the terrestrial slug *Limax*. *Neurobiol. Learn. Mem.,* 93, 337-42.

Matsuo, R., Kobayashi, S., Murakami, J., Ito, E., (2010b). Spontaneous recovery of the injured olfactory center in the terrestrial slug *Limax*. *PLoS One,* 5, e9054.

Matsuo, R., Kobayashi, S., Yamagishi, M., Ito, E., (2011). Two pairs of tentacles and a pair of procerebra: optimized functions and redundant structures in the sensory and central organs involved in olfactory learning of terrestrial pulmonates. *J. Exp. Biol.,* 214, 879-86.

Misanin, J. R., Miller, R. R., Lewis, D. J., (1968). Retrograde amnesia produced by electroconvulsive shock after reactivation of consolidated memory trace. *Science,* 160, 554-5.

Montarolo, P. G., Goelet, P., Castellucci, V. F., Morgan, J., Kandel, E. R., Schacher, S., (1986). A critical period for macromolecular synthesis in long-term heterosynaptic facilitation in *Aplysia*. *Science,* 234, 1249-54.

Nader, K., Schafe, G. E., Le Doux, J. E., (2000). Fear memories requires protein synthesis in the amygdala for reconsolidation after retrieval. *Nature,* 406, 722-26.

Nader, K., Einarsson, E. O., (2010). Memory reconsolidation: an update. *Ann. N.Y. Acad. Sci.,* 1191, 27-41.

Nikitin, E. S., Balaban, P., (2000). Optical recording of odor-evoked responses in the olfactory brain of the naïve and aversively trained terrestrial snail. *Learn. Mem.*, 7, 422-32.

Nikitin, E. S., Zakharov, I. S., Samarova, E. I., Kemenes, G., Balaban, P., (2005). Fine tuning of olfactory orientation behavior and single neuronal activity. *Eur. J. Neurosci.*, 22, 2833-44.

Odani, M., Komatsu, Y., Oda, S., Iwasaki, H., (2003). Screening of genes that respond to cryopreservation stress using yeast DNA microarray. *Cytobiology*, 47, 155-64.

Osborne, N. N., Cottrell, G. A., (1971). Distribution of biogenic amines in the slug, *Limax maximus*. *Z. Zellforsch. Mikrosk. Anat.*, 112, 15-30.

Pedreira, M. E., Pérez-Cuesta, L. M., Maldonado, H., (2002). Reactivation and reconsolidation of long-term memory in the crab *Chasmagnathus*: protein synthesis requirement and mediation by NMDA-type glutamatergic receptors. *J. Neurosci.*, 22, 8305-11.

Roobol, A., Carden, M. J., Newsam, R. J., Smales, C. M., (2009). Biochemical insights into the mechanisms central to the response of mammalian cells to cold stress and subsequent rewarming. *FEBS J.*, 276, 286-302.

Sahley, C., Gelperin, A., Rudy, J. W., (1981a). One-trial associative learning modifies food odor preferences of a terrestrial mollusc. *Proc. Natl. Acad. Sci. U.S.A.*, 78, 640-2.

Sahley, C., Rudy, J. W., Gelperin, A., (1981b). An analysis of associative learning in a terrestrial mollusc I. higher-order conditioning, blocking and a transient US pre-exposure effect. *J. Comp. Physiol.*, 144, 1-8.

Sakura, M., Kabetani, M., Watanabe, S., Kirino, Y., (2004). Impairment of olfactory discrimination by blockade of nitric oxide activity in the terrestrial slug *Limax valentianus*. *Neurosci. Lett.*, 370, 257-61.

Sangha, S., Morrow, R., Smyth, K., Cooke, R., Lukowiak, K., (2003a). Cooling blocks ITM and LTM formation and preserves memory. *Neurobiol. Learn. Mem.*, 80, 130-9.

Sangha, S., Scheibenstock, A., Lukowiak, K., (2003b). Reconsolidation of a long-term memory in *Lymnaea* requires new protein synthesis and RNA synthesis and soma of right pedal dorsal 1. *J. Neurosci.*, 23, 8034-40.

Samarova, E., Balaban, P., (2009). Changes in frequency of spontaneous oscillations in procerebrum correlate to behavioral choice in terrestrial snails. *Front. Cell. Neurosci.*, 3, 8.

Sara, S. J., (2000). Retrieval and reconsolidation: toward a neurobiology of remembering. *Learn. Mem.*, 7, 73-84.

Sekiguchi, T., Yamada, A., Suzuki, H., (1997). Reactivation-dependent changes in memory states in the terrestrial slug *Limax flavus*. *Learn. Mem.*, 4, 356-64.

Sekiguchi, T., Furudate, H., Kimura, T., (2010). Internal representation and memory formation of odor preference based on oscillatory activities in a terrestrial slug. *Learn. Mem.*, 17, 372-80.

Shimozono, S., Watanabe, S., Inoue, T., Kirino, Y., (2001). Identification and characterization of an output neuron from the oscillatory molluscan olfactory network. *Brain Res.*, 921, 98-105.

Silverman-Gavrila, L. B., Senzel, A. G., Charlton, M. P., Feng, Z. P., (2011). Expression, phosphorylation, and glycosylation of CNS proteins in aversive operant conditioning associated memory in *Lymnaea stagnalis*. *Neuroscience*, 186, 94-109.

Teyke, T., Gelperin, A., (1999). Olfactory oscillations augment odor discrimination not for identification by *Limax* CNS. *Neuroreport,* 10, 1061-8.

Toda, S., Kawahara, S., Kirino, Y., (2000). Image analysis of olfactory responses in the procerebrum of the terrestrial slug *Limax marginatus. J. Exp. Biol.,* 203, 2895-905.

Tully, T., Preat, T., Boynton, S. C., Del Vecchio, M., (1994). Genetic dissection of consolidated memory in *Drosophila. Cell,* 79, 35-47.

Watanabe, S., Kirino, Y., Gelperin, A., (2008). Neural and molecular mechanisms of microcognition in *Limax. Lean. Mem.,* 15, 633-42.

Wilcox, M., Franceschini, N. (1984). Illumination induces dye incorporation in photoreceptor cells. *Science,* 225, 851-4.

Yamada, A., Sekiguchi, T., Suzuki, H., Mizukami, A., (1992). Behavioral analysis of internal memory states using cooling-induced retrograde amnesia in *Limax flavus. J. Neurosci.,* 12, 729-35.

Yamane, T., Oestreicher, A. B., Gelperin, A., (1989). Serotonin-stimulated biochemical events in the procerebrum of *Limax. Cell. Mol. Neurobiol.,* 9, 447-59.

Yasui, K., Matsuo, R., Kirino, Y., (2004). Onset of amnesia is delayed with a decrease in inhibition of protein synthesis during odor-taste associative learning in the terrestrial slug *Limax valentianus. Zool. Sci.,* 21, 1163-6.

In: Memory Consolidation
Editors: Manabu Sakakibara and Etsuro Ito
ISBN: 978-1-63482-596-2

Chapter 4

OLFACTORY OSCILLATION AND ITS ROLE FOR LEANING AND DISCRIMINATION OF ODORS IN THE TERRESTRIAL SLUG

Midori Sakura[1,*] and Satoshi Watanabe[2]
[1]Department of Biology, Kobe University, Japan
[2]Department of Bioengineering and Robotics,
Tohoku University, Japan

ABSTRACT

Neural oscillation of the olfactory sensory system is widely observed in various animal species including mammals, insects and mollusks. Such oscillations are thought to be neural representations of odors in the brain. In the terrestrial slug *Limax*, a clear spontaneous oscillation of the local field potential is observed in the higher olfactory center of the brain, procerebral (PC) lobe. This oscillation is based on the synchronous firing of intrinsic inhibitory neurons, named bursting neurons, in the PC lobe. During the odor stimulation to the olfactory epithelium, the frequency of the oscillation is modified in an odor-specific manner. The frequency is also affected by experiences such as associative learning, indicating that the oscillation also represents the odor 'meaning' for the animal. Pharmacological blocking of oscillation in the PC lobe caused severe impairment in discrimination between conditioned and similar novel odors although it does not have any clear effects to discrimination between conditioned and different novel odors. Thus it is suggested that the olfactory oscillation play a crucial role for fine odor perception based on learning and memory.

Keywords: oscillation, olfactory learning, olfactory discrimination, *Limax*, procerebral lobe

* Corresponding Author address: E-mail: skr@port.kobe-u.ac.jp.

INTRODUCTION

Olfactory oscillation is one of the most well-known examples of temporal coding in the central nervous system. Olfactory oscillation, which is recorded as periodic modulations of the local field potencial (LFP) in the olfactory center of the brain (Gelperin, 2006), have been observed in wide range of animals, including mammals (Freeman, 1972; Kay et al., 2009), insects (Laurent and Naraghi, 1994; Stopfer et al., 1999) and mollusks (Gelperin and Tank, 1990; Delaney et al., 1994). This ubiquitousness is not surprising, considering that there are many similarities in olfactory anatomy and physiology across various animal phyla (Hildebrand and Shepherd, 1997). In all these animals, olfactory oscillation reflects coherent synchronous neural activities mediated by local inhibitory interneurons (mammals: Neville and Haberly, 2003; Lagier et al., 2007; insects: MacLeod and Laurent, 1996; mollusks: Gelperin and Tank, 1990).

Because of anatomical and physiological similarities in the olfactory system, invertebrates have been good model systems for elucidating functional mechanisms of the olfactory oscillation. In particular, the locust *Schistocerca americana* is one of the most extensively-studied animals (Laurent, 1996a,b; Laurent et al., 1998; Stopfer et al., 1999). In the locust, an odor stimulation onto the antennae elicit oscillatory LFP responses in the mushroom body, the higher olfactory center of the insect brain (Laurent and Naraghi, 1994). This oscillation is generated by spatial and temporal odor-specific responses of the projection neurons, olfactory information processing neurons connecting the antennal lobe (the first olfactory center) and the mushroom body (Laurent and Davidowitz, 1994; MacLeod and Laurent, 1996; Laurent, 1996a,b; Wehr and Laurent, 1996; Laurent and Wehr, 1996). These data suggests that the odor information is represented by this oscillatory LFP responses in the brain. The olfactory oscillation is not only odor-specific, but also experience-dependent. The power of the oscillation increases after a repeated odor stimulus (Stopfer and Laurent, 1999), indicating that the neural representations of odors are modifiable by the animal's experiences.

In spite of our great knowledge about mechanisms of oscillation, there are many unsolved questions for its functional roles (Tank et al., 1994; Kay et al., 2009), perhaps because we only have limited model systems dealing with odor-induced behavior as well as physiological phenomena. The terrestrial slug *Limax* is one of the most useful animals to tackle this issue. Not only because that it shows clear olfactory oscillation in the brain (Gelperin and Tank, 1990; Delaney et al., 1994) but also that it has excellent olfactory learning capability (Gelperin, 1975; Sahley et al., 1981a, b). Moreover, like other mollusks, an isolated nervous system of *Limax* survives in saline solution for over several hours, which greatly enhances our possibility to investigate neuronal functions electrophysiologically. In this review, we will summarize current knowledge about functional role of the olfactory oscillation in *Limax*, especially focusing on olfactory learning and discrimination.

OLFACTORY OSCILLATION IN *LIMAX*

Neural mechanisms of the olfactory oscillation in *Limax* are described in another chapter of this book in detail (see Chapter 4). Thus, we only point out some important features here. The olfactory center of the *Limax* brain, procerebral (PC) lobe, consists of two types of neurons called bursting and nonbursting neurons (B and NB neurons) (Delaney et al., 1994; Kleinfeld et al., 1994). NB neurons receive input from the tentacle nerve that carries olfactory information from the olfactory epithelium located on the tip of the tentacle (Inoue et al., 2000; Murakami et al., 2004). B neurons are inhibitory local interneurons within the PC lobe and directly make synapses with the NB neurons (Wang et al., 2001). The B neurons show spontaneous periodic bursting activities that cause inhibitory synaptic potentials in the NB neurons (Delaney et al., 1994; Kleinfeld et al., 1994) and these synaptic potentials are recorded as an oscillatory LFP (~ 0.7 Hz) in the PC lobe. The phase of the LFP oscillation is slightly different along the PC longitudinal axis, which produces wave propagation from the apical to basal end of the PC lobe (Delaney et al., 1994; Kleinfeld et al., 1994). The frequency of the PC oscillation is modified by odor stimulation to the olfactory epithelium in an odor-specific manner (Gervais et al., 1996), suggesting that, same as in the locust, the oscillation is an odor respresentation in the brain. A repellent odor stimulus, such as garlic or onion, and an attractive odor stimulus, such as rat chow, carrot or cucumber, elicits distinct modulation patterns in the PC oscillation frequency (Toda et al., 2000; Inoue et al., 2006).

OLFACTORY OSCILLATION AS AN OLFACTORY MEMORY

It has been well known that *Limax* can associate an odor with a reward or a punishment (Gelperin, 1975; Sahley et al., 1981a, b; Sahley et al., 1990). Especially, the aversive conditioning paradigm, in which an animal is trained to associate a food odor with a stimulus of bitter quinidine solution, has been used as a powerful model system to clarify neural and molecular mechanisms of olfactory learning (Watanabe et al., 2008; see also Chapter 4). Olfactory leaning capability of *Limax* is quite useful to investigate how the olfactory oscillation contributes to the odor perception because one can control the odor preference of the animal by the conditioning process, which allows us to compare the neural representations of the same odor with different cognitive meanings.

Olfactory memory modulates the frequency of the PC oscillation. The oscillation frequency during the odor stimulus is significantly different between naïve and conditioned animals (Kimura et al., 1998a, b). Moreover, the aversive and appetitive conditioning have opposite effects to the frequency modulation; the aversively-trained odor decreases the frequency whereas the appetitively-trained one increases (Kimura et al., 1998a). These results suggest that the oscillatory pattern of the PC response reflects the animal's odor preference in some way. Similar differences in the modulation of PC oscillation frequency between neutral and aversively-trained odors have been demonstrated in the terrestrial snails (Nikitin and Balaban, 2000).

More direct evidence of the correlation between neural responses and the behavior has been shown using an *in vitro* aversive-conditioning paradigm using an isolated-brain preparation (Inoue et al., 2006). In this paradigm, the number of spikes of an identified motoneuron (*p*-VN) innervated to the mantle muscles via the parietal nerve is monitored as an indicator of the odor-avoidance behavior, i.e., mantle contraction (Inoue et al., 2004). The spikes of *p*-VN are easily monitored by extracellular recordings from the parietal nerve using a glass suction electrode (Inoue et al., 2004). As an aversive paired-conditioning, odor stimulation to the olfactory epithelium (CS+) and tetanic electrical stimulation (1V; duration,

1ms; interval, 150 ms; 100 times) to the lip nerve (US), which mimics aversive bitter stimulation to the mouth, are simultaneously applied. 30 minutes after the conditioning, another odor stimulation (CS-) is applied without any US as an unpaired-conditioning. The number of *p*-VN spike responses to the CS+ and CS- are compared between before and after the conditioning (Figure 1A). Using this *in vitro* conditioning paradigm, the motor output directly related to the conditioned behavior can be monitored throughout the memory formation process, which has great advantages for functional investigation of olfactory learning and memory.

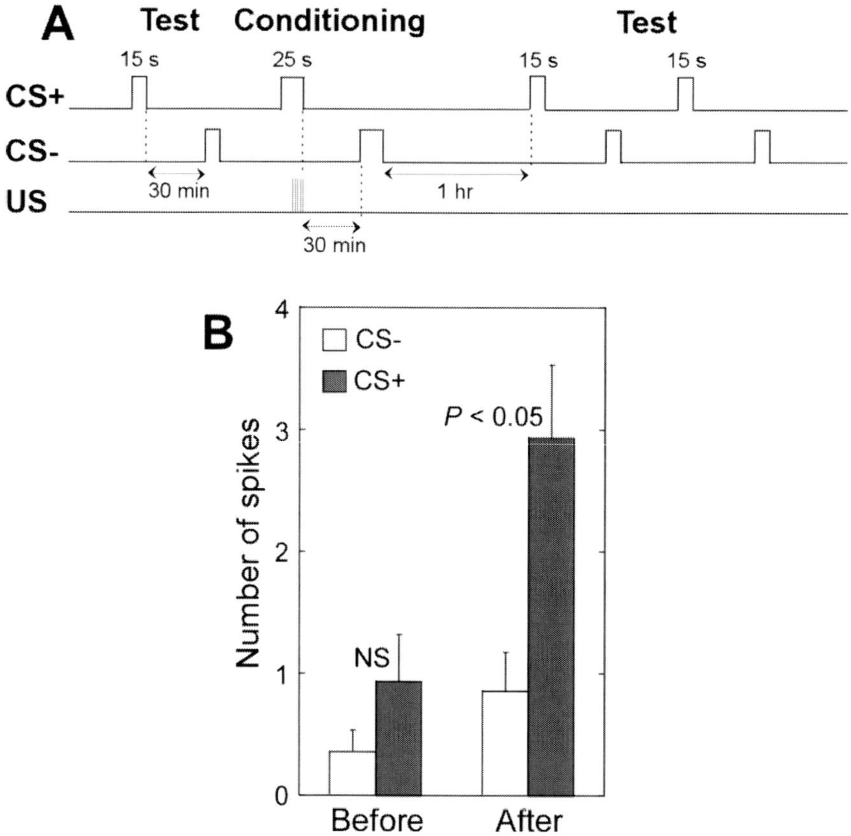

Figure 1. *In vitro* odor-aversion conditioning. A: experimental procedure of *in vitro* odor-aversion conditioning. Before conditioning, CS+ and CS- (either cucumber or carrot, respectively) are applied for 15 sec with 30 min interval. During the conditioning, CS+ is paired with US whereas CS- is applied 30 min after the US (1V; duration, 1ms; interval, 150 ms; 100 times). One hour after the conditioning, CS+ and CS- are applied again for 15 sec with 30 min interval. B: The number of spikes of *p*-VN (mean ± SE, $N = 14$) during the CS before and after conditioning. Wilcoxon's test was used for a statistical comparison between before and after the conditioning.

Before the conditioning, both CS+ and CS- to the olfactory epithelium only elicit few *p*-VN spikes (Figure 1B, Before), indicating that these odors do not elicit avoidance behavior. After the conditioning, on the other hand, CS+ elicits significantly higher number of *p*-VN spikes than CS- does (Figure 1B, After), indicating that the conditioning procedure forms CS+-specific olfactory memory in the isolated brain.

Inoue et al. (2006) also showed that, after the *in vitro* aversive-conditioning, the frequency of the PC oscillation during CS+ significantly increased compared with those before the conditioning. The frequency during CS-, by contrast, did not show any significant differences before and after the conditioning, indicating that only the odor paired with US increases the oscillation frequency. Interestingly, the fact that the oscillation frequency increases after aversive learning is not coincident with the former reports by Kimura et al. (1998a, b) (see above). It is difficult to compare these results directly because of the many differences in the experimental conditions, but both data strongly suggests that the frequency of PC oscillation represents the 'meaning' of the odor for the animal. The correlation between the PC oscillation frequency and odor-induced behavior has also been shown in the terrestrial snails (Samarova and Balaban, 2009), suggesting that this phenomenon is common in terrestrial mollusks.

Is the frequency of the PC oscillation a key factor for the behavioral switch caused by olfactory learning? There is some suggestive data to answer this question. The Local application of acetylcholine or serotonin to the PC lobe greatly increases oscillation frequency in the *Limax* (Watanabe et al., 2001; Inoue et al., 2001), but it only has little effect to the *p*-VN discharges (Kirino et al., 2005; Sakura, unpublished data). These results indicate that the high frequency of the PC oscillation itself does not cause aversive behavior. Probably, it has some modulatory effects on the odor information processing in the brain. In fact, application of acetylcholine to the PC lobe induces high *p*-VN spike rate during the odor stimulation even it is not paired with US (Kirino et al., 2005).

The next question is how the odor representation in the PC modulates by the olfactory leaning process, in other words, where the neural plasticity occurs as olfactory memory formation. It has been demonstrated that a tetanic stimulus on the tentacle nerve elicits a long-term synaptic potentiation (LTP) in the PC lobe, possibly at the synapses between NB and B neurons (Sakura et al., in preparation) although the molecular mechanisms underlying this LTP is still unclear. Further electrophysiological studies will unveil the relationship between the LTP and the PC oscillation and how it contributes to the olfactory memory.

Olfactory Oscillation Is Necessary for Fine Olfactory Discrimination

Olfactory learning capability of animals also allows us to investigate their olfactory discrimination capability. If the olfactory oscillation represents odor meaning in the brain, it must play a role for odor discrimination. In honeybees, which basically have the same olfactory oscillation mechanisms with the locusts (see above) (Stoper and Laurent, 1995; Stopfer et al., 1999), it has been reported that the the inhibition of odor-evoked oscillation in the brain by pharmacological interfering with GABAergic neurotransmission, also impairs the olfactory discrimination between similar odors but not between dissimilar odors (Stopfer et al., 1997; Hosler et al., 2000). In the insect antennal lobe, GABAergic local interneurons are involved in synchronous neural responses of the projection neurons that generate spatio-temporal oscillatory patterns in the brain (Leitch and Laurent, 1996; MacLeod and Laurent, 1996). Thus, once GABAergic transmission is blocked, the odor-specific oscillation pattern is collapsed and exercise an serious effect on the discrimination between similar odors. On the other hand, the discrimination between dissimilar odors is not affected by the GABA inhibitor, indicating that the olfactory oscillation is not always necessary for odor discrimination.

Also in *Limax*, the involvement of PC oscillation to the odor discrimination have been reported. In the PC lobe, nitric oxide (NO) regulates the oscillation frequency (Gelperin, 1994; Gelperin et al., 2000) by increasing the burst frequency of the B neurons (Gelperin et al., 2000; Inoue et al., 2001). Moreover, histochemical studies have shown that both NO-generating and NO-responsive cells exist in the PC lobe as well as in the tentacle, suggesting that the NO cascade mediates olfactory information processing (Fujie et al., 2002). Blocking NO in the PC lobe also modulates an odor-induced motor output. After the inhibition of NO synthesis by local application of N-nitro-l-arginine methyl ester ($_L$-NAME) to the PC lobe, the activity of the tentacular motor nerve to an odor becomes indistinguishable from that to another similar odor (Teyke and Gelperin, 1999), suggesting that NO serves a crucial function in olfactory discrimination by regulating oscillatory activity in the PC lobe.

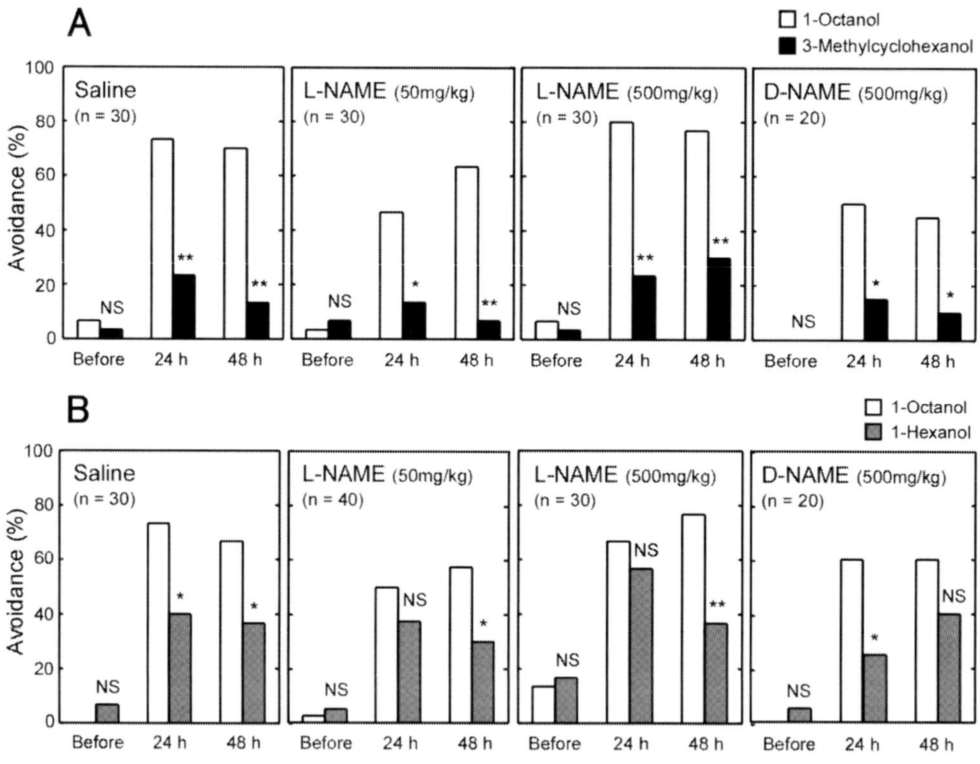

Figure 2. The effect of the injection of saline, $_L$-NAME and $_D$-NAME to odor discrimination between 1-octanol and 3-methylcyclohexanol (A) and between 1-octanol and 1-hexanol (B). Slugs were tested shortly before and 24 h and 48 h after the conditioning. The injection was made 1 h before the test, after conditioning. The vertical axis represents the relative number (%) of slugs that avoided each odor during the test ($*P < 0.05$; $**P < 0.01$ by χ^2-tests).

To clarify the effect of inhibition of NO to olfactory discrimination behaviorally, we conducted pharmacological experiments using a differential aversive conditioning paradigm, in which the animal is trained to associate one odor stimulus with quinidine solution (CS+) but not to associate another odor stimulus (CS-) (Sakura et al., 2004; Figure 2). The discrimination capability can be tested by the differences in the avoidance percentage for each odor stimulus (CS+ and CS-) after the conditioning. Two pairs of odorant, 1-octanol/3-methylcyclohexanol (i.e., structurally dissimilar pair) and 1-octanol/1-hexanol (i.e., structurally similar pair) were used for the CS, both of which they can significantly discriminate each other although the similar odor has a strong generalization effect (Sakura et al. 2004). After the L-NAME solution was injected, they could not discriminate between 1-octanol and 1-hexanol (Figure 2B, L-NAME, 24h). Injections of Saline or D-NAME, the less active optical isomer of L-NAME, did not have any effects to the discriminatory performance between 1-octanol and 1-hexanol (Figure 2B, Saline, D-NAME). In contrast, L-NAME injection did not impair the discrimination between 1-octanol and 3-methylcyclohexane, as well as Saline and D-NAME injections (Figure 2A), indicating that the L-NAME injection only impairs discrimination between structurally similar odors. These results imply the NO function of olfactory discrimination by modulating oscillation in the PC lobe. NO inhibition does not affect to the avoidance performance to the CS+, again suggesting that the oscillation frequency is not the key factor of the learning behavior (see above). Further comparison between the results of *in vivo* and *in vitro* experiments is necessary to understand the whole picture.

Conclusion and Future Perspectives

Olfactory oscillation in *Limax* has drawn attention from many researchers for the last few dacades. It is not only because the oscillation is prominent compared with other animals but also because of their many advantages as a research object. Especially, the *in vitro* brain preparation gives us great opportunities to access the neural functions during the memory formation at the cellular lebel. So far, most researchs have been focused on frequency and wave propagation (see Chapter 4) of the oscillation. However, we now aware that it is generally insufficient to understant olfactory leaning and discrinimation mechanisms. Further electrophysiological and molecular biological studies focusing on each celluler event in the PC lobe may give us deeper insight into the animals' olfactory system in general.

References

Delaney, K.R., Gelperin, A., Fee, M.S., Flores, J.A., Gervais, R., Tank, D.W., Kleinfeld, D., 1994. Waves and stimulus-modulated dynamics in an oscillating olfactory network. *Proc. Natl. Acad. Sci. U.S.A.* 91, 669-673.

Freeman, W.J., 1972. Measurement of oscillatory responses to electrical stimulation in olfactory bulb of cat. *J. Neurophysiol.* 35, 762-779.

Fujie, S., Aonuma, H., Ito, I., Gelperin, A., Ito, E., 2002. The nitric oxide/cyclic GMP pathway in the olfactory processing system of the terrestrial slug *Limax marginatus*. *Zool. Sci.* 19, 15–26.

Gelperin, A., 1975. Rapid food-aversion leaning by a terrestrial mollusk. *Science* 189, 567-570.

Gelperin, A., 1994. Nitric oxide mediates network oscillations of olfactory interneurons in a terrestrial mollusc. *Nature* 369, 61-63.

Gelperin, A., 2006. Olfactory computations and network oscillations. *J. Neurosci.* 26, 1663-1668.

Gelperin, A., Flores, J., Raccuia-Behling, F., Cooke, I.R.C., 2000. Nitric oxide and carbon monoxide modulate oscillations of olfactory interneurons in a terrestrial mollusk. *J. Neurophysiol.* 83, 116-127.

Gelperin, A., Tank, D.W., 1990. Odour-modulated collective network oscillations of olfactory interneurons in a terrestrial mollusc. *Nature* 345, 437-440.

Gervais, R., Kleinfeld, D., Delaney, K.R., Gelperin, A., 1996. Central and reflex neuronal responses elicited by odor in a terrestrial mollusk. *J. Neurophysiol.* 76, 1327-1339.

Hildebrand, J.G., Shepherd, G.M., 1997. Mechanisms of olfactory discrimination: converging evidence for common principles across phyla. *Annu. Rev. Neurosci.* 20, 595-631.

Hosler, J.S., Buxton, K.L., Smith, B.H., 2000. Impairment of olfactory discrimination by blockade of GABA and nitric oxide activity in the honey bee antennal lobes. *Behav. Neurosci.* 114, 514-525.

Inoue, T., Watanabe, S., Kawahara, S., Kirino, Y., 2000. Phase-dependent filtering of sensory information in the oscillatory olfactory center of a terrestrial mollusk. *J. Neurophysiol.* 84, 1112-1115.

Inoue, T., Watanabe, S., Kirino, Y., 2001. Serotonin and NO complementarily regulate generation of oscillatory activity in the olfactory CNS of a terrestrial mollusk. *J. Neurophysiol.* 85, 2634-2638.

Inoue, T., Inokuma, Y., Watanabe, S., Kirino, Y.,2004. In vitro study of odor-evoked behavior in a terrestrial mollusk. *J. Neurophysiol.* 91, 372-381.

Inouoe, T., Murakami, M., Watanabe, S., Inokuma, Y., Kirino, Y., 2006. In vitro odor-aversion conditioning in a terrestrial mollusk. *J. Neurophysiol.* 95, 3898-3903.

Kay, L.M., Beshel, J., Brea, J., Martin, C., Rojas-Líbano, D., Kopell, N., 2009. Olfactory oscillations: the what, how and what for. *Trends Neurosci.* 32, 207-214.

Kimura, T., Toda, S., Sekiguchi, T., Kirino, Y. 1998a. Behavioral modulation induced by food odor aversive conditioning and its influence on the olfactory responses of an oscillatory brain network in the slug *Limax marginatus*. *Learn. Mem.* 4, 365-375.

Kimura, T., Toda, S., Sekiguchi, T., Kawahara, S., Kirino, Y., 1998b. Optical recording analysis of olfactory response of the procerebral lobe in the slug brain. *Learn. Mem.* 4, 389-400.

Kirino, Y., Inoue, T., Watanabe, S., 2005. Behavioral determination of odor preference is coded by the oscillation frequency in a collective oscillating network of a terrestrial mollusk. *Chem. Senses* 30 (Suppl 1), i154-i155

Kleinfeld, D., Delaney, K.R., Fee, M.S., Flores, J.A., Tank, D.W., Gelperin, A., 1994. Dynamics of propagating waves in the olfactory network of a terrestrial mollusc: An electrical and optical study. *J. Neurophysiol.* 72, 1402-1419.

Lagier, S., Panzanelli, P., Russo, R. E., Nissant, A., Bathellier, B., Sassoè-Pognetto, M., Fritschy, J.-M., Lledo, P.-M., 2007. GABArgic inhibition at dendrodendritic synapses tunes γ oscillations in the olfactory bulb. *Proc. Natl. Acad. Sci. U.S.A.* 104, 7259-7264.

Laurent, G., 1996a. Dynamical representation of odors by oscillating and evolving neural assemblies. *Trends Neurosci.* 19, 489-496.

Laurent, G., 1996b. Olfactory processing: Maps, time and codes. *Curr. Opin. Neurobiol.* 7, 547-553.

Laurent, G., Davidowitz, H., 1994. Encoding of olfactory information with oscillating neural assemblies. *Science* 265, 1872-1875.

Laurent, G., MacLeod, K., Stopfer, M., Weher, M., 1998. Spatiotemporal structure of olfactory inputs to the mushroom bodies. *Learn Mem* 5, 124-132.

Laurent, G., Naraghi, M., 1994. Odorant-induced oscillations in the mushroom bodies of the locust. *J. Neurosci.* 14, 2993-3004.

Laurent, G., Wehr, M., 1996. Temporal representations of odors in an olfactory network. *J. Neurosci.* 16, 3837-3847.

Leitch, B., Laurent, G., 1996. GABAergic synapses in the antennal lobe and mushroom body of the locust olfactory system. J. Comp. Neurol. 372, 487-514.

MacLeod, K., Laurent, G., 1996. Distinct mechanisms for synchronization and temporal patterning of odor-encoding neural assemblies. *Science* 274, 976-979.

Murakami, M., Watanabe, S., Inoue, T., Kirino, Y., 2004. Odor-evoked responses in the olfactory center neurons in the terrestrial slug. *J. Neurobiol.* 58, 369-378.

Nikitin, E.S., Balaban, P.M., 2000. Optical recording of odor-evoked responses in the olfactory brain of the naïve and aversively trained terrestrial snail. *Learn. Mem.* 7, 422-432.

Neville, K.R., Haberly, L.B., 2003. Beta and gamma oscillations in the olfactory system of the urethane-anesthetized rat. *J. Neurophysiol.* 90, 3921-3930.

Sahley, C. Gelperin, A., Rudy, J.W., 1981a. One-trial associative learning modifies food odor preference of a terrestrial mollusc. *Proc. Natl. Acad. Sci. U.S.A.* 78, 640-642.

Sahley, C., Rudy, J.W., Gelperin, A., 1981b. An analysis of associative learning in a terrestrial mollusc I. higher-order conditioning, blocking and a transient US pre-exposure effect. *J. Comp. Physiol.* 144, 1-8.

Sahley, C., Martin, K.A., Gelperin, A., 1990. Analysis of associative learning in the terrestrial mollusc *Limax maximus*. II. Appetitive learning. *J. Comp. Physiol. A* 167, 339-345.

Sakura, M., Watanabe, S., Kirino, Y., 2004. Impairment of olfactory discrimination by blockade of nitric oxide activity in the terrestrial slug *Limax valentianus*. *Neurosci. Lett.* 370, 257-261.

Samarova, E., Balaban, P., 2009. Changes in frequency of spontaneous oscillations in procerebrum correlate to behavioural choice in terrestrial snails. *Front. Cell. Neurosci.* 3, 8.

Stopfer, M., Laurent, G., 1995. A comparative analysis of olfactory processing in insects. *Soc. Neurosci. Abstr.* 21, 134.

Stopfer, M., Laurent, G., 1999. Short-term memory in olfactory network dynamics. *Nature* 402, 664-668.

Stopfer, M., Bhagavan, S., Smith, B.H., 1997. Impaired odour discrimination on desynchronization of odour-encoding neural assemblies. *Nature* 390, 70-74.

Stopfer, M., Wehr, M., Macleod, K., Laurent, G., 1999. Neural dynamics, oscillatory synchronisation, and odour codes. In *Insect Olfaction* (ed. B. S. Hansson), pp. 163-180. Berlin Heidelberg: Springer-Verlag.

Tank, D.W., Gelperin, A., Kleinfeld, D., 1994. Odors, oscillations, and waves: does it all compute? *Science* 265, 1819-1820.

Teyke, T., Gelperin, A., 1999. Olfactory oscillations augment odor discrimination not odor identification by *Limax* CNS. *NeuroReport* 10, 1061-1068.

Toda, S., Kawahara, S., Kirino, Y., 2000. Image analysis of olfactory responses in the procerebrum of the terrestrial slug *Limax marginatus*. *J. Exp. Biol.* 203, 2895-2905.

Watanabe, S., Inoue,T., Murakami, M., Inokuma, Y., Kawahara, S., Kirino Y., 2001. Modulation of oscillatory neural activities by cholinergic activation of interneurons in the olfactory center of a terrestrial slug. Brain Res. 896, 30–35.

Watanabe, S., Kirino, Y., Gelperin, A., 2008. Neural and molecular mechanisms of microcognition in *Limax*. *Learn. Mem.* 15, 633-642.

Wang, J., Denk, W., Flores, D., Gelperin, A., 2001. Initiation and propagation of calcium-dependent action potentials in a coupled network of olfactory interneurons. *J. Neurophysiol.* 85, 977-985.

Wehr, M., Laurent, G., 1996. Odour encoding by temporal sequences of firing in oscillating neural assemblies. *Nature* 384, 162-166.

In: Memory Consolidation
Editors: Manabu Sakakibara and Etsuro Ito

ISBN: 978-1-63482-596-2

Chapter 6

MECHANISM OF MEMORY CONSOLIDATION FOR TASTE AVOIDANCE CONDITIONING IN INVERTEBRATE MODEL SYSTEM

Manabu Sakakibara*

Laboratory of Neurobiological Engineering,
Department of Biological Science and Technology,
School of High-Technology for Human Welfare and
Graduate School of Bioscience, Tokai University, Japan

ABSTRACT

"Spaced training" results in a longer lasting 'Long Term Memory (LTM)' than does "massed training" from species of invertebrates to vertebrates including mammalian. The memory enhancing mechanism is generally accepted as the cellular attributes developed in synaptic modulation, membrane excitability and cytosolic organelle. Memory is divided into 3 subclasses: Short Term Memory (STM); Intermediate Term Memory (ITM); Long Term Memory (LTM). The cellular mechanisms are assumed as followings: STM lasting minutes only requires modification of pre-existing substrate proteins, while ITM lasting a few hours is dependent on *de novo* protein synthesis, and LTM lasting more than a few hours to days, weeks and even years requires both new protein synthesis plus altered gene activity. Based on the previous studies, the critical period to form LTM in classical conditioning of gastropod model system, such as *Hermissenda* and *Lymnaea,* lies within 10 min or later than 180 min following conditioning. These 2 critical periods are correlated with the time windows of macromolecular protein synthesis and new mRNA synthesis. It is thought that there is a serial ordered transition seen behaviorally from STM to ITM then to LTM that corresponds to sequential molecular events occurring both pre- and post-synaptic sites in neurons necessary for LTM formation. In this chapter memory enhancing mechanisms was discussed from the conditioned behavioral observations obtained from the fresh pond water pulmonate *Lymnaea stagnalis (Lymnaea s.)*. First, the most efficient number of paired presentation of the conditional stimulus (CS) and the unconditional stimulus (US) was demonstrated to form

* Nishino317, Numazu 410-0321, Shizuoka, Japan: manabu@tokai.ac.jp.

STM and LTM, and then the facilitating methods to form substantial transition for longer lasting memory were presented from behavioral observation. We also examined whether the cooling induced amnesia effectively blocked the formation of STM/LTM. Finally I discussed the memory enhancing mechanisms involved the protein kinase C (PKC) phosphorylation and the necessary interval, corresponding to macro-molecular synthesis and mRNA alteration.

Keywords: short term memory, long term memory, protein kinase C, bryostatin, cooling induced amnesia, spaced training, massed training, taste avoidance conditioning, *Lymnaea stagnalis, Hermissenda crassicornis,* CS-US temporal specificity, critical period, protein synthesis, altered gene activity

List of Abbreviations

Bryo	bryostatin
CGC	cerebral giant cell
CLTM	consolidated long-term memory
CS	onditioning stimulus
CR	conditioning response
Hermissenda c.	*Hermissenda crassicornis*
IPSP	inhibitory post synaptic potential
ITM	intermediate-term memory
LTM	long-term memory
Lymnaea s.	*Lymnaea stagnalis*
MAPK	mitogen activated protein kinase
MIP	molluscan insulin related peptide
N1M	the neuron 1 medial cell
N3t	the neuron tonic 3 cell
PKA	protein kinase A
PKC	protein kinase C
RPeD11	right pedal dorsal 11 neuron
Ro-32	bis-indolylmaleimide-XI Ro-0432
STM	short-term memory
TAC	taste avoidance conditioning
US	unconditioning stimulus
WBWR	whole-body-withdrawal-response

Introduction

Classical conditioning known as Pavlovian conditioning is one of the learning paradigm capable of association between the temporal sequence of CS (conditioning stimulus) and US (unconditioning stimulus). CS was originally assumed to be a neutral stimulus for an animal not to induce any physiological response, i.e., sound of metronome in Pavlovian dog, whereas an animal responds to US, i.e., smell of meat to induce salivation in Pavlovian dog, inducible physiological response. After repeated pair of CS and US presentations, an animal can learn the temporal sequence of CS and US, thus CS induces new physiological response previously induced by US. This response is termed as conditioning response (CR). Since classical conditioning is observed among a variety of animal species ranging from nematode to human, this type of learning is considered to be one of the fundamental learning process, varieties of animals are the good models for analyzing the mechanisms of learning and memory, researchers select animals based on their aim and the favorable techniques such as molecular biology, electrophysiology and/or differential equations based computational study. The research trend shifts from macro molecular perspective, i.e., from studies of whole animal, to micro molecular perspective, such as one of cellular, molecular, and/or genetical studies. In this respect gastropod has been a favorable model for neuroethological analysis. In this chapter I discussed the plausible memory consolidation mechanisms in gastropod model systems studied in *Hermissenda crassicornis (Hermissenda c.)* and *Lymnaea stagnalis (Lymnaea s.)* based on Pavlovian conditioned behavioral studies, in addition to physiological and biophysical findings I have studied in these more than 20 years in Tokai University at Numazu Campus.

Hermissenda c. is one of the good model for Pavlovian classical conditioning from behavioral to molecular studies as we have obtained following findings; 1) After paired presentation of CS-US, snails are capable of learning the temporal sequence of CS-US, thus they did not show the innate photo-tactic behavior anymore after conditioning [1-3]; 2) The underlying neuronal mechanisms are arising from inter-sensory interaction between photoreceptors and vestibular hair cells in the statocyst [4]; 3) After acquisition of learning, the type B photoreceptors are in the long-lasting excited state due to inactivation of K^+ channels (decrease of I_A, $I_{Ca2+\text{-}K+}$ and delayed rectifier) [5-8]; 4) Inactivation of the type B photoreceptors are due to phosphorylation of the GTP-binding protein (Calexcitin) through PKC involved signaling cascade [6, 9, 10]; 5) After acquisition of learning, the morphological alternation called 'synaptic-focusing' has appeared at the type B photoreceptors synaptic ending [11]; 6) These conditioning effects are reproducible and mimicked with Hodgkin-Huxley type equations [12]; 7) TRP channels are involved in the photo-transduction in the type B photoreceptors [13, 14].

A number of classical conditioning [15-20] and operant conditioning [21, 22] paradigms have been used successfully in *Lymnaea s.*, and cellular traces of behavioral conditioning have been identified in isolated, semi-intact and simplified cultural networks [23-25]. In this model system, classical conditioning of appetitive [16] and aversive conditioning [26, 27] of the feeding behavior and aversive operant conditioning of respiratory behavior [28] are well characterized in this animal. To understand the conditioning-induced modification of the neural networks underlying feedback behaviors, four groups of neurons: sensory, modulatory, central pattern generator, and motor neurons have been studied, and each component neurons in each group have been analyzed at the cellular and molecular levels [29-32]. For the appetitive conditioning, a mechanical gentle touch to mouth area is the CS and sucrose application is the US, after conditioning procedure snails responded to a CS as starting mouth opening as if they behaved to sucrose application even they received 1 paired CS-US presentation (one-trial conditioning) [16]. On the other hand the same sucrose application is the CS and aversive stimulation such as KCl application [26], mechanical prodding to the head [18], and electrical shock [33, 34] are all the US available in the aversive conditioning. After a number of repetitive presentations of CS-US, snails responded to a CS presentation so as to reduce their feeding behavior significantly.

The spaced training, interposing a rest interval between the multiple training sessions, produces stronger and longer-lasting memory than the same number of training session with no rest interval or massed training [35, 36]. This phenomenon has been demonstrated in various invertebrate species such as *Drosophila* [37], *Apis* [38], *Aplysia* [39] and *Lymnaea* [19] in addition to vertebrate animals [40, 41]. Naive adult *Lymnaea* does not show any photo-tactic behavior, i.e., the animal does not move toward a lighter place rather the animal responds to shadow or light-off stimulus as an escape behavior [42]. This behavioral tendency shows the CS is the neutral stimulus for *Lymnaea* and of course snails hesitate to vestibular turbulence to US presentation as same as *Hermissenda* does, the snail responded with whole-body-withdrawal-response (WBWR) which is the only escape behavior available to this animal.

In our previous attempt *Lymnaea* was conditioned with at least 30 repeated pairs of flash light (CS) and vestibular turbulence (US) for 3 days. After 30 paired CS-US stimuli for 3 days conditioning resulted in escape behavior only available response for this animal, i.e., snails responded with WBWR in response to a CS presentation, flash of light. This escape behavior to a CS presentation lasted for at least a week. However their behavior to a CS presentation did not alter after training with conditioning trial of 90 paired CS-US presentations in a day nor each 45 trials for 2 days. From these observations *Lymnaea* can be conditioned by spaced training with visuo- vestibular conditioning paradigm. The training paradigm is identical as *Hermissenda* does [1, 43] though CS is not a neutral stimulus for *Hermissenda* because of their innate photo-tactic behavior. Considering the neuronal mechanism for this visuo-vestibular conditioning, I previously mentioned the involved neural pathway is different from as in *Hermissenda* and *Lymnaea* [20, 44]. As same as the visuo-vestibular conditioning in *Lymnaea s.* the snail is also available for feeding related conditioning, i.e., appetitive and/or aversive conditioning as mentioned above. Our original attempt was conducted using sucrose application as CS and mechanical prodding to the head with hand-held stick as US. However the US presentation needs skillful technique to condition animals, i.e., constant stimulus timing, strength, duration and so forth are difficult to apply consistently, because strong poking to the head induces strong escape behavior not to recover from withdrawal response for a couple of hours. In order to apply consistent aversive stimulus we recently developed the new method employed by 'electrical shock'.

In this chapter I discussed the one plausible mechanism based on the behavioral findings to consolidate the taste aversive conditioning according to the following consolidation hypothesis in terms of memory period: STM (short-term memory), ITM (intermediate-term memory) and LTM (long-term memory). STM lasting only minutes requires modification of pre-existing substrate proteins, while ITM lasting a few hours is dependent on *de novo* protein synthesis, and finally that LTM lasting more than a few hours to days, weeks and years requires both new protein synthesis and altered gene activity [45-52].

MATERIALS AND METHODS

Animals

Laboratory-reared freshwater pond snails, *Lymnaea s.* with shell lengths of 20 to 25 mm, were maintained at 22 °C in well-aerated fresh water on a 12 h light: 12 h dark cycle (on at 8:00). Snails larger than 20 mm in shell length are thought to be capable of the classical conditioning with enough matured nervous system [53]. Animals were maintained on a diet of cabbage and goldfish pellets until food-deprivation was initiated 24 h prior to experimentation.

Experimental Apparatus for Behavior

The Plexiglas experimental container (diameter: 60 mm and height: 20 mm) had a perfusion system with a single inlet and outlet that allowed the solution inside the container to be entirely replaced within 30 s (rate: 250 ml/min). The container held 10 ml fresh aquarium water in which snails were kept. The conditioning response, i.e., feeding response, was readily observed using a mirror under the container as shown in Figure 1A.

Behavioral Pharmacology-Bryostatin (Brio) or Ro-32-0432 (Ro-32)

The PKC activator Bryo was initially dissolved in 100 μl of ethanol and diluted in water to make a 1 μg/ml stock solution. This solution was further diluted in water to a final concentration of 0.1 or 0.5 ng/ml, with a final ethanol content of 0.0001/0.0005%. This was the maximum upregulating concentration of Bryo found to be effective in *Hermissenda* [54]. Animals in the experimental apparatus were immersed in a 10-ml water flow containing Bryo (0.1 or 0.5 ng/ml) for 45 min. The Bryo water was then completely replaced (in 30 s) with fresh aquarium water. To test for PKC activation, other snails were exposed to the PKC inhibitor Ro-32-0432 at a concentration of 80 nM for 20 min before or after Bryo incubation [54]. Ro-32 has a 10-fold selectivity for PKC-α and 4-fold selectivity for PKC-βI over Ca^{2+} independent PKC-ε, we selected a concentration of 80 nM as this concentration was previously used in another gastropod: even though the IC50 for PKC-ε was 108 nM for heterogeneous rat brain [55]. Control animals were incubated in fresh aquarium water containing 0.005% ethanol.

Statistics

The behavioral differences between pre- and post-conditioning (10 min, 24 h) were evaluated with two-way repeated measures analysis of variance (repeated-measure ANOVA). Scheffe's *F* post hoc test was then used to determine statistical significance between groups. Both the behavioral measurements and the analyses were performed blind to the behavioral manipulation of the animals.

Assessment of STM and LTM

Before starting conditioning procedure, snails' naive response, i.e., the number of mouth opening per minute to a CS presentation was evaluated as the control. Through this chapter we evaluated the CR at 10 min later following the repeated pairs of CS-US as STM and at 24 h later as LTM, if the CR was significantly suppressed compared to the control. A schematic illustration of conditioning apparatus and the temporal sequence of each stimulus was shown in Figure 1.

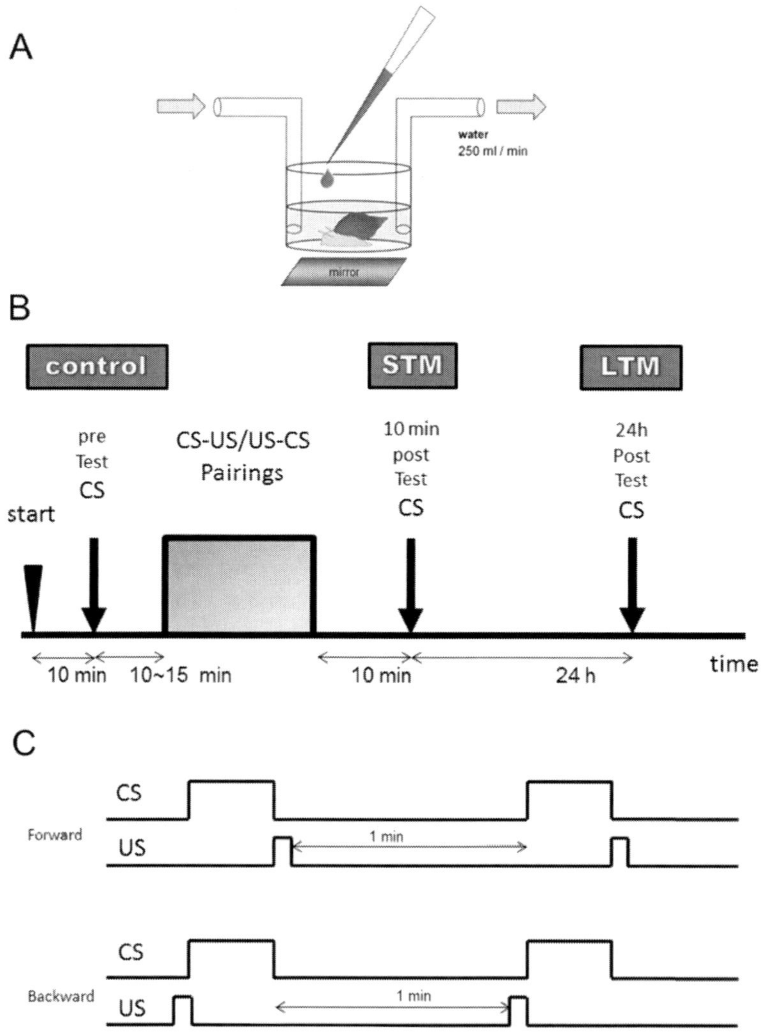

Figure 1. Schematic illustration of the experimental setup to apply the sucrose CS is shown in (A), and the temporal schedule in (B) and (C). (A) The CS, sucrose (1ml of 0.1 M solution) was applied to a mouth of *Lymnaea s.* with a 1 ml syringe. A mirror put under the test chamber allowed us to easily monitor the feeding response (i.e., the number of mouth opening per minute in response to a CS). (B) The feeding response to a CS was tested 3 times, once before the conditioning trial (i.e., the pre-test as a control) and twice after the conditioning paradigm at 10 min and 24 h as post-test. STM and LTM were evaluated in comparison with the number of bites in the pre-test at 10 min and 24 h, respectively. (C) The temporal sequence of CS-US for the forward conditioning and US-CS for the backward conditioning. Note that CS and US presentation is not overlapped. Inter-stimulus interval between each CS-US/US-CS pair is set 1 min.

RESULTS AND DISCUSSION

Classical Conditioning in *Lymnaea Stagnalis* (*Lymnaea s.*)

1. Minimum Repetition Number of CS-US to Acquire STM and/or LTM

We previously demonstrated that *Lymnaea s.* can be conditioned by visuo-vestibular conditioning paradigm with 30 paired presentation of CS-US for 3 days, i.e., total 90 CS-US trials separated by 3 D was necessary for LTM formation [19]. It is generally assumed that minimum number to acquire STM and/or LTM is dependent on the strength of nociceptivity in US employed for the aversive conditioning, e.g., under the traumatic stress such as crisis of life, the life-long memory is formed even experienced once [56]. While for the appetitive conditioning with one trial CS-US presentation snails are able to be conditioned [57]. I first examined how many trials were necessary to form STM and LTM for the aversive conditioning by the taste avoidance conditioning (TAC) with sucrose and mechanical poking as CS and US, respectively. In our previous attempt we found that 20 paired CS-US in a day was the minimal condition resulted in both STM and LTM [18].

As mentioned in Figure 1C the inter-stimulus interval was fixed 1 min, we examined the minimum repetition number to result in STM/LTM. We first examined the efficacy of 20 CS-US trials [18]. We studied 3 situations: one-day-conditioning (20 CS-US pairs; n=9); two-days-conditioning (10 trials each; n=8); 4-days-conditioning (five trials per day for 4 days; n=9). The behavioral results were shown in Figure 2.

LTM were observed only in group of 4-days-conditioning, which received 5 pairings/day for 4 days (4D × 5). That is, the number of bites in the pre-test session in response to a CS was significantly greater than the number of bites elicited by a CS in the 24 h post-test session ($p<0.05$). The 24 h post-test session in the other two groups were not significantly different, however, STM was acquired in every group. Thus, STM was observed in every groups while LTM was only observed in one of three groups, 4D × 5.

Using the 5 pairings/day for 4 days training procedure, we examined the savings that occur on each subsequent day of training that ultimately leads to the establishment of LTM (Figure 2-C). The number of bites during the immediate post-test each day was always less than the pre-test score, even if the decrease was not statistically significant. Further, the pre-test score on the following day (which is also a test for 24-h memory) was always, with the exception of day 2, higher than the 10-min post-test of the previous day. STM had appeared first on the first day and snails saved STM by the "every day conditioning paradigm" furthermore by day 4, LTM had formed. This saw-tooth decrease over the training days indicates that the snails had a day-to-day memory savings (i.e., the association between the CS-US was stronger each day) that eventually resulted in LTM formation by day 4.

Thus our experimental findings demonstrated that the TAC was acquired with the spaced conditioning of 5 trails for 4 days training. The minimum requirement to acquire STM and LTM can be modulated by various environmental change such as immediate change of physical (environmental temperature) and/or chemical environments (degree of phosphorylation).

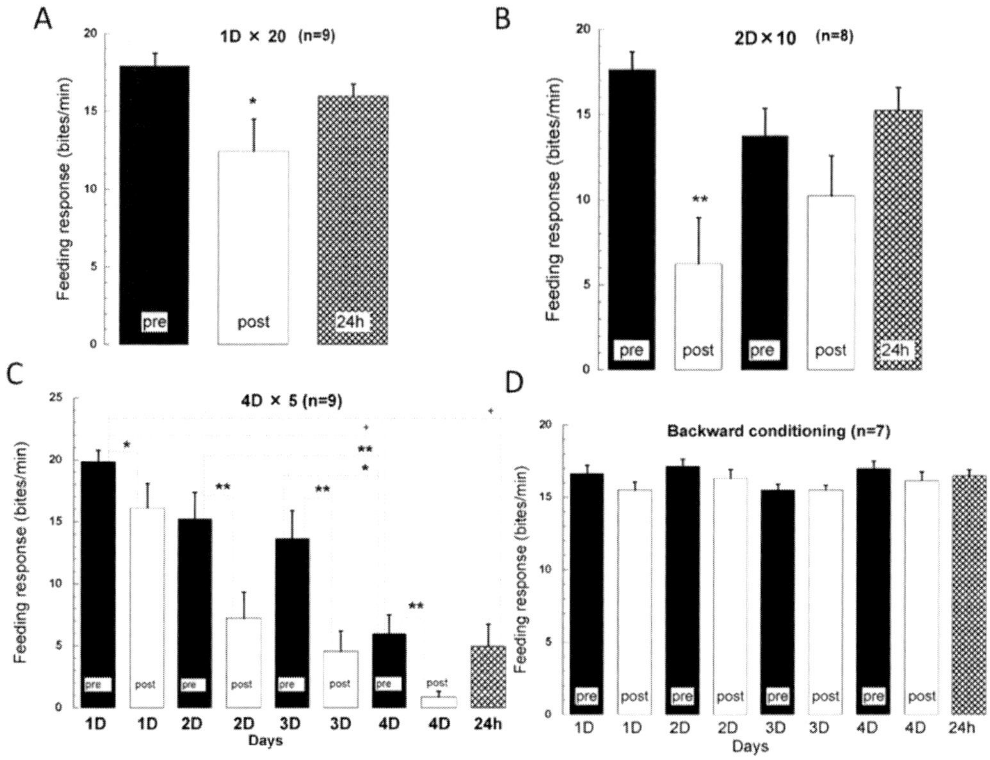

Figure 2. Feeding responses after 1-day conditioning (1D × 20 trials) (A), 2-day conditioning (2D × 10 trials) (B) and 4-day conditioning (4D × 5 trials) (C). Note that only 4-day conditioning resulted in LTM, at 24 h's memory. (D) This conditioning shows the temporal specificity as the paired US-CS (Backward conditioning) trials do not result in STM/LTM. * $p<0.05$, ** $p<0.01$, $^{+}p<0.0001$ in significant level compared to the control (pre). In this figure 'post' denotes 10 min post-test. Note that with 1-day, 2-day conditioning the conditioning effect, i.e., memory did not last until the following day. On the other hand with 4-day training procedure (C) the suppressive effect was cumulative, the feeding response decreased day by day. Modified from Takahashi et al. 2013 [58].

2. Modulation of STM and LTM

2.1. Activation of Protein Kinase C Enhances Memory Consolidation

Prior activation of the α and ε isozymes of protein kinase C (PKC), using Bryostatin (Bryo) eliminates the requirement for protein synthesis in classical conditioning of the marine mollusc *Hermissenda c.* [54, 59] and has similar efficacy in mammals [60]. These findings suggested that protein synthesis was not critical for the actual encoding of associative memory, but rather for providing essential biochemical and structural components of the synaptic apparatus required for implementation of the memory storage process. After acquisition of learning in Pavlovian conditioned *Lymnaea* their feeding and withdrawal related neurons were selectively stained with the antibody of the phosphorylated protein by PKC [61, 62]. These findings suggested that after memory formation feeding- and withdrawal-related neurons were selectively phosphorylated by PKC with TAC. In this chapter, I described a mechanism by which STM becomes consolidated into LTM by the activation of PKC-mediated phosphorylation, following a classical conditioning procedure in *Lymnaea s.* based on the recent findings. We previously studied whether PKC was involved in TAC in *Lymnaea* by the application of low doses of Bryo (sub nano molar dose) and specific PKC inhibitor bis-indolylmaleimide-XI Ro-0432 (Ro-32) before the conditioning regimen, and determined the Bryo application time-window in which ITM or LTM formation was most enhanced [63]. We examined whether a PKC activator could enhance memory formation by incubating snails with a PKC activator, Bryo, or a PKC inhibitor, Ro-32, before conditioning and compared the pre-test scores with the 10 min post-test and 24 h post-test scores as shown in Figure 3.

Five paired CS-US presentations did not significantly alter feeding behavior after conditioning (5-cont), while 10 paired CS-US presentations resulted in a significantly decreased numbers of feeding bites (10-cont). Thus, the criteria defining STM were met by the significantly reduced 10 min post-test feeding score compared with the pre-test following 10 paired CS-US training events (Figure 3A).

Snails incubated in nano-dose of Bryo (0.5 ng/ml) exhibited significantly reduced numbers of mouth openings with 5 paired CS-US presentations. These data show that Bryo augmented STM formation. That is, STM formed after 5 pairings of the CS-US. However, in Bryo 5 pairings of the CS-US were not sufficient to cause LTM formation.

However, when we incubated *Lymnaea* in Bryo and then subjected them to 10 CS-US pairings we observed both STM and LTM formation.

In summary, the feeding behavior, i.e., the number of mouth openings to CS presentation, progressively suppressed as the number of CS-US pairings increased. With the same number of pairings, however, Bryo-treated animals always showed greater reductions in the number of biting response compared to Bryo-untreated animals. LTM always formed in Bryo-treated animals, even with suboptimal numbers of CS-US pairings, i.e., snails conditioned with 10 CS-US pairings acquired LTM.

To examine and confirm the contribution of PKC to memory formation, snails were pre-incubated before the conditioning procedure an trained with a sub optimal training procedure as follows: First, the PKC activator Bryo for 45 min, followed by the PKC inhibitor Ro-32 for 25 min, or *vice versa*. Control snails simply received 10 paired CS-US training trials. The experimental results indicated that STM was formed under all conditions, including in the presence of the PKC inhibitor Ro-32. Thus, STM appeared to be formed even with minimal activation of PKC phosphorylation. Only snails treated with Bryo alone or with Bryo first followed immediately by Ro-32 developed memory that persisted for at least 24 h (LTM). Animals incubated first with Ro-32 followed by Bryo incubation also showed STM, but not LTM. The results were the same as those of the control or those with only incubation with Ro-32. We have to mention that the feeding score tested at 10 min post-test with Ro-32 was not greatly decreased as to the score of control (statistical significance in control was $p<0.0001$, while $p<0.05$ in Ro-32). In this sense there was minor effect on STM formation mechanism with Ro-32. Although the statistical probabilities for STM compared to controls were the same for Bryo and Bryo-Ro exposure ($p<0.0001$), the actual bite scores demonstrating STM were lower for Bryo alone (Bryo: 5.4 ± 1.05 bites/min; Bryo-Ro: 11.5 ± 0.99 bites/min). Importantly, LTM formation was still facilitated when snails were treated with Bryo followed by Ro-32 before the conditioning paradigm started (Bryo: 16 ± 1.19 bites/min; Bryo-Ro: 14.3 ± 1.02 bites/min). Thus, the PKC inhibitor was unable to suppress LTM formation initiated by pre-exposure to Bryo.

Memory formation in TAC under various experimental conditions in terms of PKC involvement is shown in Figure 3B.

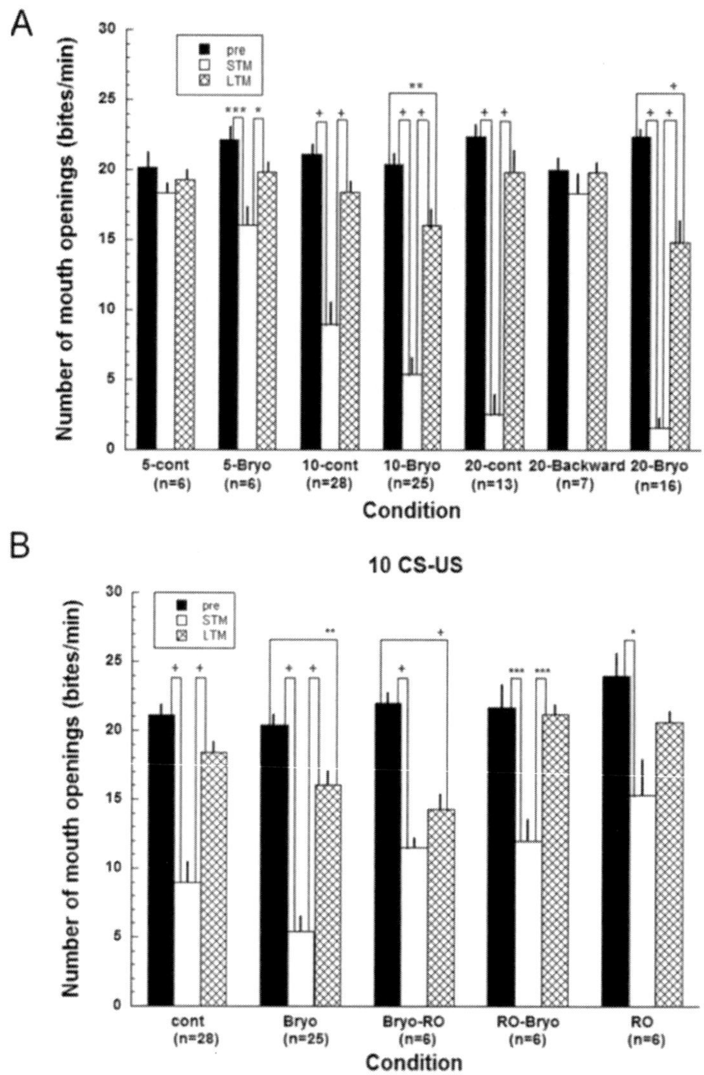

Figure 3. (A) Effects of training trial repetitions (5,10,20) of CS-US associative pairings and Bryo on memory formation. STM and LTM were evaluated from 10 min post-test and 24 h post-test, respectively. Five paired CS-US presentations did not lead to the formation of STM, unless Bryo was added. Ten and 20 paired CS-US trials alone led to STM formation, but Bryo enhanced LTM formation progressively. Snails incubated with Ro-32 prior to starting the TAC procedure for 20 min did not show STM and LTM. Twenty US-CS pairings, backward conditioning did not lead to suppressed feeding behavior. (B) Role of PKC phosphorylation on taste avoidance conditioning. Feeding behavior was examined at 10 nib post-test (STM) and 24 h post-test (LTM) following 10 paired CS-US training trials subsequent to application of Bryo and/or the PKC inhibitor, Ro-32. LTM was formed with pre-incubation with Bryo alone or Bryo followed immediately by Ro-32. Ro-32 alone or preceding Bryo, however, inhibited LTM formation. The dose of Bryo was 0.5 ng/ml in all the experiment. $*p<0.05$; $**p<0.01$; $***p<0.001$; $^{+}p<0.0001$. Modified from Takigami et al. 2014[63].

As previously mentioned, control group snails with conditioned with less than 10 paired CS-US presentations demonstrated only STM but never LTM (Figure 3A). Furthermore, snails incubated exclusively with Bryo always showed STM acquisition independent of the timing of Bryo application (Figure 4).

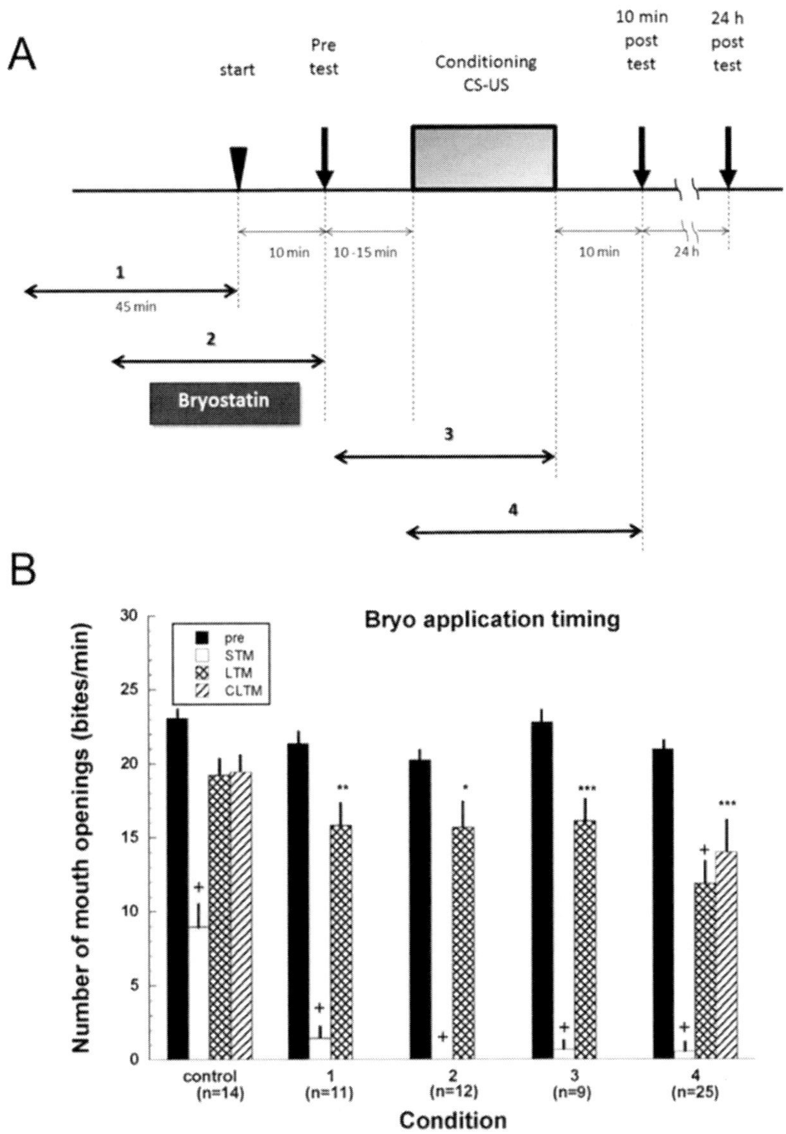

Figure 4. (A) Experimental protocols for basic taste avoidance conditioning of *Lymnaea s.* Snails were adapted for 10 min prior to sucrose administration (CS) and pre-test recording of feeding behavior (bites/min) was made, followed by a recovery period (10-15 min) and then CS-sucrose/US-head touch paired associative conditioning trails (10 or 20 pairings). Post-conditioning feeding response were subsequently measured at the 10 min post-test, 24 h post-test and compared to pre-test score. Application time-line effects of Bryo. Snails were incubated with Bryo (0.5 ng/ml; 45 min total period) at four different time-points during the basic protocol: 1, before the start of the experiment; 2, before the pre-test; 3, during the conditioning period; 4, after conditioning. (B) Memory consolidation versus application timing of Bryo. STM and LTM formation is dependent of the timing of Bryo administration; all the animals (n=71) acquired STM and LTM for condition 1, 2, 3, and 4 (see (A) for explanation). Consolidated LTM (CLTM), persisting longer than 2 days, was observed only when Bryo was administered immediately following the 20-trial training regimen (condition4). Data are presented as mean ± S.E. *$p<0.05$; **$p<0.01$; ***$p<0.001$; $^{+}p<0.0001$. Modified from Takigami et al., 2014 [63].

Conversely, animals incubated with Bryo at different time-points during the training-testing procedure demonstrated a less significant reduction in feeding behavior at the 24 h post-test (LTM) than at the pre-test ($p<0.05$), compared to their previous STM-biting rates. Snails treated with Bryo after the conditioning paradigm (i.e., post conditioning period; condition 4) showed long-lasting reduced feeding response: STM ($p<0.0001$), 24 h LTM ($p<0.0001$), and 48 h LTM ($p=0.0005$). Thus, Bryo exposure immediately after conditioning (experimental condition 4) shows that this is a critical period to enhance the transition from 24 h LTM to 48 h LTM or as it is sometimes called consolidated long-term memory (CLTM). We have not carried out 48 h post-test in condition 1-3. STM, LTM, and CLTM criteria were based on a comparison of the pre-test scores with the experimental condition results.

The fact that snails required only 10 CS-US presentations for STM formation indicates how rapid their acquisition of learning and memory are compared to other model systems. The finding that forming STM did not necessarily lead to LTM formation, is noteworthy with respect to the PKC activation levels required for the consolidation process leading to LTM formation. A minimum number of CS-US pairings are required to form STM irrespective of whether Bryo is present. However, if Bryo is present then the minimum number of pairings leads to LTM formation. Thus, the addition of nano-dose levels of Bryo was sufficient to achieve the necessary levels of PKC for LTM formation.

The effects of PKC inhibitor Ro-32 and its interaction with Bryo differed for each of the two memory states, and may reflect the time-course of their mode of actions.

LTM formation progressed normally when Bryo treatment preceded Ro-32 treatment (Bryo-RO). The degree of LTM feeding suppression compared to pretest levels was greater with Bryo-RO than Bryo treatment alone (Bryo-RO, 14.3 ± 1.02 bites/min vs. Bryo, 16 ± 1.19 bites/min; 36% vs. 20% suppression rate, respectively). This was not the case if the drug applications were reversed. Ro-32 effectively eliminated the transition of STM to LTM when administered prior to Bryo.

The results of the Bryo application timeline studies indicated that the timing of Bryo application is not critical to LTM formation: all four time-points facilitated the consolidation process from STM to LTM. LTM lasting longer than 48 h (CLTM) was observed when Bryo was applied following conditioning with 20 paired training (condition 4 in Figure 4A). However we did not examine whether LTM was present 48 h after training in experimental condition 1-3. As noted above, LTM was prevented when Ro-32 was administered prior to Bryo exposure (Figure 3B) due to the inhibitory effect of Ro-32 on the initial LTM formation.

Unlike most other known exogenous PKC activator, such as phorbol esters, Bryo is a highly specific, non-tumorigenic, and potent activator of PKC [64]. The physiologic effects observed in *Lymnaea* mimicked those of Bryo and Ro-32 in the classical conditioning in *Hermissenda*, namely: 1) PKC activation and the resultant memory enhancement conditions could be initiated solely by Bryo exposure either before training, or synergistically if given immediately after paired training; 2) Ro-32 inhibited LTM only if administered before training or before Bryo exposure; and 3) Conversely, post-training or post-Bryo administration of Ro-32 had no inhibitory effects. Bryo exposure also induced memory enhancement of TAC in *Lymnaea s.* from STM to LTM or LTM to CLTM (Figure 3). The PKC inhibitory drug application results indicated that Ro-32 does not completely inhibit learning or STM formation. This suggests to us that the initial influx of calcium through voltage dependent channels still triggers PKC autophosphorylation, leading to the initial establishment of STM or involvement of another signaling molecules. The greater degree of feeding suppression seen under experimental conditions using Bryo followed by Ro-32 compared to Bryo alone, also suggests synergistic actions other than just PKC-related mechanisms. This synergism might involve protein kinase A (PKA) or mitogen activated protein kinase (MAPK) mediated phosphorylation, as reported in *Aplysia*, *Hermissenda*, and *Lymnaea* [65-68].

All snails tested (n=71) on involvement of PKC application timing for the memory formation demonstrated a significant reduction of feeding behavior at the 10 min post-test compared to the pre-test ($p<0.001$). These findings indicate that at least 10 paired CS-US presentations are needed to cause learning and the subsequent formation of STM. The data also indicated that each of the four Bryo time-window application trails significantly enhanced LTM formation ($p<0.05$) regardless of the timing of Bryo exposure. Under conditions 1-3, significantly more snails incubated with Bryo had the ability to form LTM compared with controls. Bryo exposure immediately following a 20-trial, CS-US conditioning presentation (condition 4) however, not only led to LTM ($p<0.0001$), but facilitated its conversion to CLTM ($p=0.0005$; Figure 4B). However we could not rule out the possibility that this enhancement was exclusively due to the timeline of Bryo exposure, because we did not examine 48 h post-test in condition 1-3.

2.2. Cooling Induced Amnesia-'Cold-block'- Interfered with Formation of STM and LTM

Previously a terrestrial slug, *limax*, abolished the memory by the instantaneous cooling, and this phenomenon was called 'cooling induced amnesia' [69, 70]. After the original findings, this instantaneous amnesia induced by cooling was true for the most of invertebrate animals [71-73] and xenopus oocyte [74]. We tested whether STM/LTM was modulated by the instantaneous application of lowering environmental temperature, which was referred to 'cold-block'. The original findings in the land snail has some inherent limitation for stimulus application, it was not easy to lower the ambient temperature instantly. On the other hand as *Lymnaea s.* lives in fresh water it is easy to change their body temperature by putting snails into the chilled water.

We have established that our TAC needs 20 CS-US repeated paired presentations split into 5 trails per day for 4 days as shown in Figure 2C. This classical conditioning was temporal specific as shown in Figure 2D by the behavioral scores after the backward conditioning. The backward conditioning is one of the control paradigm because by presenting the reversed sequence, i.e., US-CS paired stimuli, does not result in feeding score suppression. The data shown in Figure 2D indicate that backward conditioning does not prevent the CS from eliciting feeding. The cold-block procedure immediate after the 5th presentation of the CS-US pairings in each day prevented any savings from occurring over days and also prevented both STM and LTM formation (Figure 5B).

The cold-block applied immediate after the conditioning trials was the reversible procedure.

The timing to apply cold-block was critical to form STM/LTM as shown in Figure 5A. We demonstrated that STM and LTM were blocked by subjecting trained snails to the cold-block procedure immediately after the last training session each day (Figure 5B). Next, we set out to determine the critical time period in the memory consolidation process during which the cold-block procedure no longer disrupted memory formation. We did this by varying the time to apply the cold-block procedure after the last CS-US pairing each day as shown in Figure 5A. We demonstrated that STM and LTM were blocked by subjecting trained snails to the cold-block procedure immediately after the last training session each day.

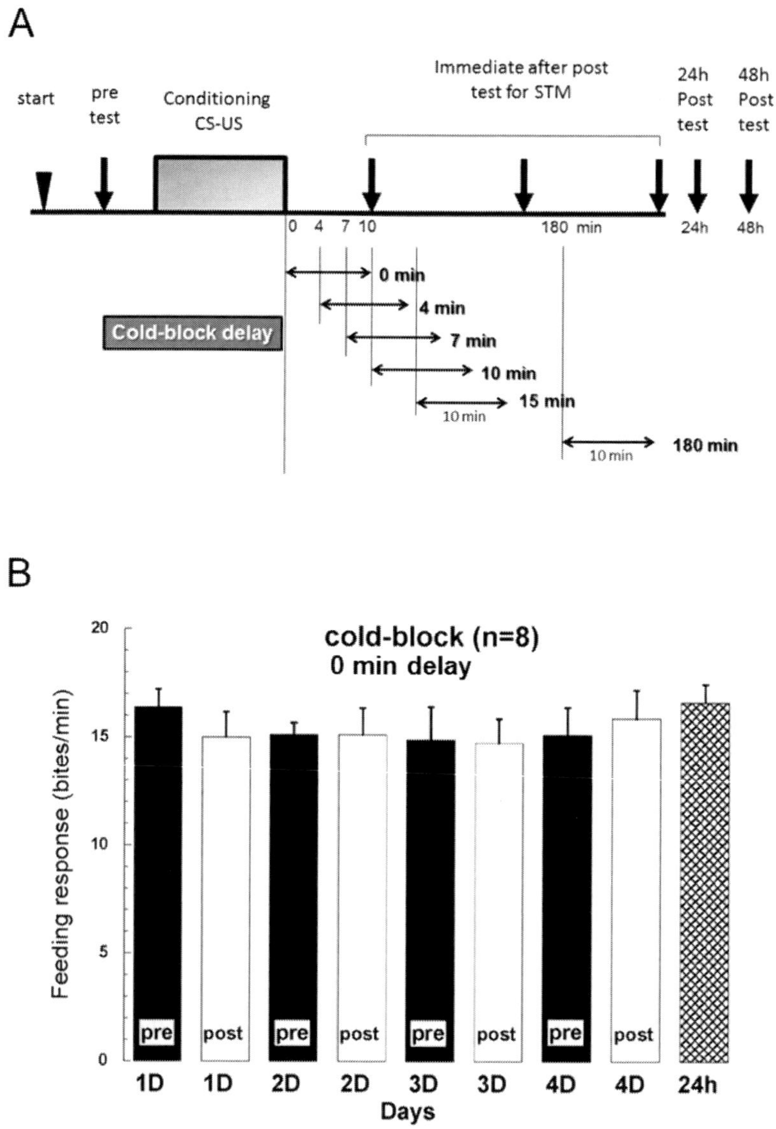

Figure 5. (A) Experimental protocols for Taste Avoidance Conditioning (TAC) of *Lymnaea s.* in a day. Snails were first adapted for 10 min prior to sucrose administration (CS) and pre-test recording of feeding behavior (bites/min) was made; then there was a 10-15 min recovery period followed by CS-sucrose/US-head touch paired associative conditioning trials (4, 10, or 20pairings). Post-conditioning feeding responses were subsequently measured at ~10 min, after cooling, at 24 h, and at 48 h, and compared to the pre-test scores. Cold-block was applied immediately after conditioning (0-min delay), 4 min, 7 min, 10 min, 15 min, and 180 min, and the effect of the cold-block was evaluated by sucrose administration (CS). STM was evaluated by an immediate post-test. LTM was evaluated by a 24 h and 48 h post-tests. (B) Dynamic changes in the feeding response of 4-days training including 'cold-block' procedure. No significant differences were detected after applying the cold-block immediately after conditioning each day. Values were compared to the pre-test value (1D pre) if there was a significant main effect evaluated by repeated measures of ANOVA. Data are presented as mean ± standard error. Modified from Takahashi et al., 2013 [58].

Next, we set out to determine the critical time period in the memory consolidation process during which the cold-block procedure no longer disrupted memory formation. We did this by varying the time to apply the cold-block procedure after the last CS-US pairing each day as shown in Figure 5A. Each group of 8 snails experienced the same number of cold-block procedure, but the cold-block was applied immediately (0 min), 4, 7, 10, 15 or 180 min after the last pairing. The daily effects of cold-block on feeding behavior are shown in Figure 6. Each graph is made up of both the mean response and individual response of each snail. On each day the pre-test score is plotted first and then the immediate post-test scores follow. The scores of the second immediate post-test applied 27 min after the last CS-US pairing in Figure 6E and 6F are not included. The groups receiving the 4 and 7 min delayed cold-block exhibited STM but not LTM. The groups receiving the cold-block 10 and 15 after the final pairing showed both STM and LTM formation. Also note that in Figure 6D and 6E, the feeding score tended to decrease day by day slowly compared to Figure 2C. Finally, LTM was not formed when the cold-block was given 180 min after the last pairing.

The detailed statistical test is available in the original study [58]. Both STM and LTM formation was prevented if the cold-block procedure was initiated without delay of the last CS-US pairing, i.e., 0 min delay. With 10-min delay STM was formed by day 1 and LTM by day 3. Following a 15-min delay after the last CS-US pairing, STM was observed on the first day, however, LTM formed by day 2. We concluded that delayed application of the cold-block procedure for 10 min or 15 min following each day's CS-US pairing allowed for the formation of both STM and LTM. The results among 6 different delays on STM and LTM were summarized in Table 1. The decreasing tendency in feeding response to CS was obvious in Figure 6D and 6E; STM formation was characterized in saw-tooth decrease over the training days; LTM formation was characterized in approaching the mean value to zero but took time in comparison with that without cold-block.

The results of the presentation of the cold-block procedure to the snails at 180 min following the last pairing each day are shown in Figure 6F. Neither STM nor LTM was formed on any day, and the variation of data was smaller than the other conditions in Figure 6. Thus, we conclude that delaying the cold-block procedure for 180 min following the last paring of the CS-US each day was sufficient to block both STM and LTM formation.

In the series of experiments, we investigated whether snails that underwent the cold-block procedure to induce amnesia could still learn and form memory. In these experiments (Figure 7A), 7 snails that had received 5 CS-US pairings for 4 days and that had received the cold-block immediately following the last pairing on each day were again trained for another 4 days, but without cold-block. These snails exhibited STM and LTM on the expected days. Animals that had been unable to to form STM because of the cold-block were now able to form both STM and LTM. Thus, the cold-block, while able to block both STM and LTM when applied at specific times, did not cause an irreversible blockage of learning and memory formation. Also, the rate of learning and memory in these snails appeared to be no faster than that in naïve snails (see Figure 2C) despite having received previous training, albeit followed by the cold-block each day.

To test whether the conditioned memory was enhanced following PKC activation, we used Bryo following training and cold-block application. Thus, a group of 6 snails received 3 days of training (5 pairings of the CS-US followed immediately by the cold-block each day). Neither STM nor LTM was observed. Before beginning the day 4 training, we placed the snails in Bryo-containing pond water at 22 °C for 45 min and then proceeded to train the snails (5 CS-US pairings).

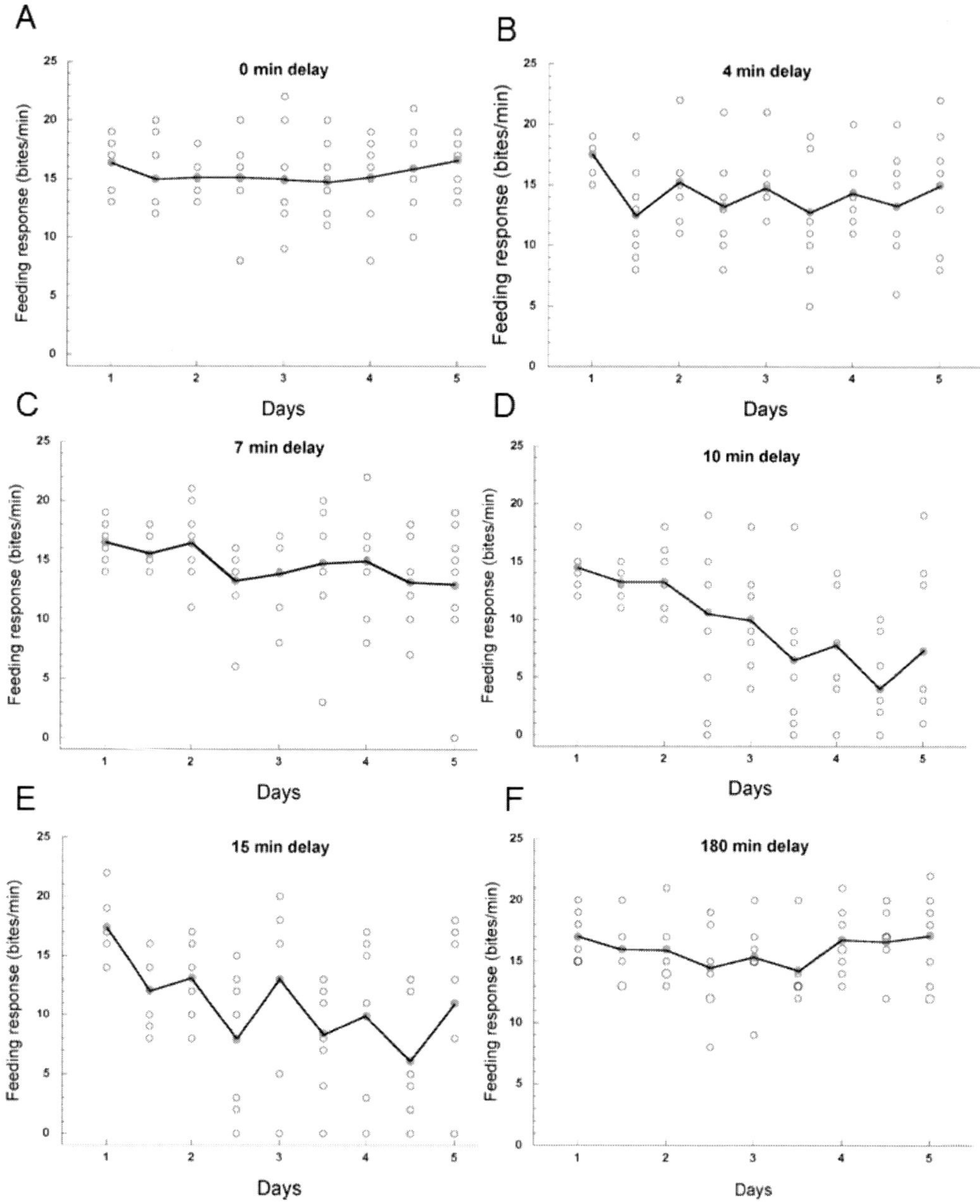

Figure 6. Effect of cold-lock on the feeding response to the CS at various time points, (A) 0-min delay (n=8), (B) 4-min delay (n=8), (C) 7-min delay (n=8), (D) 10-min delay (n=8), (E) 15-min delay (n=8), and (F) 180-min delay (n=8). Conditioning was performed with 5 paired CS-US presentations for 4 days. Each scattered graphs were composed of 8 snails' individual data, the mean of feeding scores was connected along with daily change. Within a day the scores of pre-test were shown first then immediate post-test score were followed. The immediate post-test in 15 min and 180 min delay was applied to snails twice at 10 min and 27min after the final CS-US pairing. In the case snails received 15 min delay cold-block, they were examined to CS application twice before and after the cold-block, while 180 min delayed snails were tested before the cold-block twice. The score of second immediate post-test in E and F were not included. Modified from Takahashi et al., 2013 [58].

This training procedure should not have resulted in STM, however STM (significant feeding suppression after the 5 pairings within a day) was observed even though the cold-block was applied immediately after the 5th CS-US pairing on day 4 of training as shown in Figure 7B.

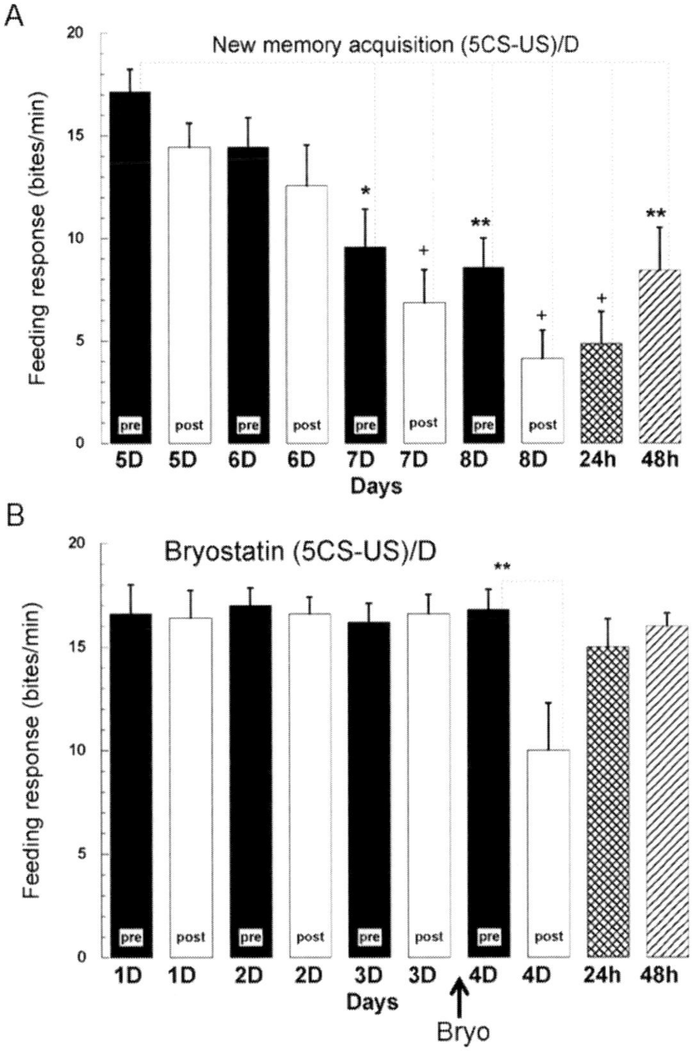

Figure 7. Taste avoidance conditioning was reversible (A) after immediate cold-block, and (B) during the amnesic state when PKC activation resulted in STM on the final day of conditioning. (A) Snails experienced 0-min delay cold-block after conditioning for 4 days, and were then conditioned without cold-block for the next 4 days (n=7). Note that the learning kinetics to obtain the new memory formation were similar to those shown in Figure 6A. (B) Snails did not exhibit memory when the cold-block was applied immediately following the CS-US pairing (i.e., days 1-3). However, on day 4 before the training procedure snails were exposed to the PKC activator Bryo (0.5 ng/ml) for 45 min, and then received the 5 CS-US pairings in 'normal water'. This resulted in STM but not LTM formation (n=5). Note that 5 paired CS-US presentation was insufficient to cause STM in Figure 2A. Bryo represents the time to soak animals in Bryo containing water on day 4 before the conditioning paradigm in the horizontal axis. **p*<0.05, ***p*<0.01, ****p*<0.001, ^{+}p<0.0001. Modified from Takahashi et al., 2013 [58].

The findings from the cold-block experiments confirmed followings; 1) A cold-block procedure prevents the formation of both STM and LTM if applied without delay of the last CS-US pairing on each day's training; 2) A cold-block procedure prevents the formation of

LTM if applied within 10 min of the last CS-US pairing on each day's training; 3) A time delay of 180 min following the last CS-US pairing each day leads to no memory saving; 4) The cold-block procedure, while sufficient to block both STM and LTM formation, did not permanently prevent snails from learning and forming both STM and LTM; 5) PKC activity appeared to enhance memory formation.

Table 1. Summary of the cold-block delay on the formation of STM and LTM, ×: no memory formation; 1D: appearance at the first day; 2D: appearance at the second day; 3D: appearance at the third day; 4D: appearance at the 4th day; 5D: appearance at the 5th day

Delay (min)	STM	LTM
0	×	×
4	1D, 2D	×
7	2D	×
10	1D	3D, 4D, 5D
15	1D	2D, 4D
180	×	×

2.3. Effective Methods of TAC by Spaced Training Procedure Resulted in STM and LTM in a Single Day

As mentioned above, we developed TAC in *Lymnaea s.* to study the neuronal basis of associative learning and memory formation. To accomplish this, we applied a sucrose solution to the lips, which elicits feeding, as the CS and we applied a tactile stimulus to the head, which stops feeding, as the US. After paired CS-US presentations, the snails learn and form memory for TAC. Importantly, however, this method of conditioning requires manual skill to consistently apply the US to the head of the animal. If the US is too week, feeding does not cease, and if the US is too strong, the snails respond with WBWR, from which recovery took 10 min to hours. To establish a more consistently effective conditioning procedure, we developed new methodology for use as the US. We used a high electrical voltage field around the animal, which we refer to as US2, while the original tactile stimulus is referred to as US1 hereafter. The variables to apply US are the stimulus strength, duration and the timing in relation to preceding CS. It was difficult to consistently apply the constant US1, i.e., the mechanical head poking stimulus should apply at the same timing, same strength and constant duration. On the other hand, the electrical stimulation is controllable for every parameter mentioned above we can look forward to obtaining the consistent conditioning animals [33].

US2 in our new procedure serves as an enough aversive stimulus that causes the snail to start withdrawing, which is incompatible with feeding. Thus, the feeding response elicited by the CS is immediately terminated. However, possibly because there is little current flow (i.e., micro ampere order), the snails quickly cease withdrawal. Thus, paired presentations of the CS-US2 can occur, and with successful conditioning, the CS no longer elicits feeding. However, if the order of the pairings is reversed, i.e., US2-CS did not elicit feeding suppression in response to CS presentation. Previously, 20 paired CS-US1 presentations were required to obtain learning and LTM formation [18]. We found that with the new US2, successful TAC and LTM formation occurred sooner and persisted longer.

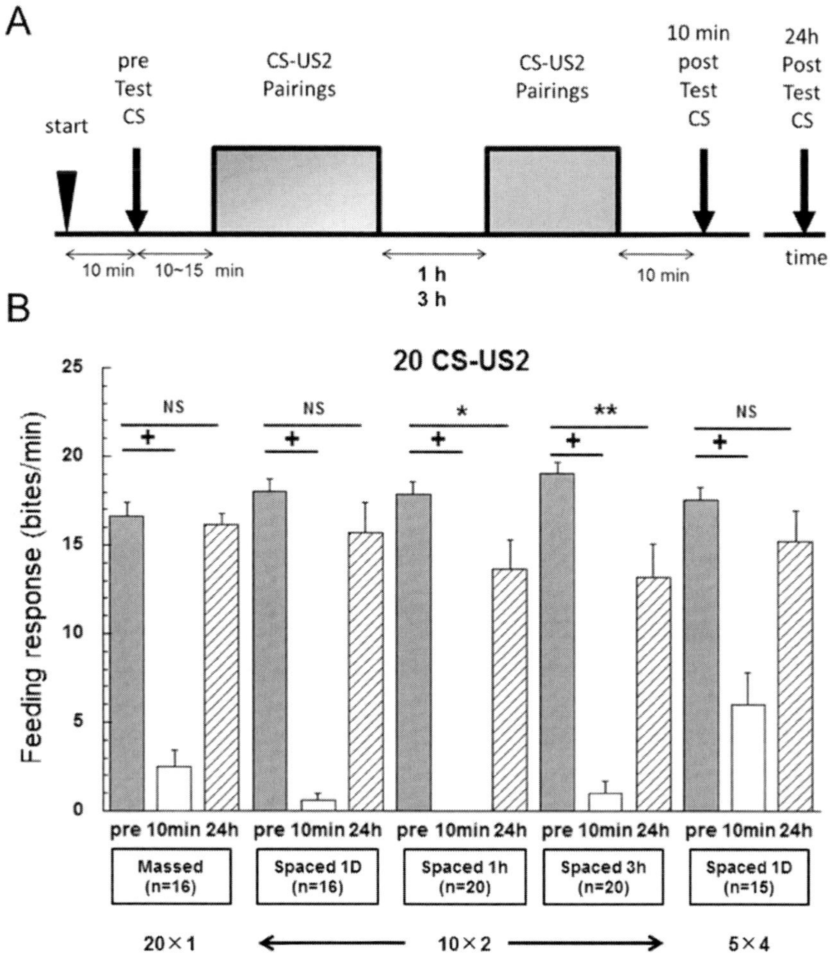

Figure 8. (A) Temporal sequence of the spaced training procedure is illustrated. In this conditioning procedure US was high voltage with little current (~80 μA) as US2. In the conditioning trial, the first and second sets of the CS-US2 presentation was separated by 1 or 3 h interval. In case of 20/8 CS-US2 presentations, the first trial was composed of 10/5 repeated pairs of CS-US2, while the second trail composed of 10/3 CS-US2 presentations. (B) Twenty paired CS-US2 presentations in various conditions: massed conditioning (20 CS-US2 in a day); spaced conditioning (10 CS-US2 in a day over 2 days); spaced conditioning (10 CS-US2 pairs interposing 1 h interval in a day); spaced conditioning (10 CS-US2 pairs interposing 3 h interval in a day); spaced conditioning (5 CS-US2 pairs in a day over 4 days). Note that the spaced conditioning procedure consisting of 10 pairs presentations in a day with interposing interval resulted in both STM and LTM. NS: not significant, $*p<0.05$, $**p<0.01$, $^{+}p<0.0001$. Modified from Takigami et al., 2014 [34].

The minimal number of pairings to form LTM for TAC with sucrose and mechanical stimulation of the snail's head was 20 repeated CS-US1 presentations over the course of 4 days as shown in Figure 2 [58]. Here we tested whether a similar number of pairings using sucrose as CS and US2, was necessary to produce LTM. These data are presented in Figure 8B: Twenty presentations of CS-US2 on a single day (i.e., the massed training procedure in the leftmost of Figure 8B), also resulted in STM but did not form LTM; Ten pairs of the CS-US2 on a day for 2 days, which we termed 'spaced 1D' training also resulted in STM but not LTM; On the other hand interposing a 1 h or 3 h interval, which were designated as 'spaced 1 h' or 'spaced 3 h', between the first 10 pairs presentations and second 10 pairs resulted in both STM and LTM; The 'original' spaced training of 5 pairs of CS-US1 over 4 days training procedure resulted in STM but not LTM as shown in Figure 8B of the last panel.

We then examined using the massed training procedure how many trials were necessary to form both STM and LTM. We subjected different naïve cohort of snails to either 5, 8, 10, 15 or 20 paired presentations of CS-US2 on a single day. We found that each pairing procedure resulted in STM but we only observed LTM ($p<0.001$) with the 15 paired presentation procedure. These data are shown in Figure 10A. In Figure 10B we plotted for each of the pairing groups in their response to both STM (10 min post-test) and LTM (24 h post-test). As can be best seen there is a U-shape curve for the 24 h post-test response. While LTM was only demonstrated for the 15 paired presentations it can be seen that the 'dose-response' curve for the number of pairings needed to result in LTM is not a 'straight' line. When we plot the 10 min post-test results it can be seen that while all numbers of pairings produce STM, STM gets 'better' with increased pairings.

With the massed training procedure we found that 15 paired CS-US presentations was the optimal number. That is, 15 pairings on a single day was the only procedure that resulted in LTM formation. Next we asked whether allowing a 'rest' interval of 1 h or 3 h between 'sets' of pairings would result in 'better' LTM formation. For example, would a rest interval of 1 h or 3 h following the first 10 CS-US pairings cause better LTM? These data are presented in Figure 11A and show that there is more significant suppression in terms of the feeding score to a CS. Both scores of post-test at 10 min and 24 h by the spaced training showed more significant suppression than that by the massed training procedure. Though the total paired presentations are the same, inserting a longer rest interval after the first 10 pairings resulted in a greater degree of feeding suppression. Interestingly, the degree of suppressed feeding response at 10 min after the conditioning was almost the same in massed or spaced conditioning. Neither STM nor LTM was observed with backward conditioning as shown in Figure 11A in the rightmost panel. This experiment confirmed that the decreased response to the CS following our conditioning procedure was a true example of associative behavior and not the result of just the presentation of the US. Having shown that the interposition of a rest interval between the first 10 pairings of a 15 paired training procedure resulted in better LTM formation, we tested whether such a rest interval would also result in LTM with reduced numbers of pairings. We therefore used a 8 paired presentation sequence in which we inserted a rest interval following the first 5 paired CS-US2 presentations. These are shown in Figure 11B. Interposition of a 1 h interval following the fifth CS-US2 pairing resulted in STM but not LTM. However, when a 3 h rest interval was interposed LTM was observed. Furthermore, if we first exposed snails to the PKC activator, Bryo also greatly suppressed the feeding response at the 10 min post-test.

Having shown that pre-treatment with Bryo significantly enhanced LTM formation we asked whether Bryo if applied during the pairing sequence (8 paired CS-US2 trials) would also enhance LTM. We termed this application of Bryo as 'Bryo-Mid'. The rest interval in this series of experiments occurred following the fifth paired CS-US2 presentation. In the Bryo-Mid exposure group snails were exposed to Bryo during 1 h-interval for 40 min before the start of the experiment. The Bryo-Mid exposure group had stronger feeding suppression in the LTM post-test session (Bryo pre: $p<0.01$; Bryo Mid: $p<0.001$).

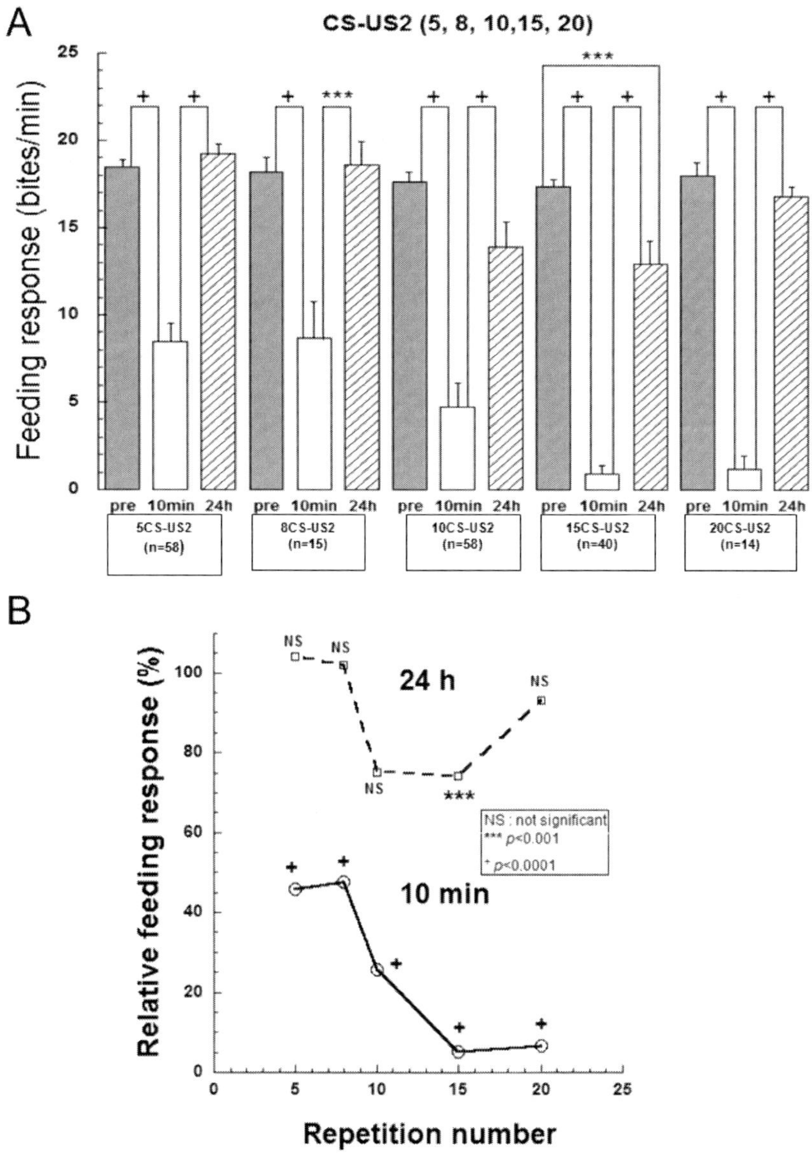

Figure 10. Massed training. (A) Comparison of the feeding response following 5, 8, 10, 15 and 20 paired presentations of the CS-US2 in massed training procedure. Note that all condition resulted in STM formation, however LTM was only formed with 15 paired CS-US2 presentations. $^{***}p<0.001$, $^{+}p<0.0001$. (B) Relationship between the number of CS-US2 presentations and memory formation. Data are depicted from A. The ordinate, relative feeding response (%) represented the relative suppression score: pre-test score vs. 10 min post-test; pre-test score vs. 24 h post-test. Note that STM and LTM were formed only with 15 paired of CS-US2 presentations. NS: not significant, $^{***}p<0.001$, $^{+}p<0.0001$. Modified from Takigami et al., 2014 [34].

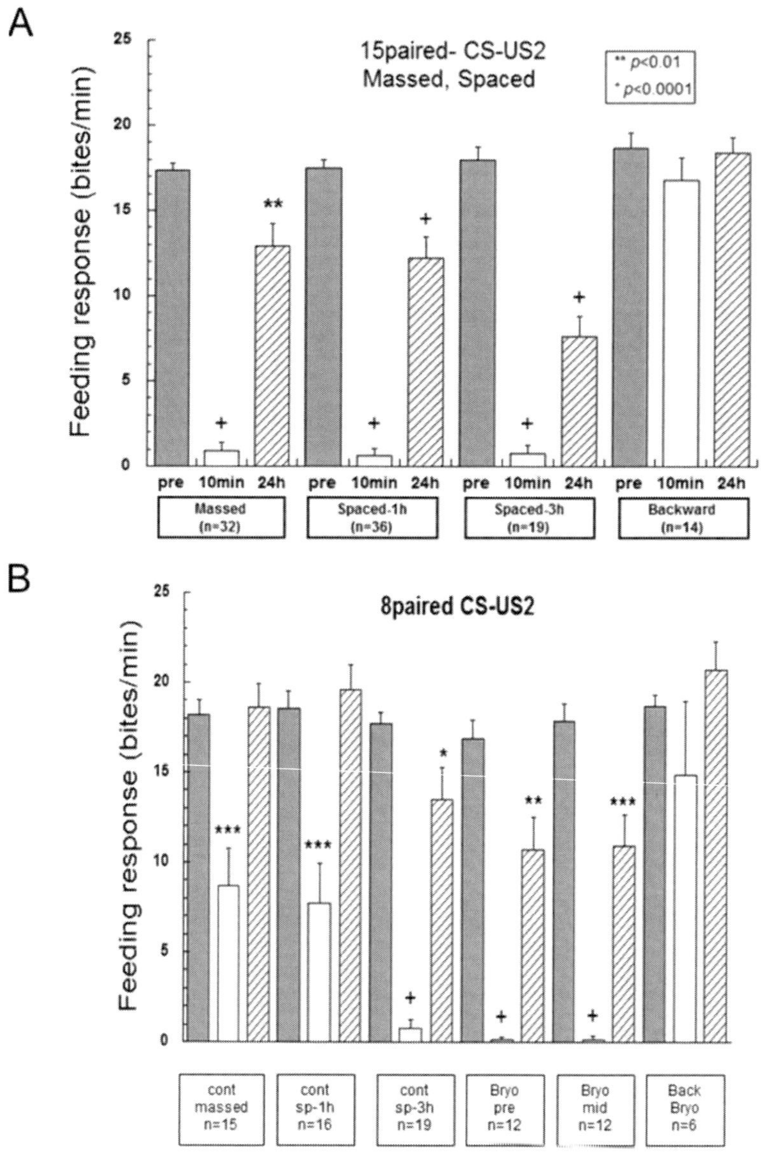

Figure 11. (A) The rest interval enhances memory formation. Comparison of the memory formation with of 15 paired presentations of CS-US2/US2-CS presentations. Spaced training was performed in the following manner. We first presented a set of 10 paired of CS-US2 presentations, following the 10th pair a rest interval (1 h or 3 h) was interposed. Then the remaining 5 paired presentations commenced. The reversed sequence of US2-CS was performed (i.e., backward conditioning) and this did not result in either STM or LTM. Note that with 15 paired CS-US2 presentations both massed and spaced training resulted in STM and LTM. However, the magnitude of feeding suppression was greater with the spaced training procedure and a 3 h interval was more effective than the 1 h interval. (B) Bryo enhances LTM formation. Comparison of the suppression of the feeding response following of 8 paired of CS-US2 or US2-CS presentations. Note that the Bryo concentration was five times diluted (0.1 ng/ml) than that shown in Figure 4. The left side panels show the results of the control pairings; while the right hand panels show the results when snails were exposed to Bryo either before (Bryo-pre) or during the training (Bryo-mid). Even with Bryo exposure backward conditioning did not result in either STM or LTM. Modified from Takigami et al., 2014 [34].

We have made use of a new high voltage electrical stimulus as the US2 in place of the tactile stimulus or chemical to evoke the WBWR that causes termination of the feeding response [33]. This US2 has the chief advantage in consistency. The US evokes the WBWR, but as we have found snails begin to emerge from their shells and locomote within 30 s of receiving the US2 [34]. Additionally, they also begin to feed again within the same period. We showed using the US2 when paired in a forward manner (CS-US2) with the sucrose CS it is possible to establish associative learning. By the new conditioning procedure with US2, the massed training as shown in Figure 10A, the CS-US2 lead to the establishment of STM. That is, the CS evoked significantly fewer biting movements 10 min after the last CS-US2 pairing. However, in this 'massed' training only with 15 pairings did we observe LTM. That is, the CS elicited statistically fewer bites when presented in the 24 h post-test session. It is uncertain why 20 paired presentations did not lead to LTM formation. Concerning the minimum number to produce LTM formation we can think about the Yerkes-Dorson Law which predict the optimal stress level for establishing LTM [75]. From this view point 15 paired presentations of CS-US may be less stressful than 20 pairings as shown in the U-shape curve in Figure 10B. This will be discussed below in the section 3.3 in this chapter.

Activation of PKC with low dose of Bryo (0.25 ng/ml) on days before associative learning was found to be effective in memory enhancement in the type B photoreceptor of sea snail *Hermissenda* [54, 59]. This memory enhancing effects was not merely occurred in invertebrate model systems, such as *Hermissenda*, and *Lymnaea* [58, 63, 76] but also true for the water maze-trained rat [60]. Activation of Bryo further induced the morphological modification of hippocampal pyramidal ells with increasing in the number of mushroom spines in the dendrite [77]. Activity of PKC isoforms is essential for many types of learning and memory and is crucial for memory disorders in human Alzheimer's dementia [78]. In the mammalian brain PKC isozymes play a regulatory role in amyloid production and accumulation, furthermore restoration of the impaired PKC signaling pathway pharmacologically results in therapeutic effects on Alzheimer's disease [79] and enhanced memory capacity, synaptic remodeling, repair and synaptogenesis [80].

3. Memory Formation

3.1. Mechanisms of STM Formation

At the cellular level, memory is evidenced by a change in synaptic activity at the pre-and/or post-synaptic neurons [81-83]. Changes in intrinsic membrane properties, however, also contribute to memory formation [84, 85].

In *Hermissenda*, a marine mollusk related species to *Lymnaea*, classical conditioning (a photic stimulus paired with a rotational turbulence) results in a change in the excitability of type B photoreceptors due to long-lasting K^+ channels inactivation [5, 86]. In *Hermissenda*, bryostatin enhances memory formation via a PKC pathway [87]. We demonstrated that Bryo enhances memory formation in *Lymnaea* as shown in Figure 3. These data are consistent with data from operant conditioning studies in *Lymnaea* showing that Bryo enhances long-lasting memory formation [76]. Future studies in *Lymnaea* should focus on examining changes in the excitability of neurons such as RPeD11 [25, 42, 88] to determine whether the changes in excitability induced by Bryo are required for LTM formation.

The data shown in this chapter are also consistent with the finding reported by Epstein et al. in 2003 that the protein synthesis required for ITM and LTM occurs within 15 min following conditioning and again after 60 min, suggesting that a second round of protein synthesis is required for memory following classical conditioning in *Hermissenda* [89]. In this chapter we showed that memory formation could be disrupted by applying a cold-block 180 min following completion of the conditioning. Thus, it appears that different forms of memory (e.g., ITM and LTM) following training. These findings are in contrast with the previous simpler suggestion that a single period of protein synthesis occurring immediately following conditioning underlies memory formation [57].

3.2. Mechanisms of LTM Formation

It is general assumed that LTM requires altered gene activity. Altering the molecular processes following learning during the consolidation period blocks the formation of memory. A critical step in the consolidation process leading to long-lasting forms of synaptic plasticity and LTM is the activation of immediate early genes such as activity-related cytoskeleton association protein, early growth response gene 1, CCAAT/enhancer binding protein, *c-fos*, *c-jun*, and the recently identified target gene insulin-like growth factor 2 [90-96].

One of the known immediate early genes cloned and sequenced in TAC in *Lymnaea* is CCAAT/enhancer binding protein [65, 97] and recent findings revealed that molluscan insulin-related peptide is involved in consolidation of TAC in *Lymnaea.* [98].

Protein synthesis is essential for LTM formation. Two periods of sensitivity are assumed; one occurring immediately after training [51] and the other occurring several hours later [89]. In *Lymnaea s.*, Fulton et al. 2005 published that one-trial appetitive conditioning data suggested that there is a single period of sensitivity for the protein synthesis necessary for LTM formation that lasts from between 10 min to 1 h after conditioning, based on studies using pharmacological agents (the translation inhibitor anisomycin and he transcription blocker actinomycin D) [57]. Fulton et al. 2008 showed the quite contrast results with the same appetitive conditioning that a cold-block procedure disrupted LTM only when applied immediately after conditioning whereas delaying the treatment by 10 min left the 24 h memory trace intact [71]. Pharmacological treatments, such as those used in the Fulton studies of 2005, have long-lasting irreversible effects on *Lymnaea* [57]. As demonstrated in our previous study, the cold-block induced amnesia was totally reversible and instantaneous [58, 72].

Cooling snails after operant conditioning within 30 s, reactivation of memory, or extinction does not adversely affect the health of the snails [72, 99, 100]. In our study, to block LTM formation, the cold-block had to be applied immediately within 10 min following the last CS-US presentation. Cooling snails immediately after the first bout of ITM training also prevents the establishment of the residual molecular trace [101].

We demonstrated that the formation of both STM and LTM was blocked by cooling immediately after conditioning. The group overall at certain time points did statistically exhibit STM (4, 7, 10 and 15 min delay), but not meet LTM formation in 4 and 7 min delay. We therefore believe that process of consolidation may proceed at earlier period by 10 min. We thought that the effect of the cold-block was due to cold-block induced halting or slowing down of the metabolic process and/or macromolecular synthesis necessary for LTM. Actually this cooling induced slow down macromolecular process was apparent comparing with Figure 6F and Figure 2A suggesting the protein synthesis for LTM formation may be interfered by a long delayed cold-block.

To further complicate matters, we found that applying the cold-block 180 min after the last CS-US pairing each day blocked memory formation entirely, which was un expected finding. It could be argued that STM was not observed because once formed it only persisted a short period of time and had dissipated prior to testing 180 min after the last CS-US pairing and after the cold-block procedure. Why LTM was not apparent on the last day following this procedure remains an important question because it appears that the consolidation period should have been completed within 10 min after each day's training. A somewhat similar phenomenon was recently been reported in various gastropod model systems such as *Lymnaea, Hermissenda*, and *Aplysia*. In the *Lymnaea* study, which examined appetitive one-trial conditioning, memory could be demonstrated at 10 min, 1 h, and 3 h, but not at 30 min or 2 h after a single training trials [102]. Moreover, the authors demonstrated that memory could be perturbed by a tactile stimulus, which initiated the WBWR, when applied 30 min and 2 h after the training trial. These memory lapses and susceptibility corresponded to transition between different phases of memory with different molecular requirements. In the *Hermissenda* study by Craw and Xue-Bian, changes in the protein levels at two different times following one-trial in vitro conditioning were examined using two-dimensional difference gel electrophoresis. Significant protein regulation was detected at 30 min and 3 h post-conditioning. These proteins were involved in diverse cellular function, such as translational regulation, cell signaling, cytoskeletal regulation, metabolic activity, and protein degradation [103]. In the *Aplysia* studies, memory at the behavioral level and its synaptic correlates were not observed at various times after training [104, 105]. Interestingly, the Botzer et al. 1988 also examined feeding aversion. In their study, LTM only appeared 12 to 24 h after training. Memory was not apparent between 4 h and 12 h after training. These authors also used a cold-block procedure that differed from ours in that it involved placing animals in a freezer, but the behavioral data do not radically differ from what we observed here. Cooling in the aforementioned *Aplysia* study, however, did not block STM as we showed [58].

Finally there is another possibility. We cannot rule out the possibility that the experimental procedure used in the "180 min delay cold-block" induced extinction of the learned response. With the "180 min delay cold-block" the CS alone was applied twice (10min and 27 min) after the last CS-US in order to test for STM before the cold-block. The extra CS application may initiate an extinction process and this process might be enhanced during a cold-block period, thus no LTM was observed.

3.3. How Many Repetition of CS-US Requires to Form LTM

Based on the data by Takigami et al. 2014 and previously obtained paring a sucrose CS with the high electric voltage stimulus as the US2, 15 repeated pairings of the CS-US2 on a single day is sufficient to result in LTM formation [33, 34]. Interestingly, using KCl as an aversive stimulus it only took 10 paired CS-US presentations to cause LTM formation that persisted for 1 month [26]. It may be that the KCl stimulus, which causes complete withdrawal of the snail into its shell, is much more stressful stimulus than either of the stimuli we have used (i.e., US1 and US2) and that stress associated with the KCl stimuli causes enhanced LTM formation [106]; while if a non-invasive mechanical tactile stimulus to animal's head was used as the US it took 20 paired presentations to result in LTM [58]. Finally, if rotational turbulence was used as the US it required 30 paired presentations to result in LTM [19]. Thus, we can conclude that the effectiveness of the US is very important in causing LTM formation. What is not clear however, is that while each of the stimuli used as the US causes the WBWR into its shell, why different USs require more pairings to produce LTM. Thus, it is not certain why the different stimuli have differences in the ability to lead to LTM formation. Besides the 'strength' or salience of the US the developmental state of the nervous system governs if LTM will be formed following training. This is especially true with regards to conditioning where an aversive stimulus in *Lymnaea s.* is used [53, 107-109].

The persistence of a memory is one index of how strong a memory is. Another metric in *Lymnaea* is to determine how resistant the memory is to extinction [110, 111]. With food avoidance it becomes difficult to determine exactly how long the memory persists because as we showed in this chapter snails had to be food-deprived for 1 week prior to conditioning. Thus following the formation of LTM snails have strong motivation for food and therefore tend to show an enhanced, homeostatic driven response to the CS which interferes with the memory phenotype. We may be able to circumvent this appetitive problem to the CS by training snails with sucrose as the CS and having a small amount of cabbage (i.e., different food substance) in the training beaker. Then we could feed snails cabbage after LTM has established and better determine how long the memory persist.

In our attempt to determine the persistence of LTM following the 15 paired presentations of the CS-US2, we found that memory persisted for at least 1 week. We believe this may be an underestimation of how long the LTM can persist [33]. There are two reasons for this assumption. The first is that snails during this period were not fed, thus, increasing the motivation for them to eat. Therefore, the 'effectiveness' of the CS was enhanced by the lack of food, which would result in a snails that was more motivated to eat. Secondly the CS alone was presented to the snails on a number of associations during the testing period (i.e., 10 min, 24 h, 3 days, and 7 days) and thus extinction of the learned response could have occurred. In this respect, *Lymnaea s.* could be differentially conditioned with two different appetitive foods, e.g., sucrose and cabbage as mentioned above [112].

We showed in the massed training experiments that 15 paired presentations of the CS-US2 resulted in learning, STM, and LTM. However, both 8 and 20 paired presentations only resulted in learning and STM and not LTM. We hypothesize therefore that the 20 paired presentations may be too stressful for the snails, while the 8 paired presentations may not be stressful enough. As previously mentioned Yerkes-Dodson Law indicated that there is an optimal level of stress, below which little attention is paid to training, while above the optimum level the stimulus is just too stressful for animals to pay attention to training [113].

4. Neuronal Correlates of TAC

In both classical and operant conditioning procedure that employ aversive conditioning, the noxious stimulus elicits in whole or in part the WBWR, which is the only available defensive behavior for this animal. Many different noxious sensory inputs including: 1) chemosensory input (e.g., KCl); 2) tactile stimulation; 3) electrical; 4) rotational turbulence; and 5) light-off visual stimulus (i.e., shadow response) mediated via either ocular or dermal photoreceptors [18-20, 26, 42, 114-117]. All these stimuli are sufficient to elicit the WBWR. The neurons that mediate the WBWR alter activity of neurons in the feeding circuit (e.g., cerebral giant cells, CGCs). The CGCs play a key role in conditioning taste aversion [118]. Whether memory formation culminating in LTM is a serial or a parallel process (e.g., whether LTM requires the prior formation of ITM) is a subject currently under investigation. However, it has been shown in both *Lymnaea* and *Aplysia* that following the cessation of training, there may be 'memory lapses' [51, 102]. It is unclear if these lapses are the result of transitions from one form of memory to another (i.e., from ITM to LTM), during which the organism cannot access the memory. Our present working hypothesis is that the macromolecular protein synthesis processes necessary to form ITM and LTM occur within the neural circuit that enables TAC, the cerebral giant cells (CGC), which play a key role as a neuromodulator. Specifically, we hypothesize that CGC activity is modified by the repeated paired presentations of the CS-US2. The change in CGC activity alters the activity of the neuron tonic 3 cell (N3t), ultimately resulting in an increase in the amplitude in the inhibitory post-synaptic potential (IPSP) to one of the feeding central pattern generator neurons, the neuron 1 medial cell (N1M). This finally, results in the suppression of the feeding response to CS [118].

Right Pedal Dorsal 11 (RPeD11), the neuron which was shown to mediate the WBWR [119] also contributed in establishment of TAC [42, 88, 115]. This neuron sends an important and effective inhibitory input to the CGCs [25]. There is also a one-trial training which is very effective for establishing appetitive food conditioning [57, 120, 121]. The important difference between aversive and appetitive food conditioning is in the neuron circuit involved. In the case of appetitive conditioning changes occur within the feeding circuit itself and it is not necessary to invoke any other auxiliary circuit. However, in case of aversive conditioning, neural circuits extrinsic to the feeding circuit are necessary involved. That is, a different neuronal circuit is involved if forming the association between the CS and US. In the naïve state this circuit is not functional. It takes the pairing of the CS and US in close temporal contiguity to cause the circuit to function in a manner such that the CS now leads to food aversion. How this actually occurs at the circuit and neuronal level is not yet known.

5. Critical Period for Memory Consolidation

We have previously demonstrated that there is a critical period during which it is possible to enhance LTM formation, and this period occurred approximately 3 h after the final CS-US2 presentation [58]. A similar critical period following training to enhance memory formation was found in *Hermissenda* [89]. In our most recent study we did not observe any memory enhancing effects in the massed training procedure as shown in Figure 11A of massed training. However when we interposed a 'rest' interval of 1 h or 3 h between 'sets' of paired presentations of the CS-US2 (Figure 11B left half of 3 control graphs) both STM and LTM formation was enhanced [34]. These data are consistent with previous findings in *Hermissenda* by Epstein et al. 20003. The interposition of a 3 h interval between the 'sets' of paired CS-US2 presentations results in more consistent memory forming process similar to what is seen in 'spaced' training procedures.

We found that exposure to Bryo and 3 h rest interval resulted in equivalent memory formation that resulted following the interposition of a 3 h 'rest' interval between 'sets' of CS-US2 pairings (Figure 11B). We used a dose of Bryo that was 5 times lower (0.1 ng/ml) [34] than what we had previously used [58]. We interpret these data as showing that the altered gene activity necessary for LTM formation leading to new mRNA production occurred during this period with the aid of PKC phosphorylation. Thus, the data shown here suggests that the memory consolidation process necessary for long-lasting LTM requires both the involvement of PKC-activated protein synthesis and post-conditioning period necessary to allow for modification of post-transcriptional mRNA.

Studies by Ito's group has shown that the molecular events underlying conditioned taste aversion in *Lymnaea* involve both the cyclic AMP response binding protein (CREB) and CREB has a phosphorylation cite by PKC [27, 122]. Finally we have also recently found that molluscan insulin related peptides (MIPs) and their receptors play a key role in this conditioning process [98, 123, 124]. We still do not know the complete sequence of molecular events that are causal to the formation of LTM following the initial formation of STM. Our study may provide one clue for molecular event during memory consolidation.

CONCLUSION

Classical conditioning procedure has established with various combination of CS (flash of light irradiation, sucrose application) and US (rotational turbulence, mechanical tactile stimulation, electrical high voltage stimulus) in *Lymnaea s.* These USs attribute to invasive and vigilance stimulus eliciting whole-body-withdrawal-response which is the only escape behavior available to this animal. Among them taste avoidance conditioning was successfully achieved by the paired presentation of CS, sucrose application, and US, invasive stimulus which elicits escape behavior. Originally we applied mechanical tactile stimulus to the head with a hand-held stick as US, however successful conditioning was dependent on the examiner's skill because it is difficult to apply US consistently. Contrary, it is easy to condition snails with sucrose application (CS) and high voltage electrical stimulation (US). We first examined the efficacy between the nature of US, the mechanical prodding (US1) and the electrical stimulation (US2) by comparison with the minimal number of repetition to obtain STM and LTM. STM and LTM is defined by the significant feeding suppression of mouth opening (bites/min) to a CS presentation between the pre-test score and 10 min or 24 h post-test score, respectively. If we trained snails in a day, the combination of CS-US1 required more than 20 times repetition, while with CS-US2 combination we need 15 repetition. However if the repeated CS-US presentations are interposed with a rest interval, the training effects was maximally improved with the 3 h rest interval in a day. These finding suggest that the critical period to form LTM should be around 3 h during which it may correspond to mRNA alteration time-point. Concerning the crucial time line of 3 h following the conditioning trials for STM and LTM formation, we found that cooling induced amnesia provided us a good measure refereed as 'cold-block' to estimate the temporal sequence of intracellular molecular switching events. If we applied the cold-block immediate after the training trial no STM nor LTM was formed, furthermore 180 min following the training session 'cold-block' also interfered the formation of both STM and LTM. These findings suggested that formation of STM was critical soon after the training session, while for the formation of LTM 3 h should be critical. In addition to these temporal time-point corresponding to intracellular molecular events hypothesized, activation of PKC irrespective of conditioning experience facilitates memory formation with less repeated presentations of CS-US, i.e., from STM to ITM or STM into LTM. These findings suggested that protein synthesis was not crucial for the actual encoding of associative memory, but rather for providing essential biochemical and structural components of the synaptic apparatus required for implementation of the memory storage process.

ACKNOWLEDGMENTS

The original studies to in this chapter were conducted in collaboration with Prof. Dan Alkon of Blanchette Rockefeller Neuroscience Institute, Dr. Alan Kuzirian of Marine Biological Laboratory, Prof. Tohru Yoshioka of Waseda University, Prof. Ken Lukowiak of University of Calgary, Prof. Etsuro Ito of Tokushima Bunri University, Prof. Tetsuro Horikoshi of Tokai University, Dr. Nobuko Katagiri of Tokyo Women's Medical University, Dr. Dai Hatakeyama of Hokkaido University, and Dr. Sabrina Pankey of University of California at Santa Barbara; and my graduate students, Mr. Takeshi Kouchi, Dr. Ryo Kawai, Ms. Megumi Ono, Ms. Satoko Inamura, Ms. Noriko Tsubata, Mr. Futoshi Okuda, Ms. Tomoyo Aritaka, Dr. Akira Iizuka, Mr. Hiroyuki Suzuki, Ms. Chiye Toyoshima, Mr. Yasutaka Nomura, Dr. Hiroshi Sunada, Dr. Satoshi Takigami, and Ms. Tomoyo Takahashi. I would like to express my thanks to all of them. Some of the studies were financially supported by KAKENHI from JSPS to M.S.

REFERENCES

[1] Alkon, D.L.,(1974). Associative training of *Hermissenda. J. Gen. Physiol.* 64, 70-84.

[2] Alkon, D.L.,(1983). Learning in a marine snail. *Sci. Am.* 249, 70-84.

[3] Alkon, D.L.,(1980). Cellular analysis of a gastropod (*Hermissenda crassicornis*) model of associative learning. *The Biological Bulletin.* 159, 505-560.

[4] Alkon, D.L.,(1973). Intersensory interactions in *Hermissenda. J. Gen. Physiol.* 62, 185-202.

[5] Alkon, D.L., Sakakibara, M., Forman, R., Harrigan, J., Lederhendler, I., and Farley, J.,(1985). Reduction of two voltage-dependent K^+ currents mediates retention of a learned association. *Behav. Neural. Biol.* 44, 278-300.

[6] Sakakibara, M., Alkon, D.L., Neary, J.T., Heldman, E., and Gould, R.,(1986). Inositol trisphosphate regulation of photoreceptor membrane currents. *Biophys. J.* 50, 797-803.

[7] Sakakibara, M., Collin, C., Kuzirian, A., Alkon, D.L., Heldman, E., Naito, S., and Lederhendler, I.,(1987). Effects of alpha 2-adrenergic agonists and antagonists on photoreceptor membrane currents. *J. Neurochem.* 48, 405-16.

[8] Alkon, D.L. and Sakakibara, M.,(1985). Calcium activates and inactivates a photoreceptor soma potassium current. *Biophys. J.* 48, 983-95.

[9] Alkon, D.L., Nelson, T.J., Zhao, W., and Cavallaro, S.,(1998). Time domains of neuronal Ca^{2+} signaling and associative memory: steps through a calexcitin, ryanodine receptor, K^+ channel cascade. *Trends in Neurosciences.* 21, 529-537.

[10] Sakakibara, M., Alkon, D.L., DeLorenzo, R., Goldenring, J.R., Neary, J.T., and Heldman, E.,(1986). Modulation of calcium-mediated inactivation of ionic currents by Ca^{2+}/calmodulin-dependent protein kinase II. *Biophys J.* 50, 319-27.

[11] Kawai, R., Horikoshi, T., and Sakakibara, M.,(2004). Involvement of the ryanodine receptor in morphologic modification of *Hermissenda* type B photoreceptors after in vitro conditioning. *J. Neurophysiol.* 91, 728-35.

[12] Sakakibara, M., Ikeno, H., Usui, S., Collin, C., and Alkon, D.L.,(1993). Reconstruction of ionic currents in a molluscan photoreceptor. *Biophys. J.* 65, 519-27.

[13] Sakakibara, M., Inoue, H., and Yoshioka, T.,(1998). Evidence for the involvement of inositol trisphosphate but not cyclic nucleotides in visual transduction in *Hermissenda* eye. *J. Biol. Chem.* 273, 20795-801.
[14] Sakakibara, M., Alkon, D.L., Kouchi, T., Inoue, H., and Yoshioka, T.,(1994). Induction of photoresponse by the hydrolysis of polyphosphoinositides in the *Hermissenda* type B photoreceptor. *Biochem. Biophys. Res. Commun.* 202, 299-306.
[15] Audesirk, T.E., Alexander, J.E., Jr., Audesirk, G.J., and Moyer, C.M.,(1982). Rapid, nonaversive conditioning in a freshwater gastropod. I. Effects of age and motivation. *Behav. Neural Biol.* 36, 379-90.
[16] Kemenes, G. and Benjamin, P.R.,(1989). Appetitive learning in snails shows characteristics of conditioning in vertebrates. *Brain Res.* 489, 163-6.
[17] Kojima, S., Nakamura, H., Nagayama, S., Fujito, Y., and Ito, E.,(1997). Enhancement of an inhibitory input to the feeding central pattern generator in *Lymnaea stagnalis* during conditioned taste-aversion learning. *Neurosci. Lett.* 230, 179-82.
[18] Kawai, R., Sunada, H., Horikoshi, T., and Sakakibara, M.,(2004). Conditioned taste aversion with sucrose and tactile stimuli in the pond snail *Lymnaea stagnalis*. *Neurobiol. Learn Mem.* 82, 164-168.
[19] Sakakibara, M., Kawai, R., Kobayashi, S., and Horikoshi, T.,(1998). Associative learning of visual and vestibular stimuli in *Lymnaea*. *Neurobiol. Learn Mem.* 69, 1-12.
[20] Sakakibara, M.,(2006). Comparative study of visuo-vestibular conditioning in *Lymnaea stagnalis*. *Biol. Bull.* 210, 298-307.
[21] Lukowiak, K., Cotter, R., Westly, J., Ringseis, E., and Spencer, G.,(1998). Long-term memory of an operantly conditioned respiratory behaviour pattern in *lymnaea stagnalis*. *J. Exp. Biol.* 201 (Pt 6), 877-82.
[22] Lukowiak, K., Adatia, N., Krygier, D., and Syed, N.,(2000). Operant conditioning in *Lymnaea*: evidence for intermediate- and long-term memory. *Learn Mem.* 7, 140-50.
[23] Syed, N.I., Bulloch, A.G., and Lukowiak, K.,(1990). In vitro reconstruction of the respiratory central pattern generator of the mollusk *Lymnaea*. *Science*. 250, 282-5.
[24] Syed, N.I., Roger, I., Ridgway, R.L., Bauce, L.G., Lukowiak, K., and Bulloch, A.G.,(1993). Identification, characterisation and in vitro reconstruction of an interneuronal network of the snail *Helisoma trivolvis*. *J. Exp. Biol.* 174, 19-44.
[25] Sunada, H., Lukowiak, K., and Sakakibara, M.,(2012). In vitro aversion conditioning in *Lymnaea*. Short communication. *Acta Biol. Hung.* 63 Suppl 2, 190-3.
[26] Kojima, S., Yamanaka, M., Fujito, Y., and Ito, E.,(1996). Differential neuroethological effects of aversive and appetitive reinforcing stimuli on associative learning in *Lymnaea stagnalis*. *Zoolog. Sci.* 13, 803-812.
[27] Ito, E., Kojima, S., Lukowiak, K., and Sakakibara, M.,(2013). From likes to dislikes: conditioned taste aversion in the great pond snail (*Lymnaea stagnalis*). *Canadian Journal of Zoology*. 91, 405-412.
[28] Lukowiak, K., Ringseis, E., Spencer, G., Wildering, W., and Syed, N.,(1996). Operant conditioning of aerial respiratory behaviour in *Lymnaea stagnalis*. *J. Exp. Biol.* 199, 683-91.
[29] Benjamin, P.R., Staras, K., and Kemenes, G.,(2000). A systems approach to the cellular analysis of associative learning in the pond snail *Lymnaea*. *Learn. Mem.* 7, 124-31.

[30] Sakakibara, M., *Cellular and molecular aspect of short-term and long-term memory from molluscan system*, in *Novel Trends in Brain Science*, C.-T.Y. Minoru Onozuka, Editor. 2008, Springer: Tokyo. p. 131-148.

[31] Sakakibara, M., Aritaka, T., Iizuka, A., Suzuki, H., Horikoshi, T., and Lukowiak, K.,(2005). Electrophysiological responses to light of neurons in the eye and statocyst of *Lymnaea stagnalis*. *J. Neurophysiol.* 93, 493-507.

[32] Tsubata, N., Iizuka, A., Horikoshi, T., and Sakakibara, M.,(2003). Photoresponse from the statocyst hair cell in *Lymnaea stagnalis*. *Neurosci. Lett.* 337, 46-50.

[33] Takigami, S., Sunada, H., Lukowiak, K., and Sakakibara, M.,(2013). High voltage with little current as an unconditional stimulus for taste avoidance conditioning in *Lymnaea stagnalis*. *Neurosci. Lett.* 555, 149-153.

[34] Takigami, S., Sunada, H., Lukowiak, K., and Sakakibara, M.,(2014). Spaced taste avoidance conditioning in *Lymnaea*. *Neurobiol. Learn Mem.* 107, 79-86.

[35] Hintzman, D.L., *Theoretical implications of the spacing effect.*, in *Theories in cognitive psychology: The loyola symposium*, R.L. Solso, Editor. 1974, Lawrence Erlbaum Associates: Potomac, MD. p. 77-99.

[36] Adams, J.A.,(1987). Historical review and appraisal of research on the learning, retention, and transfer of human motor skills. *Psychological Bulletin.* 101, 41.

[37] Tully, T., Preat, T., Boynton, S.C., and Del Vecchio, M.,(1994). Genetic dissection of consolidated memory in *Drosophila*. *Cell.* 79, 35-47.

[38] Bitterman, M.E., Menzel, R., Fietz, A., and Schafer, S.,(1983). Classical conditioning of proboscis extension in honeybees (*Apis mellifera*). *J. Comp. Psychol.* 97, 107-19.

[39] Frost, W.N., Castellucci, V.F., Hawkins, R.D., and Kandel, E.R.,(1985). Monosynaptic connections made by the sensory neurons of the gill- and siphon-withdrawal reflex in *Aplysia* participate in the storage of long-term memory for sensitization. *Proc. Natl. Acad. Sci. USA.* 82, 8266-9.

[40] Kogan, J.H., Frankland, P.W., Blendy, J.A., Coblentz, J., Marowitz, Z., Schutz, G., and Silva, A.J.,(1997). Spaced training induces normal long-term memory in CREB mutant mice. *Curr. Biol.* 7, 1-11.

[41] Ishida, M. and Papini, M.R.,(1997). Massed-trial overtraining effects on extinction and reversal performance in turtle (*Geoclemys reevesii*). *Quarterly Journal of Experimental Psychology.* 50B, 1-16.

[42] Sunada, H., Sakaguchi, T., Horikoshi, T., Lukowiak, K., and Sakakibara, M.,(2010). The shadow-withdrawal response, dermal photoreceptors and their input to a higher order interneuron, RPeD11 in the pond snail *Lymnaea stagnalis*. *J. Exp. Biol.* 213, 3409-15.

[43] Lederhendler, II, Gart, S., and Alkon, D.L.,(1986). Classical conditioning of *Hermissenda*: origin of a new response. *J. Neurosci.* 6, 1325-31.

[44] Suzuki, H., Horikoshi, T., and Sakakibara, M.,(2008). Neurophysiological analysis of visuo-vestibular conditioning in *Lymnaea stagnalis*. *Acta Biol. Hung.* 59 Suppl, 93-5.

[45] Abraham, W.C. and Williams, J.M.,(2008). LTP maintenance and its protein synthesis-dependence. *Neurobiol. Learn Mem.* 89, 260-8.

[46] Castellucci, V.F., Blumenfeld, H., Goelet, P., and Kandel, E.R.,(1989). Inhibitor of protein synthesis blocks long-term behavioral sensitization in the isolated gill-withdrawal reflex of *Aplysia*. *J. Neurobiol.* 20, 1-9.

[47] DeZazzo, J. and Tully, T.,(1995). Dissection of memory formation: from behavioral pharmacology to molecular genetics. *Trends Neurosci.* 18, 212-8.
[48] Ramirez, R.R., Gandhi, C.C., Muzzio, I.A., and Matzel, L.D.,(1998). Protein synthesis-dependent memory and neuronal enhancement in *Hermissenda* are contingent on parameters of training and retention. *Learn Mem.* 4, 462-77.
[49] Sangha, S., Scheibenstock, A., McComb, C., and Lukowiak, K.,(2003). Intermediate and long-term memories of associative learning are differentially affected by transcription versus translation blockers in *Lymnaea. J. Exp. Biol.* 206, 1605-13.
[50] Schafe, G.E. and LeDoux, J.E.,(2000). Memory consolidation of auditory pavlovian fear conditioning requires protein synthesis and protein kinase A in the amygdala. *J. Neurosci.* 20, RC96.
[51] Sutton, M.A., Masters, S.E., Bagnall, M.W., and Carew, T.J.,(2001). Molecular mechanisms underlying a unique intermediate phase of memory in *aplysia. Neuron.* 31, 143-54.
[52] Wustenberg, D., Gerber, B., and Menzel, R.,(1998). Short communication: long- but not medium-term retention of olfactory memories in honeybees is impaired by actinomycin D and anisomycin. *Eur. J. Neurosci.* 10, 2742-5.
[53] Ono, M., Kawai, R., Horikoshi, T., Yasuoka, T., and Sakakibara, M.,(2002). Associative Learning Acquisition and Retention Depends on Developmental Stage in *Lymnaea stagnalis. Neurobiol. Learn Mem.* 78, 53-64.
[54] Alkon, D.L., Epstein, H., Kuzirian, A., Bennett, M.C., and Nelson, T.J.,(2005). Protein synthesis required for long-term memory is induced by PKC activation on days before associative learning. *Proc. Natl. Acad. Sci. U S A.* 102, 16432-7.
[55] Wilkinson, S.E., Parker, P.J., and Nixon, J.S.,(1993). Isoenzyme specificity of bisindolylmaleimides, selective inhibitors of protein kinase C. *Biochem. J.* 294 (Pt 2), 335-7.
[56] Sugai, R., Azami, S., Shiga, H., Watanabe, T., Sadamoto, H., Kobayashi, S., Hatakeyama, D., Fujito, Y., Lukowiak, K., and Ito, E.,(2007). One-trial conditioned taste aversion in *Lymnaea*: good and poor performers in long-term memory acquisition. *J. Exp. Biol.* 210, 1225-37.
[57] Fulton, D., Kemenes, I., Andrew, R.J., and Benjamin, P.R.,(2005). A single time-window for protein synthesis-dependent long-term memory formation after one-trial appetitive conditioning. *Eur. J. Neurosci.* 21, 1347-58.
[58] Takahashi, T., Takigami, S., Sunada, H., Lukowiak, K., and Sakakibara, M.,(2013). Critical Period of Memory Enhancement during Taste Avoidance Conditioning in *Lymnaea stagnalis. PloS one.* 8, e75276.
[59] Kuzirian, A.M., Epstein, H.T., Gagliardi, C.J., Nelson, T.J., Sakakibara, M., Taylor, C., Scioletti, A.B., and Alkon, D.L.,(2006). Bryostatin enhancement of memory in *Hermissenda. Biol. Bull.* 210, 201-14.
[60] Sun, M.K. and Alkon, D.L.,(2008). Synergistic effects of chronic bryostatin-1 and alpha-tocopherol on spatial learning and memory in rats. *Eur. J. Pharmacol.* 584, 328-37.
[61] Nomura, Y., Hatakeyama, D., Horikoshi, T., and Sakakibara, M.,(2008). Immunohistological studies on the distribution of learning-related peptides in the central nervous system of conditioned *Lymnaea. Acta Biol. Hung.* 59 Suppl, 81-92.

[62] Hatakeyama, D., Inamura, S., Ito, E., Sakakibara, M., Nelson, T., and Alkon, D.L.,(2004). Calexcitin-like immunoreactivity in the pond snail *Lymnaea stagnalis*. *Neuroscience Research Communications*. 35, 32-40.

[63] Takigami, S., Sunada, H., Lukowiak, K., Kuzirian, A., Alkon, D.L., and Sakakibara, M.,(2014). Protein kinase C mediates memory consolidation of taste avoidance conditioning in *Lymnae stagnalis*. *Neurobiol. Lern. and Mem.* 111, 9-18.

[64] Smith, J.B., Smith, L., and Pettit, G.R.,(1985). Bryostatins: potent, new mitogens that mimic phorbol ester tumor promoters. *Biochem. Biophys. Res. Commun.* 132, 939-45.

[65] Azami, S., Wagatsuma, A., Sadamoto, H., Hatakeyama, D., Usami, T., Fujie, M., Koyanagi, R., Azumi, K., Fujito, Y., Lukowiak, K., and Ito, E.,(2006). Altered gene activity correlated with long-term memory formation of conditioned taste aversion in *Lymnaea*. *J. Neurosci. Res.* 84, 1610-20.

[66] Crow, T., Xue-Bian, J.J., Siddiqi, V., Kang, Y., and Neary, J.T.,(1998). Phosphorylation of mitogen-activated protein kinase by one-trial and multi-trial classical conditioning. *J. Neurosci.* 18, 3480-7.

[67] Guan, Z., Kim, J.H., Lomvardas, S., Holick, K., Xu, S., Kandel, E.R., and Schwartz, J.H.,(2003). p38 MAP kinase mediates both short-term and long-term synaptic depression in aplysia. *J. Neurosci.* 23, 7317-25.

[68] Michel, M., Kemenes, I., Muller, U., and Kemenes, G.,(2008). Different phases of long-term memory require distinct temporal patterns of PKA activity after single-trial classical conditioning. *Learn Mem.* 15, 694-702.

[69] Yamada, A., Sekiguchi, T., Suzuki, H., and Mizukami, A.,(1992). Behavioral analysis of internal memory states using cooling-induced retrograde amnesia in *Limax flavus*. *J. Neurosci.* 12, 729-35.

[70] Sekiguchi, T., Suzuki, H., Yamada, A., and Mizukami, A.,(1994). Cooling-induced retrograde amnesia reflexes Pavlovian conditioning associations in *Limax flavus*. *Neurosci. Res.* 18, 267-75.

[71] Fulton, D., Kemenes, I., Andrew, R.J., and Benjamin, P.R.,(2008). Time-window for sensitivity to cooling distinguishes the effects of hypothermia and protein synthesis inhibition on the consolidation of long-term memory. *Neurobiol. Learn Mem.* 90, 651-4.

[72] Sangha, S., Morrow, R., Smyth, K., Cooke, R., and Lukowiak, K.,(2003). Cooling blocks ITM and LTM formation and preserves memory. *Neurobiol. Learn. Mem.* 80, 130-9.

[73] Xia, S.Z., Feng, C.H., and Guo, A.K.,(1999). Temporary amnesia induced by cold anesthesia and hypoxia in *Drosophila*. *Physiol. Behav.* 65, 617-23.

[74] Sommerville, J.,(1999). Activities of cold-shock domain proteins in translation control. *Bioessays*. 21, 319-25.

[75] Yerkes, R.M. and Dodson, J.D.,(1908). The relation of strength of stimulus to rapidity of habit-formation. *J. Comp. Neurol. Psychol.* 18, 459-482.

[76] Rosenegger, D., Parvez, K., and Lukowiak, K.,(2008). Enhancing memory formation by altering protein phosphorylation balance. *Neurobiol. Learn Mem.* 90, 544-52.

[77] Hongpaisan, J. and Alkon, D.L.,(2007). A structural basis for enhancement of long-term associative memory in single dendritic spines regulated by PKC. *Proc. Natl. Acad. Sci. U S A.* 104, 19571-6.

[78] Sun, M.K. and Alkon, D.L.,(2005). Dual effects of bryostatin-1 on spatial memory and depression. *Eur. J. Pharmacol.* 512, 43-51.

[79] Etcheberrigaray, R., Tan, M., Dewachter, I., Kuipéri, C., Van der Auwera, I., Wera, S., Qiao, L., Bank, B., Nelson, T.J., Kozikowski, A.P., Van Leuven, F., and Alkon, D.L.,(2004). Therapeutic effects of PKC activators in Alzheimer's disease transgenic mice. *Proc. Natl. Acad. Sci. U S A.* 101, 11141-11146.

[80] Sun, M.K. and Alkon, D.L.,(2009). Protein kinase C activators as synaptogenic and memory therapeutics. *Arch. Pharm. Chem. Life Sci.* 342, 689-698.

[81] Kandel, E.R.,(2001). The molecular biology of memory storage: a dialogue between genes and synapses. *Science.* 294, 1030-8.

[82] Staras, K., Kemenes, G., and Benjamin, P.R.,(1998). Neurophysiological correlates of unconditioned and conditioned feeding behavior in the pond snail *Lymnaea stagnalis. J. Neurophysiol.* 79, 3030-40.

[83] Fulton, D., Condro, M.C., Pearce, K., and Glanzman, D.L.,(2008). The potential role of postsynaptic phospholipase C activity in synaptic facilitation and behavioral sensitization in *Aplysia. J. Neurophysiol.* 100, 108-16.

[84] Braun, M.H. and Lukowiak, K.,(2011). Intermediate and long-term memory are different at the neuronal level in *Lymnaea stagnalis* (L.). *Neurobiol. Learn Mem.* 96, 403-16.

[85] Mozzachiodi, R., Lorenzetti, F.D., Baxter, D.A., and Byrne, J.H.,(2008). Changes in neuronal excitability serve as a mechanism of long-term memory for operant conditioning. *Nat. Neurosci.* 11, 1146-8.

[86] Alkon, D.L., *Memory Traces in the Brain.* 1987, New York: Cambridge University Press.

[87] Nelson, T.J., Cavallaro, S., Yi, C.L., McPhie, D., Schreurs, B.G., Gusev, P.A., Favit, A., Zohar, O., Kim, J., Beushausen, S., Ascoli, G., Olds, J., Neve, R., and Alkon, D.L.,(1996). Calexcitin: a signaling protein that binds calcium and GTP, inhibits potassium channels, and enhances membrane excitability. *Proc. Natl. Acad. Sci. U S A.* 93, 13808-13.

[88] Sunada, H., Horikoshi, T., Lukowiak, K., and Sakakibara, M.,(2010). Increase in excitability of RPeD11 results in memory enhancement of juvenile and adult of *Lymnaea stagnalis* by predator-induced stress. *Neurobiol. Learn Mem.* 94, 269-277.

[89] Epstein, H.T., Child, F.M., Kuzirian, A.M., and Alkon, D.L.,(2003). Time windows for effects of protein synthesis inhibitors on Pavlovian conditioning in *Hermissenda*: behavioral aspects. *Neurobiol. Learn Mem.* 79, 127-31.

[90] Alberini, C.M., Ghirardi, M., Metz, R., and Kandel, E.R.,(1994). C/EBP is an immediate-early gene required for the consolidation of long-term facilitation in *Aplysia. Cell.* 76, 1099-114.

[91] Alberini, C.M. and Chen, D.Y.,(2012). Memory enhancement: consolidation, reconsolidation and insulin-like growth factor 2. *Trends Neurosci.* 35, 274-83.

[92] Anokhin, K.V. and Rose, S.P.,(1991). Learning-induced Increase of Immediate Early Gene Messenger RNA in the Chick Forebrain. *Eur. J. Neurosci.* 3, 162-167.

[93] Chen, D.Y., Stern, S.A., Garcia-Osta, A., Saunier-Rebori, B., Pollonini, G., Bambah-Mukku, D., Blitzer, R.D., and Alberini, C.M.,(2011). A critical role for IGF-II in memory consolidation and enhancement. *Nature.* 469, 491-7.

[94] Guzowski, J.F., Setlow, B., Wagner, E.K., and McGaugh, J.L.,(2001). Experience-dependent gene expression in the rat hippocampus after spatial learning: a comparison of the immediate-early genes Arc, c-fos, and zif268. *J. Neurosci.* 21, 5089-98.

[95] Jones, M.W., Errington, M.L., French, P.J., Fine, A., Bliss, T.V., Garel, S., Charnay, P., Bozon, B., Laroche, S., and Davis, S.,(2001). A requirement for the immediate early gene Zif268 in the expression of late LTP and long-term memories. *Nat. Neurosci.* 4, 289-96.

[96] Taubenfeld, S.M., Milekic, M.H., Monti, B., and Alberini, C.M.,(2001). The consolidation of new but not reactivated memory requires hippocampal C/EBPbeta. *Nat. Neurosci.* 4, 813-8.

[97] Hatakeyama, D., Sadamoto, H., Watanabe, T., Wagatsuma, A., Kobayashi, S., Fujito, Y., Yamashita, M., Sakakibara, M., Kemenes, G., and Ito, E.,(2005). Requirement of New Protein Synthesis of a Transcription Factor for Memory Consolidation: Paradoxical Changes in mRNA and Protein Levels of C/EBP. *J. Mol. Biol.*

[98] Murakami, J., Okada, R., Sadamoto, H., Kobayashi, S., Mita, K., Sakamoto, Y., Yamagishi, M., Hatakeyama, D., Otsuka, E., Okuta, A., Sunada, H., Takigami, S., Sakakibara, M., Fujito, Y., Awaji, M., Moriyama, S., Lukowiak, K., and Ito, E.,(2013). Involvement of insulin-like peptide in long-term synaptic plasticity and long-term memory of the pond snail *Lymnaea stagnalis. J. Neurosci.* 33, 371-383.

[99] Sangha, S., McComb, C., and Lukowiak, K.,(2003). Forgetting and the extension of memory in *Lymnaea. J. Exp. Biol.* 206, 71-7.

[100] Sangha, S., Scheibenstock, A., Morrow, R., and Lukowiak, K.,(2003). Extinction requires new RNA and protein synthesis and the soma of the cell right pedal dorsal 1 in *Lymnaea stagnalis. J. Neurosci.* 23, 9842-51.

[101] Parvez, K., Stewart, O., Sangha, S., and Lukowiak, K.,(2005). Boosting intermediate-term into long-term memory. *J. Exp. Biol.* 208, 1525-36.

[102] Marra, V., O'Shea, M., Benjamin, P.R., and Kemenes, I.,(2013). Susceptibility of memory consolidation during lapses in recall. *Nat. Commun.* 4, 1578.

[103] Crow, T. and Xue-Bian, J.J.,(2011). Proteomic analysis of short- and intermediate-term memory in *Hermissenda. Neuroscience.* 192, 102-11.

[104] Botzer, D., Markovich, S., and Susswein, A.J.,(1998). Multiple memory processes following training that a food is inedible in *Aplysia. Learn Mem.* 5, 204-19.

[105] Sherff, C.M. and Carew, T.J.,(2004). Parallel somatic and synaptic processing in the induction of intermediate-term and long-term synaptic facilitation in *Aplysia. Proc. Natl. Acad. Sci. U S A.* 101, 7463-8.

[106] Lukowiak, K., Martens, K., Rosenegger, D., Browning, K., de Caigny, P., and Orr, M.,(2008). The perception of stress alters adaptive behaviours in *Lymnaea stagnalis. J. Exp. Biol.* 211, 1747-56.

[107] McComb, C., Rosenegger, D., Varshney, N., Kwok, H.Y., and Lukowiak, K.,(2005). Operant conditioning of an in vitro CNS-pneumostome preparation of *Lymnaea. Neurobiol. Learn Mem.* 84, 9-24.

[108] Orr, M., Hittel, K., and Lukowiak, K.,(2010). Predator detection enables juvenile *Lymnaea* to form long-term memory. *J. Exp. Biol.* 213, 301-307.

[109] Yamanaka, M., Sadamoto, H., Hatakeyama, D., Nakamura, H., Kojima, S., Kimura, T., Yamashita, M., Urano, S., and Ito, E.,(1999). Developmental changes in Conditioned taste aversion in *Lymnaea stagnalis. Zoolog Sci.* 16, 9-16.

[110] Kennedy, C.D., Houmes, S.W., Wyrick, K.L., Kammerzell, S.M., Lukowiak, K., and Sorg, B.A.,(2010). Methamphetamine enhances memory of operantly conditioned respiratory behavior in the snail Lymnaea stagnalis. *J. Exp. Biol.* 213, 2055-65.
[111] Fruson, L., Dalesman, S., and Lukowiak, K.,(2012). A flavanol present in cocoa ((-)epicatechin) enhances snail memory. *J. Exp. Biol.* 215, 3566-3576.
[112] Sugai, R., Shiga, H., Azami, S., Watanabe, T., Sadamoto, H., Fujito, Y., Lukowiak, K., and Ito, E.,(2006). Taste discrimination in conditioned taste aversion of the pond snail *Lymnaea stagnalis. J. Exp. Biol.* 209, 826-33.
[113] Calabrese, E.J.,(2008). Converging concepts: adaptive response, preconditioning, and the Yerkes-Dodson Law are manifestations of hormesis. *Ageing Res. Rev.* 7, 8-20.
[114] Chono, K., Fujito, Y., and Ito, E.,(2002). Non-ocular dermal photoreception in the pond snail *Lymnaea stagnalis. Brain Res.* 951, 107-12.
[115] Pankey, S., Sunada, H., Horikoshi, T., and Sakakibara, M.,(2010). Cyclic nucleotide-gated channels are involved in phototransduction of dermal photoreceptors in *Lymnaea stagnalis. J. Comp. Physiol. B.* 180, 1205-11.
[116] Sakakibara, M., Okuda, F., Nomura, K., Watanabe, K., Meng, H., Horikoshi, T., and Lukowiak, K.,(2005). Potassium currents in isolated statocyst neurons and RPeD1 in the pond snail, *Lymnaea stagnalis. J. Neurophysiol.* 94, 3884-92.
[117] Stoll, C.J.,(1972). Sensory systems involved in the shadow response of *Lymnaea stagnalis. Proc. K Ned. Akad. Wet.* C75, 342-351.
[118] Kobayashi, S., Kojima, S., Yamanaka, M., Sadamoto, H., Nakamura, H., Fujito, Y., Kawai, R., Sakakibara, M., and Ito, E.,(1998). Operant Conditioning of Escape Behavior in the Pond Snail, Lymnaea stagnalis. *Zoological Science.* 15, 683-690.
[119] Inoue, T., Takasaki, M., Lukowiak, K., and Syed, N.I.,(1996). Identification of a putative mechanosensory neuron in *Lymnaea*: characterization of its synaptic and functional connections with the whole-body withdrawal interneuron. *J. Neurophysiol.* 76, 3230-8.
[120] Kemenes, I., Kemenes, G., Andrew, R.J., Benjamin, P.R., and O'Shea, M.,(2002). Critical time-window for NO-cGMP-dependent long-term memory formation after one-trial appetitive conditioning. *J. Neurosci.* 22, 1414-25.
[121] Straub, V.A., Styles, B.J., Ireland, J.S., O'Shea, M., and Benjamin, P.R.,(2004). Central localization of plasticity involved in appetitive conditioning in Lymnaea. *Learn Mem.* 11, 787-93.
[122] Sadamoto, H., Sato, H., Kobayashi, S., Murakami, J., Aonuma, H., Ando, H., Fujito, Y., Hamano, K., Awaji, M., Lukowiak, K., Urano, A., and Ito, E.,(2004). CREB in the pond snail Lymnaea stagnalis: cloning, gene expression, and function in identifiable neurons of the central nervous system. *J. Neurobiol.* 58, 455-66.
[123] Hatakeyama, D., Okuta, A., Otsuka, E., Lukowiak, K., and Ito, E.,(2013). Consolidation of long-term memory by insulin in Lymnaea is not brought about by changing the number of insulin receptors. *Commun. Integr. Biol.* 6, e23955.
[124] Ito, E., Okada, R., Sakamoto, Y., Otshuka, E., Mita, K., Okuta, A., Sunada, H., and Sakakibara, M.,(2012). Insulin and memory in Lymnaea. *Acta Biol. Hung.* 63 Suppl 2, 194-201.

In: Memory Consolidation
Editors: Manabu Sakakibara and Etsuro Ito
ISBN: 978-1-63482-596-2

Chapter 7

TRANSCRIPTION MECHANISM IN A SINGLE NEURON TO CONTROL CONDITIONED TASTE AVERSION IN THE POND SNAIL LYMNAEA STAGNALIS

Etsuro Ito[1,*], Miki Yamagishi[1] and Yusuke Ikemoto[2]
[1]Laboratory of Functional Biology,
Kagawa School of Pharmaceutocal Sciences,
Tokushima Bunri University, Sanuki, Japan
[2]Department of Mechanical and Intellectual Systems Engineering,
Faculty of Engineering, University of Toyama, Toyama, Japan

ABSTRACT

Conditioned taste aversion (CTA) can be classically conditioned in the pond snail *Lymnaea stagnalis* and subsequently be consolidated into long-term memory (LTM). The neural mechanism that subserves CTA-LTM are summarized as follows: A polysynaptic inhibitory postsynaptic potential recorded in the neuron 1 medial (N1M) cell in the buccal ganglion of the conditioned snails as a result of activation of the cerebral giant cell (CGC) is larger and lasts longer than that in control snails. The N1M cell is the member of feeding central pattern generator, whereas the CGC is a multimodal interneuron thought to play a key role in the feeding behavior. The CGC is serotonergic and the secretion of setotonin from the CGC is controlled by cAMP, PKA and CREB, showing that these signaling cascades are involved in CTA. In this chapter we review the transcription mechanism, partuculary the cascades of cAMP-PKA-CREB and CCAAT/enhancer binding protein, in CTA-LTM.

Keywords: cAMP, CCAAT/enhancer binding protein, cerebral giant cell, conditioned taste aversion, CREB, PKA, *Lymnaea*

* Corresponding Author address. Kagawa School of Pharmaceutocal Sciences, Tokushima Bunri University, Sanuki 769-2193, Japan, E-mail: eito@kph.bunri-u.ac.jp.

INTRODUCTION

The causal mechanisms underlying the neuronal and molecular processes by which an associatively learned behavior is transformed into long-term memory (LTM) are still subjects to be examined. As a way of elucidating the mechanisms involved in the formation of memory, investigators have turned to molluscan systems (Alkon and Nelson, 1990; Gelperin et al., 1996; Kandel, 2001; Matsuo et al., 2011). One such good system is that of the pond snail *Lymnaea stagnalis* (Figure 1). An advantage that this system has over the many other excellent systems is that robust associative learning and LTM have been shown by a number of different laboratories with a number of different training procedures. Not only can *Lymnaea* be classically conditioned (Alexander et al., 1982; Audesirk et al., 1982; Kemenes and Benjamin, 1989a, 1989b; Kojima et al., 1996, 1998; Kemenes et al., 1997; Sakakibara et al., 1998; Ito et al., 1999, 2013), but they are also amenable to one-trial learning (Kemenes et al., 2002; Sugai et al., 2007) as well as being suitable for operant conditioning studies (Kobayashi et al., 1998; Spencer et al., 1999; Lukowiak et al., 2000). In all cases, the associative learning can be committed to LTM that persists for days, weeks, and months depending on the training procedure used. Their neuronal mechanisms have also been investigated for the simplicity of the central nervous system (CNS) of *Lymnaea*. Therefore, *Lymnaea* may be one of the best systems to use in the studies of how associative learning is remembered at the cellular and molecular levels. In addition, CNS preparations have also been utilized to study neural analogues of associative learning in vitro (Veprintsev and Rozanov, 1967; Kemenes et al., 1997; Sunada et al., 2012).

The Ito group has so far noted one remarkable learning ability in *Lymnaea*. This is the capacity to establish taste aversion and consolidate it into LTM. This phenomenon is referred to as conditioned taste aversion (CTA) (Kojima et al., 1996; Kita et al., 2011). To produce CTA in *Lymnaea*, an appetitive stimulus (e.g., sucrose) is used as the conditioned stimulus (CS). Application of the CS to the lips increases the feeding response (i.e., the number of bites) in snails. An aversive stimulus (e.g., KCl) is used as the unconditioned stimulus (US).

Application of the US to the snails inhibits feeding behavior. In the taste aversion training procedure, the CS is paired with the US. After repeated temporal contingent presentations of the CS and US, the CS no longer elicits a feeding response (Figure 2), and this taste aversion persists for more than a month (Kojima et al., 1996).

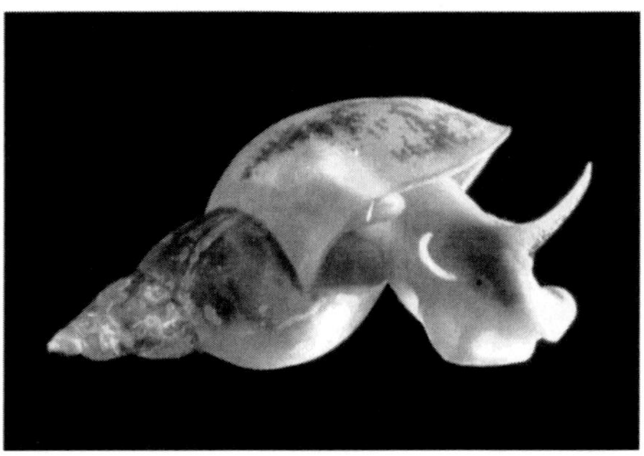

Figure 1. The pond snail *Lymnaea stagnalis*. The snails with 20 mm shell length were used.

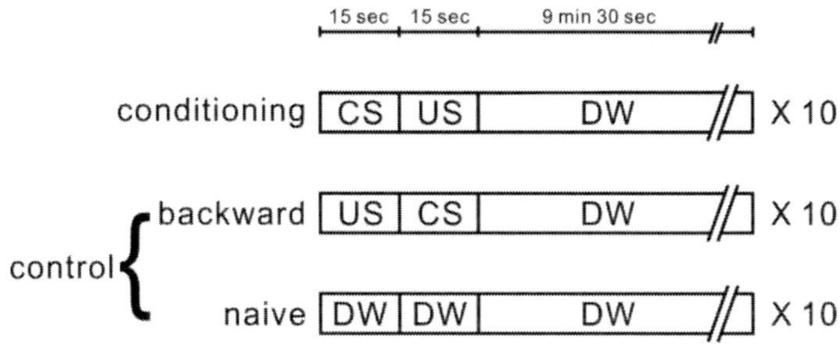

Figure 2. Training paradigm. CS is a 10 mM sucrose solution; US is a 10 mM KCl solution; DW is distilled water. As the pretest, the number of feeding responses was counted in distilled water following a 15-s application of the 10 mM sucrose CS to the lips. In the taste aversion training procedure as shown conditioning, we paired the 10 mM sucrose CS with the 10 mM KCl US. Snails received 10 paired CS-US trials on a single day. Controls included a backward-conditioned (US-CS) group and a naive group to validate associative learning. For the naive control group, only distilled water was applied to the lips instead of the CS and US. In the posttest sessions, snails were again challenged with the CS, and the number of bites was recorded in the 1-min interval in distilled water after a 15-s application of the CS. All tests were performed blindly.

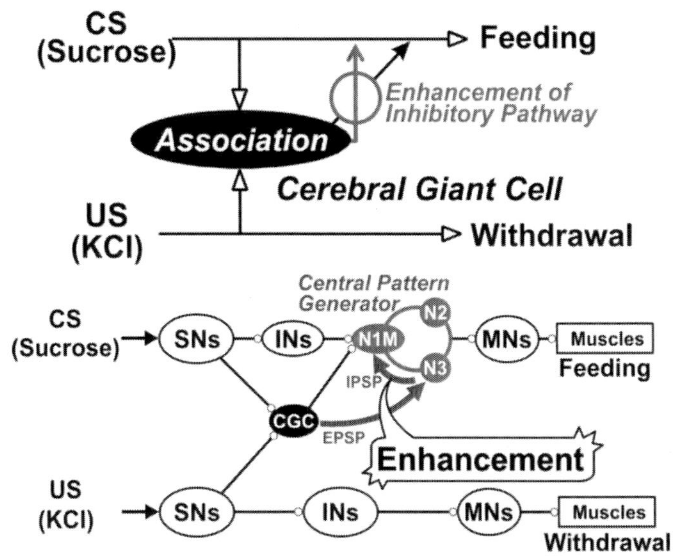

Figure 3. Our working hypothesis for *Lymnaea* CTA. Upper: Neuromodulatory model. When sucrose (CS) is followed by KCl (US) in the training session, the association of these stimuli occurs at one or more loci in the CNS. Then this association enhances an inhibitory pathway, resulting in suppression of the feeding response to sucrose (CS). Lower: Neural circuitry model. The sensory neurons (SNs) sensitive to sucrose excite the interneurons (INs), including the feeding CPG neurons, and the motor neurons (MNs) to induce a feeding response, whereas the sensory neurons to KCl excite the interneurons (INs) and the motor neurons (MNs) in the withdrawal pathway, resulting in a withdrawal response. Based on previous observations by many researchers, we hypothesized that a pair of CGCs receive the information of the above two stimuli and that the CGC exerts a strong polysynaptic inhibitory influence on one of the feeding CPG neurons (N1M cell) via the N3t cell.

On the basis of the above behavioral experiments, we proposed a working hypothesis for CTA in *Lymnaea* (Figure 3). We hypothesized that when the CS (sucrose) is followed by the US (KCl) in the training session, the association of the CS and US potentiates an inhibitory neuronal pathway, resulting in suppression of the feeding response to the CS (Kojima et al., 1996). Taking into account the underlying neural circuits worked out by the Benjamin group (See Benjamin et al., 2000 as a review), our model further proposes that sensory neuron(s) activated by the appetitive sucrose (CS) excite the feeding central pattern generator (CPG) neurons that drive motor neurons in the CS pathway to induce a feeding response. Similarly, sensory neuron(s) activated by the aversive KCl stimulus (US) excite withdrawal interneurons that activate motor neurons in the US pathway, resulting in a withdrawal response. The withdrawal response takes precedence over the feeding response. With the pairing of CS-US, the CS is no longer capable of eliciting feeding. It is the association of the CS and US in the key interneurons that results in the CS no longer being able (i.e., while LTM persists) to elicit the feeding response.

Previous studies have shown that the cerebral giant cells (CGCs) exert both a weak excitatory monosynaptic influence and a strong inhibitory polysynaptic influence on the neuron 1 medial (N1M) cells of the feeding CPG, and that the repetitive firing of the CGCs.

This results in inhibitory influences on the N1M cells (Yeoman et al., 1996). The CGCs act as a pair of interneurons, one located in the right and the other in the left cerebral ganglia. We showed that the CS (sucrose) and US (KCl) are associated in the CGCs and alter the activity of the CGCs (Nakamura et al., 1999a, 1999b). Because we applied CS and US only to the lips, but not the neurons in the CNS, these solutions were used as tastes for the lips, such as sweet and bitter, respectively. We have in control experiments shown that if these solutions are directly applied to the CNS, no responses are recorded at the CGCs. Thus, the CGCs were a logical site for further explanation to elucidate the neuronal mechanisms of CTA (Ito et al., 2012). With further experiments, we found that a polysynaptic inhibitory postsynaptic potential (IPSP) recorded in the N1M cells by activation of the CGCs was larger and lasted longer in the taste-aversion trained snails than that in the control snails (Kojima et al., 1997). These data suggested to us that an enhanced IPSP in the N1M cells underlies the suppression of feeding response in *Lymnaea* CTA (Kojima et al., 2001). Interestingly, when the amplitude of the IPSP recorded in the N1M cells in the CNS taken from good memory performers was compared with the IPSP amplitude recorded in the poor memory performers, we found that there was a much greater variance in the amplitude of the IPSP from the poor performers. This suggested to us that this greater variance in IPSP amplitude in the poor performers corresponds to the instability of the input elicited by the US in those key target neurons. Thus, the polysynaptic IPSP from the CGCs to the N1M cells in memory-poor performers is not able to suppress the feeding response.

In this chapter, we note that CGC plays a key role in *Lymnaea* CTA and thus review the transcription mechanism, partuculary the cascade of cAMP-PKA-CREB, in the CGC as well as that of CCAAT/enhancer binding protein in the other neuron in CTA-LTM.

Cyclic AMP and Protein Kinase A

As described above, the CGC plays a crucial role in *Lymnaea* CTA. We thus attempted to clarify the signal transduction cascades from the CGC to the following buccal motor neurons (Nakamura et al., 1999c). We examined whether cAMP-dependent protein kinase (PKA) contributes to enhancement of a monosynaptic connection between the presynaptic CGC and the postsynaptic buccal motor neuron 1 (B1 motor neuron). Injection of cAMP into the CGC or inhibition of phosphodiesterase by isobutylmethylxanthine in the CGC increased the amplitude of excitatory postsynaptic potential (EPSP) in the B1 motor neutron, whereas no changes were detected in the electrical properties of the CGC. The synaptic enhancement in the B1 motor neuron was completely blocked by inhibition of PKA in the CGC. The increase in the EPSP amplitude of B1 motor neuron was associated with the increase in the amount of serotonin release from the CGC.

We next injected cAMP into the CGC and monitored the potentials of the B3 motor neuron activated by the CGC (Otsuka et al., 2013). The B3 activity is used as an index for the synaptic inputs from the N3t cell to the N1M cell, both of which consist of the feeding CPG. We found that the B3 potentials were transiently enlarged. Thus, when the cAMP concentration is increased in the CGC by taste aversion training, cAMP-induced changes may play a key role in the establishment of a memory trace in the N3t cell.

On the other hand, the CGC is serotonergic (Hatakeyama and Ito, 1999). We examined the serotonin receptors to identify their receptor subtypes functioning in the synapses between the CGC and the B1 motor neuron by recording the B1 EPSP evoked by stimulation of the CGC in the presence of antagonists: cinanserin for metabotropic serotonin (5-HT_2) receptors and/or MDL72222 for ionotropic (5-HT_3) receptors (Kawai et al., 2011). The EPSP amplitude was partially suppressed by the application of these antagonists. The rise time of the compound EPSP was longer in the presence of MDL72222 than in that of cinanserin. These results suggested that these two subtypes of serotonin receptors are involved in the CGC-B1 synapses, and that these receptors contribute to EPSP. That is, the fast component of compound EPSP is mediated by 5-HT_3-like receptors, and the slow component is generated via 5-HT_2-like receptors.

CREB

Isoforms

LTM formation is universally accepted to depend on altered gene activity (i.e., transcription) and new protein synthesis (i.e., translation) (McGaugh, 2000). Altered gene activity is initiated by various transcription factors (Abel and Kandel, 1998), one of which is the cAMP response element binding protein (CREB). Evidence from *Drosophila*, *Aplysia*, and mice demonstrated that specific transcription mediated by CREB is necessary for the establishment of behavioral LTM or neural analogue of LTM (Silva et al., 1998; Lonze and Ginty, 2002).

There are at least two different members of CREB/activating transcription factor (ATF) family, which play the roles in LTM formation: one known as the activator leads to an increase in transcription activity, whereas the other known as the repressor inhibits transcription activity (Bartsch et al., 1995, 1998). The ratio of activator/repressor CREBs has been proposed to act as a "molecular switch" in determining whether LTM is formed. In *Aplysia*, CREB1a homodimers bind to a DNA target site, cAMP response element (CRE), and activate transcription by phosphorylation of a particular serine residue (Bartsch et al., 1998), whereas *Aplyisa* CREB2 acts as a repressor in transcription via conformation of heterodimers with *Aplyisa* CREB1 (Bartsch et al., 1995). Similar findings have been obtained in *Drosophila* and in mammals (Karpinski et al., 1992; Yin et al., 1994).

CREB can activate the expression of target genes in a phosphorylation-dependent manner via dimerization by binding to a CRE in their promoter regions (Montminy and Bilezikjian, 1987). The CREB family gene is also known to be expressed in a variety of spliced isoforms, giving rise to functionally different CREB family proteins with the potential to either activate or repress target gene expression (Ruppert et al., 1992; Mayr and Montminy, 2001; Blöcher et al., 2003). CREB activator isoforms contain the three functional domains, phosphorylation box (P-box) or kinase inducible domain containing several phosphorylation sites for various kinases, transactivation domain (Q-domain) interacting with a component of the TFIID complex for basal transcription, and leucine zipper (bZIP) domain for DNA binding and dimerization.

We cloned and analyzed the cDNA sequences of homologues of CREB1 and CREB2 and determined the presence of these mRNAs in identified neurons of the *Lymnaea* CNS (Sadamoto et al., 2004). The deduced amino acid sequence of *Lymnaea* CREB1 is homologous to transcriptional activators, mammalian CREB1 and *Aplysia* CREB1a, in the C-terminal bZIP and P-box, whereas the deduced amino acid sequence of *Lymnaea* CREB2 is homologous to transcriptional repressors, human CREB2, mouse activating transcription factor-4, and *Aplysia* CREB2 in the bZIP domain (Figure 4). In situ hybridization revealed that only a relatively few neurons showed strongly positive signals for *Lymnaea* CREB1 mRNA, whereas all the neurons in the CNS contained *Lymnaea* CREB2 mRNA. Using one of the neurons (the CGCs) containing *Lymnaea* CREB1 mRNA, we showed that the injection of a CRE oligonucleotide inhibited a cAMP induced, long-lasting synaptic plasticity. We therefore concluded that CREBs are present in *Lymnaea* neurons and may function as necessary players in behavioral plasticity (Ribeiro et al., 2003; Sadamoto et al., 2004).

On the other hand, CREB repressor isoforms, lacking each one of P-box or Q-domain, have inhibitory effects on gene expression (Bartsch et al., 1998; Behr and Weinbauer, 2000; Blöcher et al., 2003, 2005; Poels and Vanden Broeck, 2004). Spliced isoforms of CREB also have opposite effects on learning behavior. For example, mutant mice lacking the activator isoforms of CREB have deficiencies in long-term but not in short-term memory (Bourtchuladze et al., 1994). In *Drosophila*, seven spliced isoforms of CREB protein were identified: dCREB2-a, -b, -c, -d, -q, -r, and -s (Yin et al., 1995b). Olfactory memory formation was enhanced by overexpression of transcription activator dCREB2-a (Yin et al., 1995a), and impaired by overexpression of transcription repressor dCREB2-b lacking Q-domain (Yin et al., 1994; Perazzona et al., 2004). In *Aplysia*, at least three different CREB isoforms played competing roles in LTM formation (Bartsch et al., 1998; Upadhya et al., 2004). Mohamed et al. (2005) and Liu et al. (2008) reported that inducing CREB1 gene expression by a direct CREB1 feedback loop leads to increased transcriptional activity for long-term synaptic facilitation in *Aplysia*. These studies clearly showed that spliced isoforms of CREB regulate the gene induction for memory formation, and thus the ratio between the levels of CREB activator and repressor isoforms is important for memory formation.

We thus characterized *Lymnaea* CREB1 mRNA isoforms of spliced variants in the *Lymnaea* CNS (Sadamoto et al., 2010). Among these spliced variants, the three isoforms that code a whole *Lymnaea* CREB1 protein are considered to be the activators for gene regulation (Figure 5).

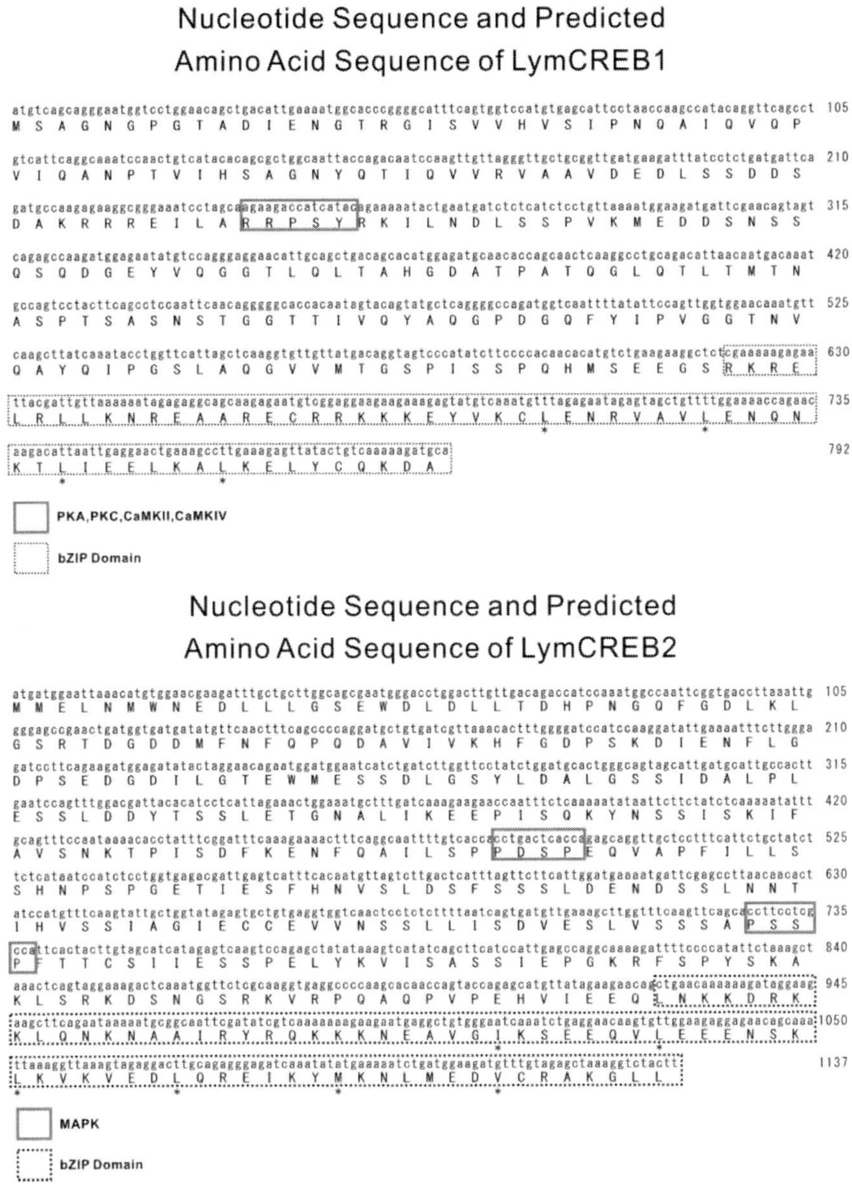

Figure 4. Nucleotide sequences and predicted amino acid sequences of *Lymnaea* CREB1 and CREB2. For CREB1, the solid square region shows the phosphorylation site by PKA, PKC, CaMK2 and CaMK4. The dashed square region shows the b-ZIP domain. For CREB2, it can be phosphorylated by MAP kinase. The dashed square region is the b-ZIP domain.

The other four isoforms, which code truncated *Lymnaea* CREB1 proteins with no kinase inducible domain, are the repressors. For a better understanding of the possible roles of different *Lymnaea* CREB1 isoforms, the expression level of these isoform mRNAs was investigated by real-time PCR method. Further, we examined the changes in gene expression for all the isoforms in the CNS after taste aversion training or backward conditioning as a control. The results showed that taste aversion training increased *Lymnaea* CREB1 gene expression, but it did not change the activator/repressor ratio. Our findings showed that the repressor isoforms, as well as the activator ones, are expressed in large amounts in the CNS, and the gene expression of CREB1 isoforms appeared to be specific for the given stimulus. This was the first quantitative analysis of the expression patterns of CREB1 isoforms at the mRNA level and their association with learning behavior.

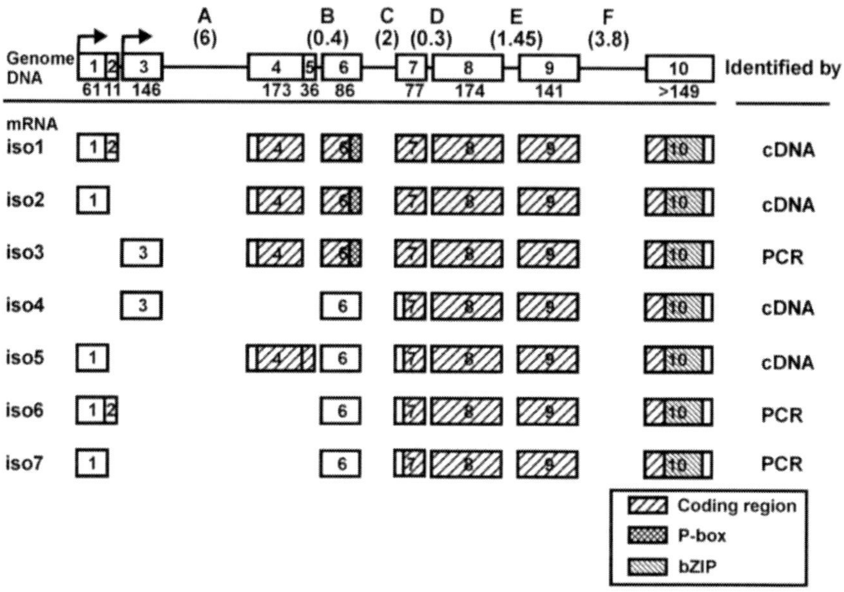

Figure 5. Genomic organization of *Lymnaea* CREB1 gene and its spliced mRNA isoforms. The genomic DNA and seven identified isoforms produced by alternative splicing are shown as iso1 to iso7. Upper: Squares represent exons 1-10 and lines represent introns A-F. The numbers below the exons indicate the exon lengths (bp). The numbers in the parentheses above the introns indicate the intron lengths (kb). Lower: Open boxes indicate non-coding regions; hatched boxes indicate coding regions. The identifying methodologies are shown on the right (Sadamoto et al., 2010).

As descrived above, the activator isoforms, possessing the functional domains for kinase induction and for interaction with other transcriptional regulators, act as transcriptional activators. On the other hand, some isoforms, lacking those functional domains, are reported to be repressors that make heterodimers with activator isoforms. The complex and ingenious function for CREB1 arises in part from the variation in their spliced isoforms, which allows them to interact with each other. To demonstrate direct observation of dimerization between CREB1 activator and repressor in living cells, we applied fluorescence cross-correlation spectroscopy (FCCS) (Sadamoto et al., 2011). The FCCS is a well established spectroscopic method to determine the interaction between the different fluorescent molecules in the aqueous condition. Using differently labeled CREB1 isoforms, we successfully observed the interaction of CREB1 activator and repressor via dimerization in the nuclei of cultured cells. As a result, we confirmed the formation of heterodimer between CREB1 activator and repressor isoforms in living cells.

Amounts and RNA Interference in the CGC

Gene expression is differently regulated in every cell even though the cells are included in the same tissue. For this reason, we need to measure the amount of mRNAs in a single neuron to understand transcription mechanism better. However, there have been no accurate, rapid and appropriate methods to determine the exact copy numbers of particular mRNAs in a single cell. We therefore developed a procedure for isolating a single, identified neuron and determining the exact copy numbers of mRNAs within it (Wagatsuma et al., 2005). We first isolated the CGC. We then determined the copy numbers of mRNAs for CREBs. The protocol uses two techniques in concert with each other: a technique for isolating a single neuron with newly developed micromanipulators coupled to an assay of mRNAs by real-time PCR. The molecular assay determined the mRNA copy numbers, each of which was compared with a standard curve prepared from cDNA solutions corresponding to the serially diluted solutions of *Lymnaea* CREB mRNA. The standard curves were linear within a range of 10 to 10^5 copies, and the intra-assay variation was within 15%. Each neuron removed from the ganglia was punctured to extract the total RNA directly and was used for the assay without further purification. Using this two-step procedure, we found that the mRNA copy number of CREB repressor (CREB2) was 30-240 in a single cerebral giant cell, whereas that of CREB activator (CREB1) was below the detection limits of the assay (< 25). These results suggest that the CREB cascade is regulated by an excess amount of CREB2 in the CGCs. Our procedure is the only quantitative analysis for elucidation of the dynamics of gene transcription in a single cell.

We still did not know whether the constitutively expressed forms of CREB were enough or the newly synthesized forms were required for the synaptic enhancement. In addition, if the newly synthesized forms were needed, we must have determined the time for translation of CREB from its mRNA. We thus applied the methods of RNA interference and real-time PCR to CREB in the CGCs. Repeatedly, the CGCs play an important role in associative learning and employ a CREB cascade for the synaptic enhancement to neurons such as the B1 motor neurons. We injected the small interfering RNA (siRNA) of CREB1 or CREB2 into the CGCs and examined the changes in amplitude of EPSP recorded in the B1 motor neurons (Wagatsuma et al., 2006). The changes in the amounts of CREB1 and CREB2 mRNAs were also examined in the CGCs. The EPSP amplitude recorded in the B1 motor neurons was suppressed 15 min after injection of CREB1 siRNA, whereas that was augmented 60 min after injection of CREB2 siRNA. In the latter case, the decrease in the amount of CREB2 mRNA was confirmed by real-time PCR. Our results showed that the de novo synthesized forms of CREB are required within tens of minutes for the synaptic enhancement in memory consolidation.

CCAAT/Enhancer Binding Protein

Other than CREB, some specific transcription factors are also thought to be essential for memory consolidation across species. One of the important transcription factors that regulate learning-induced gene expression is the CCAAT/enhancer binding protein (C/EBP) (Ramji and Foka, 2002; Hatakeyama et al., 2004). In *Aplysia*, the induction of C/EBP is regulated by the early activation of CREB1, and this process is required for long-term facilitation (Guan et al., 2002). In addition, overexpression of *Aplysia* C/EBP (ApC/EBP) in sensory neurons induced long-term facilitation after the application of a single pulse of serotonin, which normally induces only short-term facilitation (Lee et al., 2001).

C/EBP thus appears to be critical for the formation of longterm memory. We examined the changes in the mRNA and protein expression levels of C/EBP in the B2 motor neurons of *Lymnaea* (Hatakeyama et al., 2006).

These buccal motor neurons are involved in the motor control of feeding behavior, with a potentially important role in CTA. Single-cell real-time PCR revealed a significant decrease in *Lymnaea* C/EBP mRNA copy numbers in the B2 motor neurons during memory consolidation after taste aversion training (Figure 6).

By contrast, isoelectric focusing and immunoblotting of extracts of the buccal ganglia showed that translation and phosphorylation levels of *Lymnaea* C/EBP significantly increased during memory consolidation. The C/EBP-like immunoreactivity in the B2 motor neurons, which are the major immunopositive component in the buccal ganglia, also significantly increased during memory consolidation, suggesting that the main source of increase in the level of protein in the buccal ganglia are the B2 motor neurons. Thus, early memory consolidation after taste aversion training in *Lymnaea* involves both the rapid synthesis and phosphorylation of *Lymnaea* C/EBP as well as the rapid breakdown of *Lymnaea* C/EBP mRNA in the neural network controlling feeding, suggesting that all of these processes play a role in the function of C/EBP in memory consolidation.

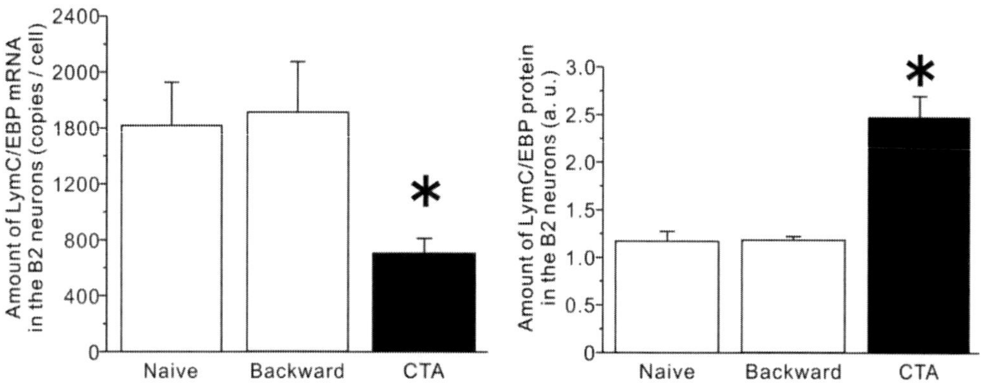

Figure 6. Changes in C/EBP mRNA and protein levels in single B2 motor neurons. Left: The copy number of *Lymnaea* C/EBP mRNA in the B2 motor neurons of taste-aversion trained snails (CTA) was less than that in B2 motor neurons from control snails (*$P < 0.01$, one-way ANOVA, Schaffé's post-hoc test). Right: The ratio of average immunocytochmistry intensity in the cytoplasm and nucleus of the taste-averion trained group (CTA) was significantly higher than in the control group (*$P < 0.01$, one-way ANOVA, Bonferroni's post-hoc test) (Hatakeyama et al., 2006).

DISCUSSION

Our nest study target is to build a mathematical model based on a molecular action described in this chapter and its corresponding behavior with respect to learning and momory in *Lymnaea*. The mathematical model emploies the Hill equations with multiple inputs/outputs (Alon, 2006) on describing chemical reactions in gene regulatory networks, and maintains compatibility with our real measurement data in *Lymnaea*. The more correctness of the model is expected to be improved through interactive cross-schemes between the computational and the biological data collations. Thease modeling studies will enable us to attempt to capture the phenomenon of learning and momory in *Lymnaea* from the viewpoint of control theorem by means of dynamical systems analyses. Particularly, we will develop the mathematical expressions for transitions of the dynamics, which determine the steady states of transcription modulating factor quantity and the second messenger titer. These theoretic approaches will provide the quantative aspects of how the transitions of the genetic dynamics can regulate the learning estarblishment and the momory consolidation. Moreover, the given control model will be constantly confirmed by computer simulations. That means, in short, the mathematical modeling is treated exactly the same way as the framework of control system designs or system identification techniques with real data.

We consider that the control theoretic approaches should be substantiative and imperative procedures to capture the invariant things of the system structure in our target phenomenon. By these cross-schemes between the computational and the biological approaches, we will try to perform reverce-engineering in biology, that is to elucidate the principle of learning and memory in biology by control theorm views.

Finally, we would like to add some comments on the function of the postsynaptic neuron for the CGC. The postsynaptic neuron may be controlled by an action of insulin.

Previous studies have shown that some molluscan insulin-related peptide (MIP) genes were upregulated in *Lymnaea* exhibiting CTA (Azami et al., 2006). We thus hypothesized that MIPs play an important role in neurons underlying the CTA-LTM consolidation process. To examine this hypothesis, we observed the distribution of MIP II, a major peptide of MIPs, and MIP receptor and determined the amounts of their mRNAs in the CNS (Murakami et al., 2013a). MIP II was only observed in the light green cells in the cerebral ganglia, but the MIP receptor was distributed throughout the entire CNS, including the buccal ganglia.

Next, when we applied exogenous mammalian insulin, secretions from MIP-containing cells or partially purified MIPs, to the isolated CNS, we observed a long-term change in synaptic efficacy (i.e., enhancement) of the synaptic connection between the CGC and the B1 motor neuron (Murakami et al., 2013a). This synaptic enhancement was blocked by application of an insulin receptor antibody to the isolated CNS. Injection of the insulin receptor antibody into *Lymnaea* before taste aversion training, while not blocking the acquisition of taste aversion learning, blocked the memory consolidation process; thus, LTM was not observed. These data suggested that MIPs trigger changes in synaptic connectivity that may be correlated with the consolidation of taste aversion learning into CTA-LTM in the *Lymnaea* CNS.

Further, we examined whether or not taste aversion training changes the mRNA expression level of MIP receptor in the *Lymnaea* CNS and found that it does not (Hatakeyama et al., 2013). This result suggested that the MIPs' effect on synaptic function in the CNS is attributable to a change in the MIP concentration, and not to a change in the mRNA expression level of MIP receptor, which is thought to reflect the number of MIP receptors. We then examined whether the observed changes in synaptic plasticity were the result of presynaptic and/or postsynaptic alterations using the paired pulse procedure. The paired pulse ratio was unaltered following insulin application, suggesting that insulin's effects on synaptic plasticity are mediated postsynaptically in the B1 motor neuron (Murakami et al., 2013b). Thus, it was suggested that postsynaptic changes need to be considered when insulin's actions on synaptic plasticity are examined.

In this chapter, we have described the transcription mechanism of cAMP-PKA-CREB in the CGC, which plays a key function in *Lymnaea* CTA. We also have reviewed the transcription mechanism of CCAAT/enhancer binding protein during CTA-LTM in *Lymnaea*.

The advantage of the use of *Lymnaea* is that we are capable of studying the transcription mechanism underlying learning and memory at the single cell level.

ACKNOWLEDGMENTS

This work was partly supported by KAKENHI from JSPS to E.I.

REFERENCES

Abel, T., Kandel, E. R., (1998). Positive and negative regulatory mechanisms that mediate long-term memory storage. *Brain Res. Rev.* 26, 360-378.

Alexander, J. E. Jr, Audesirk, T. E., Audesirk, G. J., (1982) Rapid, nonaversive conditioning in a freshwater gastropod. II. Effects of temporal relationships on learning. *Behav. Neural Biol.* 36, 391-402.

Alkon, D. L., Nelson, T. J., (1990). Specificity of molecular changes in neurons involved in memory storage. *FASEB J.* 4, 1567-1576.

Alon, U., (2006). An Introduction to Systems Biology: Design Principles of Biological Circuits. CRC, Boca Raton.

Audesirk, T. E., Alexander, J. E. Jr, Audesirk, G. J., Moyer, C. M., (1982). Rapid, nonaversive conditioning in a freshwater gastropod. I. Effects of age and motivation. *Behav. Neural Biol.* 36, 379-390.

Azami, S., Wagatsuma, A., Sadamoto, H., Hatakeyama, D., Usami, T., Fujie, M., Koyanagi, R., Azumi, K., Fujito, Y., Lukowiak, K., Ito, E., (2006). Altered gene activity correlated with long-term memory formation of conditioned taste aversion in *Lymnaea*. *J. Neurosci. Res.* 84, 1610-1620.

Bartsch, D., Ghirardi, M., Skehel, P. A., Karl, K. A., Herder, S. P., Chen, M., Bailey, C. H., Kandel, E. R., (1995). *Aplysia* CREB2 represses long-term facilitation: relief of repression converts transient facilitation into long-term functional and structural change. *Cell* 83, 979-992.

Bartsch, D., Casadio, A., Karl, K. A., Serodio, P., Kandel, E. R. (1998). CREB1 encodes a nuclear activator, a repressor, and a cytoplasmic modulator that form a regulatory unit critical for long-term facilitation. *Cell* 95, 211-223.

Benjamin, P. R., Staras, K., Kemenes, G., (2000). A systems approach to the cellular analysis of associative learning in the pond snail *Lymnaea*. *Learn. Mem.* 7, 124-131.

Behr, R., Weinbauer, G. F., (2000). CREM activator and repressor isoforms in human testis: sequence variations and inaccurate splicing during impaired spermatogenesis. *Mol. Hum. Reprod.* 6, 967-972.

Blöcher, S., Behr, R., Weinbauer, G. F., Bergmann, M., Steger, K., (2003). Different CREM-isoform gene expression between equine and human normal and impaired spermatogenesis. *Theriogenology* 60, 1357–1369.

Blöcher, S., Fink, L., Bohle, R. M., Bergmann, M., Steger, K., (2005). CREM activator and repressor isoform expression in human male germ cells. *Int. J. Androl.* 28, 215-23.

Bourtchuladze, R., Frenguelli, B., Blendy, J., Cioffi, D., Schutz, G., Silva, A. J., (1994). Deficient long-term memory in mice with a targeted mutation of the cAMP-responsive element-binding protein. *Cell* 79, 59-68.

Gelperin, A., Kleinfeld, D., Denk, W., Cooke, I. R., (1996). Oscillations and gaseous oxides in invertebrate olfaction. *J. Neurobiol.* 30, 110-122.

Guan, Z., Giustetto, M., Lomvardas, S., Kim, J. H., Miniaci, M. C., Schwartz, J. H., Thanos, D., Kandel, E. R, (2002). Integration of long-term-memory-related synaptic plasticity involves bidirectional regulation of gene expression and chromatin structure. *Cell* 111, 483-493.

Hatakeyama, D., Ito, E., (1999). Three-dimensional reconstruction and mapping of serotonin-like immunoreactive neurons in the central nervous system of the pond snail, *Lymnaea stagnails*, with the confocal laser scanning microscope. *Bioimages* 7, 1-12.

Hatakeyama, D., Fujito, Y., Sakakibara, M., Ito, E., (2004). Expression and distribution of transcription factor CCAAT/enhancer-binding protein in the central nervous system of *Lymnaea stagnalis*. *Cell Tissue Res.* 318, 631-641.

Hatakeyama, D., Sadamoto, H., Watanabe, T., Wagatsuma, A., Kobayashi, S., Fujito, Y., Yamashita, M., Sakakibara, M., Kemenes, G., Ito, E., (2006). Requirement of new protein synthesis of a transcription factor for memory consolidation: Paradoxical changes in mRNA and protein levels of C/EBP. *J. Mol. Biol.* 356, 569-577.

Hatakeyama, D., Okuta, A., Otsuka, E., Lukowiak, K., Ito, E., (2013). Consolidation of long-term memory by insulin in *Lymnaea* is not brought about by changing the number of insulin receptors. *Commun. Integr. Biol.* 6, e23955.

Ito, E., Kobayashi, S., Kojima, S., Sadamoto, H., Hatakeyama, D., (1999). Associative learning in the pond snail, *Lymnaea stagnalis*. *Zool. Sci.* 16, 711-723.

Ito, E., Otsuka, E., Hama, N., Aonuma, H., Okada, R., Hatakeyama, D., Fujito, Y., Kobayashi, S., (2012). Memory trace in feeding neural circuitry underlying conditioned taste aversion in *Lymnaea*. *PLoS ONE* 7, e43151.

Ito, E., Kojima, S., Lukowiak, K., Sakakibara, M., (2013). From likes to dislikes: conditioned taste aversion in the pond snail *Lymnaea stagnalis*. *Can. J. Zool.* 91, 405-412.

Kandel, E. R., (2001). The molecular biology of memory storage: a dialogue between genes and synapses. *Science* 294, 1030-1038.

Karpinski, B. A., Morle, G. D., Huggenvik, J., Uhler, M. D., Leiden, J. M., (1992). Molecular cloning of human CREB-2: an ATF/CREB transcription factor that can negatively regulate transcription from the cAMP responsive element. *Proc. Natl. Acad. Sci. USA* 89, 4820-4824.

Kawai, R., Kobayashi, S., Fujito, Y., Ito, E., (2011). Multiple subtypes of serotonin receptors in the feeding circuit of a pond snail. *Zool. Sci.* 28, 517-525.

Kemenes, G., Benjamin. P. R., (1989a). Appetitive learning in snails shows characteristics of conditioning in vertebrates. *Brain Res.* 489, 163-166.

Kemenes, G., Benjamin, P. R., (1989b). Goal-tracking behavior in the pond snail, *Lymnaea stagnalis*. *Behav. Neural Biol.* 52, 260-270.

Kemenes, G., Staras, K., Benjamin, P. R., (1997). In vitro appetitive classical conditioning of the feeding response in the pond snail *Lymnaea stagnalis*. *J. Neurophysiol.* 78, 2351-2362.

Kemenes, I., Kemenes, G., Andrew, R. J., Benjamin, P. R., O'Shea, M., (2002). Critical time-window for NO-cGMP-dependent long-term memory formation after one-trial appetitive conditioning. *J. Neurosci.* 22, 1414-1425.

Kita, S., Hashiba, R., Ueki, S., Kimoto, Y., Abe, Y., Gotoda, Y., Suzuki, R., Uraki, E., Nara, N., Kanazawa, A., Hatakeyama, D., Kawai, R., Fujito, Y., Lukowiak, K., Ito, E., (2011). Does conditioned taste aversion learning in the pond snail *Lymnaea stagnalis* produce conditioned fear? *Biol. Bull.* 220, 71-81.

Kobayashi, S., Kojima, S., Yamanaka, M., Sadamoto, H., Nakamura, H., Fujito, Y., Kawai, R., Sakakibara, M., Ito, E., (1998). Operant conditioning of escape behavior in the pond snail, *Lymnaea stagnalis*. *Zool. Sci.* 15, 683-690.

Kojima, S., Yamanaka, M., Fujito, Y., Ito, E., (1996). Differential neuroethological effects of aversive and appetitive reinforcing stimuli on associative learning in *Lymnaea stagnalis*. *Zool. Sci.* 13, 803-812.

Kojima, S., Nakamura, H., Nagayama, S., Fujito, Y., Ito, E., (1997). Enhancement of an inhibitory input to the feeding central pattern generator in *Lymnaea stagnalis* during conditioned taste-aversion learning. *Neurosci. Lett.* 230, 179-182.

Kojima, S., Kobayashi, S., Yamanaka, M., Sadamoto, H., Nakamura, H., Fujito, Y., Kawai, R., Sakakibara, M., Ito, E., (1998). Sensory preconditioning for feeding response in the pond snail, *Lymnaea stagnalis*. *Brain Res.* 808, 113-115.

Kojima, S., Hosono, T., Fujito, Y., Ito, E., (2001). Optical detection of neuromodulatory effects of conditioned taste aversion in the pond snail *Lymnaea stagnalis*. *J. Neurobiol.* 49, 118-128.

Lee, J. A., Kim, H. K., Kim, K. H., Han, J. H., Lee, Y. S., Lim, C. S., Chang, D. J., Kubo, T., Kaang, B. K., (2001). Overexpression of and RNA interference with the CCAAT enhancer-binding protein on long-term facilitation of *Aplysia* sensory to motor synapses. *Learn. Mem.* 8, 220-226.

Liu, R. Y., Fioravante, D., Shah, S., Byrne, J. H., (2008). cAMP response element-binding protein 1 feedback loop is necessary for consolidation of long-term synaptic facilitation in *Aplysia*. *J. Neurosci.* 28, 1970-1976.

Lonze, B. E., Ginty, D. D., (2002). Function and regulation of CREB family transcription factors in the nervous system. *Neuron* 35, 605-623.

Lukowiak, K., Adatia, N., Krygier, D., Syed, N., (2000). Operant conditioning in *Lymnaea*: evidence for intermediate- and long-term memory. *Learn. Mem.* 7, 140-150.

Matsuo, R., Kobayashi, S., Yamagishi, M,, Itom E., (2011). Two pairs of tentacles and a pair of procerebra: optimized functions and redundant structures in the sensory and central organs involved in olfactory learning of terrestrial pulmonates. *J. Exp. Biol.* 214, 879-886.

Mayr, B., Montminy, M., (2001). Transcriptional regulation by the phosphorylation-dependent factor CREB. *Nat. Rev. Mol. Cell. Biol.* 2, 599-609.

McGaugh, J. L. (2000). Memory—a century of consolidation. *Science* 287, 248-251.

Mohamed, H. A., Yao, W., Fioravante, D., Smolen, P. D., Byrne, J. H., (2005). cAMP-response elements in *Aplysia* creb1, creb2, and Ap-uch promoters: implications for feedback loops modulating long term memory. *J. Biol. Chem.* 280, 27035-27043.

Montminy, M. R., Bilezikjian, L. M., (1987). Binding of a nuclear protein to the cyclic-AMP response element of the somatostatin gene. *Nature* 328, 175-178.

Murakami, J., Okada, R., Sadamoto, H., Kobayashi, S., Mita, K., Sakamoto, Y., Yamagishi, M., Hatakeyama, D., Otsuka, E., Okuta, A., Sunada, H., Takigami, S., Sakakibara, M., Fujito, Y., Awaji, M., Moriyama, S., Lukowiak, K., Ito, E., (2013a). Involvement of insulin-like peptide in long-term synaptic plasticity and long-term memory of the pond snail *Lymnaea stagnalis*. *J. Neurosci.* 33, 371-383.

Murakami, J., Okada, R., Fujito, Y., Sakakibara, M., Lukowiak, K., Ito, E., (2013b). Paired pulse ratio analysis of insulin-induced synaptic plasticity in the snail brain. *J. Exp. Biol.* 216, 1771-1773.

Nakamura, H., Ito, I., Kojima, S., Fujito, Y., Suzuki, H., Ito, E., (1999a). Histological characterization of lip and tentacle nerves in *Lymnaea stagnalis*. *Neurosci. Res.* 33: 127-136.

Nakamura, H., Kojima, S., Kobayashi, S., Ito, I., Fujito, Y., Suzuki, H., Ito, E., (1999b). Physiological characterization of lip and tentacle nerves in *Lymnaea stagnalis*. *Neurosci. Res.* 33, 291-298.

Nakamura, H., Kobayashi, S., Kojima, S., Urano, A., Ito, E., (1999c). PKA-dependent regulation of synaptic enhancement between a buccal motor neuron and its regulatory interneuron in *Lymnaea stagnalis*. *Zool. Sci.* 16, 387-394.

Otsuka, E., Matsunaga, M., Okada, R., Yamagishi, M., Okuta, A., Lukowiak, K., Ito, E., (2013). Increase in cyclic AMP concentration in a cerebral giant interneuron mimics part of a memory trace for conditioned taste aversion of the pond snail. *BIOPHYSICS* 9, 161-166.

Perazzona, B., Isabel, G., Preat, T., Davis, R. L., (2004). The role of cAMP response element-binding protein in *Drosophila* long-term memory. *J. Neurosci.* 24, 8823-8828.

Poels, J., Vanden Broeck, J., (2004). Insect basic leucine zipper proteins and their role in cyclic AMP-dependent regulation of gene expression. *Int. Rev. Cytol.* 241, 277-309.

Ramji, D. P., Foka, P., (2002). CCAAT/enhancerbinding proteins: structure, function and regulation. *Biochem. J.* 365, 561-575.

Ribeiro, M. J., Serfőző, Z., Papp, A., Kemenes, I., O'Shea, M., Yin, J. C., Benjamin, P. R., Kemenes, G., (2003). Cyclic AMP response element-binding (CREB)-like proteins in a molluscan brain: cellular localization and learning-induced phosphorylation. *Eur. J. Neurosci.* 18, 1223-1234.

Ruppert, S., Cole, T. J., Boshart, M., Schmid, E., Schutz, G., (1992). Multiple mRNA isoforms of the transcription activator protein CREB: generation by alternative splicing and specific expression in primary spermatocytes. *EMBO J.* 11, 1503-1512.

Sadamoto, H., Sato, H., Kobayashi, S., Murakami, J., Aonuma, H., Ando, H., Fujito, Y., Hamano, K., Awaji, M., Lukowiak, K., Urano, A., Ito, E., (2004). CREB in the pond snail *Lymnaea stagnalis*. Cloning, gene expression and function in identifiable neurons of the central nervous system. *J. Neurobiol.* 58, 455-466.

Sadamoto, H., Kitahashi, T., Fujito, Y., Ito, E., (2010). Learning-dependent gene expression of CREB1 isoforms in the molluscan brain. *Front. Behav. Neurosci.* 4, 25.

Sadamoto, H., Saito, K., Muto, H., Kinjo, M., Ito, E., (2011). Direct observation of dimerization between different CREB1 isoforms in a living cell. *PLoS ONE* 6, e20285.

Sakakibara, M., Kawai, R., Kobayashi, S., Horikoshi, T., (1998). Associative learning of visual and vestibular stimuli in *Lymnaea*. *Neurobiol. Learn. Mem.* 69, 1-12.

Silva, A. J., Kogan, J. H., Frankland, P. W., Kida, S., (1998). CREB and memory. *Annu. Rev. Neurosci.* 21, 127-148.

Spencer, G. E., Syed, N. I., Lukowiak, K. (1999). Neural changes after operant conditioning of the aerial respiratory behavior in *Lymnaea stagnalis*. *J. Neurosci.* 19, 1836-1843.

Sugai, R., Azami, S., Shiga, H., Watanabe, T., Sadamoto, H., Kobayashi, S., Hatakeyama, D., Fujito, Y., Lukowiak, K., Ito E., (2007). One-trial conditioned taste aversion in *Lymnaea*: good and poor performers in long-term memory acquisition. *J. Exp. Biol.* 210, 1225-1237.

Sunada, H., Lukowiak, K., Sakakibara, M., (2012). In vitro aversion conditioning in *Lymnaea*. *Acta Biol. Hung*. 63 (Suppl. 2), 190-193.

Upadhya, S. C., Smith, T. K., Hegde, A. N., (2004). Ubiquitin-proteasomemediated CREB repressor degradation during induction of long-term facilitation. *J. Neurochem.* 91, 210-219.

Veprintsev, B. N., Rozanov, S. I. (1967). Training of the isolated snail brain [in Russian]. *Biofizika*. 12, 943-947.

Wagatsuma, A., Sadamoto, H., Kitahashi, T., Lukowiak, K., Urano, A., Ito, E., (2005). Determination of the exact copy numbers of particular mRNAs in a single cell by quantitative real-time RT-PCR. *J. Exp. Biol.* 208, 2389-2398.

Wagatsuma, A., Azami, S., Sakura, M., Hatakeyama, D., Aonuma, H., Ito, E., (2006). De novo synthesis of CREB in a presynaptic neuron is required for synaptic enhancement involved in memory consolidation. *J. Neurosci. Res.* 84, 954-960.

Yeoman, M. S., Brierley, M. J., Benjamin, P. R., (1996). Central pattern generator interneurons are targets for the modulatory serotonergic cerebral giant cells in the feeding system of *Lymnaea*. *J. Neurophysiol.* 75, 11-25.

Yin, J. C., Wallach, J. S., Del Vecchio, M., Wilder, E. L., Zhou, H., Quinn, W. G., Tully, T., (1994). Induction of a dominant negative CREB transgene specifi cally blocks long-term memory in *Drosophila*. *Cell* 79, 49–58.

Yin, J. C., Del Vecchio, M., Zhou, H., Tully, T., (1995a). CREB as a memory modulator: induced expression of a dCREB2 activator isoform enhances long-term memory in *Drosophila*. *Cell* 81, 107–115.

Yin, J. C., Wallach, J. S., Wilder, E. L., Klingensmith, J., Dang, D., Perrimon, N., Zhou, H., Tully, T., Quinn, W. G., (1995b). A *Drosophila* CREB/CREM homolog encodes multiple isoforms, including a cyclic AMP dependent protein kinase-responsive transcriptional activator and antagonist. *Mol. Cell. Biol.* 15, 5123–5130.

In: Memory Consolidation
Editors: Manabu Sakakibara and Etsuro Ito

ISBN: 978-1-63482-596-2

Chapter 8

DYNAMIC MOLECULAR MECHANISMS OF MEMORY CONSOLIDATION AFTER SINGLE-TRIAL FOOD-REWARD CLASSICAL CONDITIONING IN *LYMNAEA*

György Kemenes*
Sussex Neuroscience, School of Life Sciences,
University of Sussex, Brighton, UK

ABSTRACT

Lymnaea provides highly valuable experimental models for top-down analyses of the evolutionarily conserved molecular mechanisms of the consolidation of associative memory. Transcription and translation-dependent long-term memory (LTM) forms after single-trial classical conditioning, which allows the analysis of the role of key molecular pathways in memory consolidation in a sharply timed manner. The consolidation of LTM after single-trial food reward learning requires the activation of highly conserved signaling pathways (NO/cGMP, PACAP/cAMP/PKA, PKC, PKM, MAPK, NMDA and AMPA receptors, CaMKII), activation of gene transcription by CREB and new mRNA and protein synthesis. Importantly, these molecular mechanisms of memory consolidation are very dynamic, with different mechanisms being activated in different phases, from early through intermediate to late ('lingering') consolidation either sequentially or in a parallel and overlapping manner. Furthermore, different kinase enzymes are involved in the consolidation of early (6 h post-training) versus late (24 h post-training) phases of LTM again underscoring the highly dynamic and complex nature of the process of memory consolidation.

Keywords: feeding; reward classical conditioning; associative long-term memory; top-down approach; PACAP, cAMP, PKA, PKC, PKM, MAPK, NOS, NO, cGMP, NMDA

* Corresponding author: Professor György Kemenes: E-mail: g.kemenes@sussex.ac.uk; Telephone: +44 (0)1273 872762.

receptors, AMPA receptors, CaMKII, CREB, RNA synthesis; protein synthesis; invertebrates; molluscs; *Lymnaea*

INTRODUCTION

In the 1980's seminal behavioral work established the ability of the pond snail *Lymnaea stagnalis* to form long-term associative memory after classical conditioning with just a single pairing of a non-food chemical stimulus, amyl acetate, and a food stimulus, sucrose. [1] By the late 1990's the wealth of knowledge about the behavioral features of reward classical conditioning of feeding [2-8], together with a detailed understanding of the feeding circuitry (for a recent review see [9]) and how its neurons change after food-reward classical conditioning [7, 8, 10, 11] had made *Lymnaea* a very attractive and highly productive experimental model for top-down analyses of associative learning and memory.

Detailed analyses of molecular mechanisms of associative memory started in the early 2000's after sufficient information had been obtained on both the behavioral and cellular aspects of associative learning after classical conditioning. By the mid-2000's *Lymnaea* had emerged as the most widely used molluscan model system for the top-down analysis of both the molecular and cellular mechanisms of associative learning, including both classical and operant conditioning. This chapter reviews the findings on the molecular mechanisms of memory consolidation obtained using single-trial food-reward classical conditioning in *Lymnaea*. Other chapters of this book provide reviews of the results based on other conditioning paradigms in *Lymnaea* and other mollusks.

1. SINGLE-TRIAL FOOD-REWARD CLASSICAL CONDITIONING IN *LYMNAEA*

To date, the most detailed information on the molecular mechanisms of the consolidation of associative LTM in *Lymnaea* has been gained from experiments using single-trial food-reward (also known as appetitive) classical conditioning. In this paradigm snails are subjected to a conditioning protocol using a single pairing of amyl acetate ('pear drops') as the CS with a highly salient food stimulus, sucrose, as the US. As early as 30 min after training, the explicitly paired (CS + US) experimental group shows significantly greater feeding responses to amyl acetate over their own naïve responses and all the standard control groups used in classical conditioning experiments (Random, Explicitly Unpaired, CS alone, US alone, Handled). [1, 12-16]

The most important finding from the original and later behavioral studies was that the LTM that formed after a single pairing of CS and US lasted for several weeks. [1, 12] This is a remarkable example of single trial learning, which is now being used very effectively for analyses of the time course of the molecular mechanisms underlying memory processes. The use of this single-trial paradigm for the analysis of the molecular mechanisms of memory consolidation has two main advantages. First, after single-trial conditioning, translation- and transcription dependent memory emerges in a matter of hours [13] allowing this type of memory to be studied on a timescale of a few hours to several weeks. This early emergence of

LTM was also utilized in *in vitro* conditioning experiments investigating the cellular and molecular mechanisms of memory formation in semi-intact preparations, which are only viable for up to 6 hours. [17] Second, unlike multi-trial paradigms, single-trial conditioning allows the analyses of the amnestic effects of sharply timed manipulations of key molecular pathways during memory consolidation. [12-16, 19, 20].

2. Requirement for Macromolecular Synthesis for the Consolidation of LTM After Single-Trial Food-Reward Classical Conditioning

At the molecular level, the most important question to be addressed was whether or not LTM forming after single-trial conditioning (similar to 'flash-bulb' memory known in other systems, including humans, cf. [21]) is based on the same conserved pathways that were originally described using multi-trial paradigms in other systems. An important clue to suggest that this might be the case is that the broad-spectrum protein synthesis inhibitor anisomycin (ANI) and the transcriptional inhibitor actinomycin-D (Act-D) injected in an early time window (10 min to 1 h) after conditioning blocked the 24 h memory trace (conventionally regarded as LTM). [13, 20] Earlier (4 h and 5 h) memory traces were also blocked by both ANI and Act-D, confirming that LTM, defined as memory dependent on early post-training translation and transcription, is present as early as 4 h after single-trial conditioning. [13, 20] This observation indicates that single-trial reward conditioning triggers molecular cascades that are involved in the rapid consolidation of long-lasting memory traces. Interestingly, there is only a single early time window of sensitivity to transcription- and translation blockers [13], unlike in a number of other studies, which described a temporally distinct second window of protein synthesis-dependent LTM in both vertebrates and invertebrates. [22-25] It has been suggested that 'strong' training protocols lead to rapid memory consolidation involving only a single wave of protein synthesis dependent events, whereas 'weak' training protocols are followed by a more prolonged consolidation phase, containing two or more windows of requirement for new protein synthesis. [24] According to this categorization, single-trial reward conditioning in *Lymnaea* qualifies as a 'strong' training protocol, explaining both the rapid consolidation and persistence of the associative memory trace, characteristic of 'flash-bulb' memory.

The possibility that cooling induced blockade of LTM stems from the disruption of protein synthesis was also investigated after single-trial food-reward conditioning in *Lymnaea*. [26] By varying the timing of post-training hypothermia the critical period during which cooling disrupts the consolidation of appetitive long-term memory was determined. Post-training hypothermia was found to disrupt LTM only when applied immediately after conditioning, while delaying the treatment by 10 min left the 24 h memory trace intact. This brief (<10 min) window of sensitivity differs from the time-window previously described for the protein synthesis inhibitor anisomycin, which was effective during at least the first 30 min after conditioning. [13] Therefore, hypothermia and protein synthesis inhibition exhibit distinct time-windows of effectiveness in *Lymnaea*, a fact that is inconsistent with the hypothesis that cooling induced amnesia occurs through the direct disruption of macromolecular synthesis. It was therefore suggested that the very early blocking of the 24 h memory trace by cooling is interfering with enzymic cascades including nitric oxide synthase and kinases that are known to be activated by chemical conditioning (see sections 3, 4 and 6).

3. NO and Memory Consolidation After Single-Trial Reward Conditioning

The first ever study to investigate a specific molecular mechanism of LTM in *Lymnaea* established that consolidation of LTM in *Lymnaea* is dependent on the nitric oxide (NO)-cGMP signalling pathway. [12] There is a critical period of sensitivity up to 5 h after conditioning when blocking this pathway by drug injection prevents LTM formation (Figure 1), which was established at both the level of whole animals and the neuronal network responsible for generating the feeding behavior. [12]

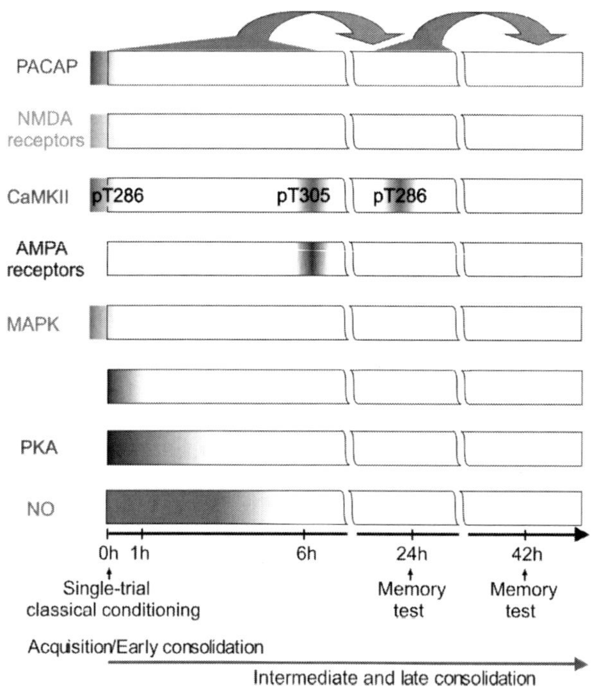

Figure 1. Time windows of molecular requirements for the consolidation of LTM after single-trial food-reward classical conditioning in *Lymnaea*. The large curved arrows indicate the requirement for the activation or synthesis of molecules in the 0 h to 6 h (acquisition/early and intermediate consolidation) and 24 h time window (late consolidation) for later memory, tested at 24 h and 42 h post-training, respectively. In the late memory consolidation stage (24 h post-training), of the molecules investigated in our studies only the intrinsic activation of CaMKII is required for memory expression at 42 h post-training. Data sources: [12-16, 18, 48].

Further evidence for a role of NO in LTM came from experiments on single CGCs (Cerebral Giant Cells, key modulatory neurons of the feeding network). [27, 28] These neurons express mRNA transcripts from two related nNOS genes (*Lym-nNOS1* and *Lym-nNOS2*) and they also express NO-sensitive soluble guanylyl cyclase (sGC). Six hours after one-trial conditioning *Lym-nNOS1* is upregulated compared with controls. This upregulation of the NOS coding transcript may be due to an earlier down-regulation of the *Lymnaea anti-nNOS* transcript at 4 h that is known to be inhibitory on NOS transcript production. Electrophysiological experiments on isolated CGCs also showed that the CGCs responded to NO by generating a prolonged depolarization of the membrane potential, similar to what was found earlier after single-trial classical conditioning. [29] The NO-induced depolarization was blocked by ODQ [28], supporting the hypothesis that it is mediated by sGC. This is a rare example of analysis where the role of a signalling molecule in memory consolidation could be traced from the behavioral to the network and single neuronal level.

4. PKA, MAPK and Memory Consolidation After Single-Trial Reward Conditioning

Other highly conserved molecular pathways that have been implicated in the consolidation of LTM after single-trial reward conditioning in *Lymnaea* are the PKA and MAPK-dependent signalling pathways. Inhibition of PKA catalytic subunit activity or MAPK phosphorylation blocked the consolidation of 24 h LTM without blocking sensory or motor pathways. [14, 18] When measured thirty min after conditioning, increased levels of both PKA activity and MAPK phosphorylation were found [14, 18], with increased PKA activation also detected when measured in earlier (5 min and 10 min) and a later (1 h) time window [14], indicating a more prolonged dependence of 24 h LTM on PKA compared to protein synthesis (Figure 1). A detailed analysis of the time windows of requirement for PKA however revealed an important difference between memory at 6 h and 24 h after training. For memory to be expressed at 6 h there is a very early (within 5 min) post-acquisition requirement for PKA with a narrow time window (around 10 min). If a PKA inhibitor is injected > 10 min after training the 6 h memory trace remains unaffected. By contrast, the 24 h memory trace is impaired by injection of the PKA inhibitor for up to 3 h post-training. This latter finding shows that the same very early PKA-mediated events that are necessary for 6 h memory are not sufficient for 24 h memory trace to form, which is dependent on much more prolonged PKA activity compared to the 6 h memory trace.

5. CREB and Memory Consolidation After Single-Trial Reward Conditioning

Detailed molecular analyses identified highly conserved cyclic AMP-responsive element binding protein (CREB) genes (LymCREB1 and LymCREB2) and CREB-like proteins in *Lymnaea*. [30, 31] These were important findings because regulation of gene expression during memory consolidation is known to involve a variety of transcription factors, with CREB playing a particularly important role in the switch between short-term to long-term memory storage in a variety of different species and paradigms. [21, 32-43] Consistent with a role for CREB in *Lymnaea* LTM is the observation that levels of phosphorylated CREB1 are increased in neurons of the feeding network following single-trial food-reward conditioning. [30].

6. NMDA and AMPA Receptors, CaMKII and Memory Consolidation After Single-Trial Reward Conditioning

The involvement of NMDA and AMPA receptors and CaM-kinase II (CaMKII) in different stages of associative LTM induced by single-trial reward classical conditioning also has been investigated in *Lymnaea*. [15, 16] First, by using a general CaM-kinase inhibitor, KN-62, it was established that CaM-kinase activation by autophosphorylation at T286 (pT286) was necessary for acquisition and late consolidation (Figure 1), but not early or intermediate consolidation or retrieval of LTM [15]. Then, western blot assays of the buccal and cerebral ganglia from classically conditioned animals and treatment of intact animals with CaMKIINtide identified CaMKII as the main CaM-kinase required for learning and also for late memory consolidation in a critical time window (around 24 h after learning).[15] Acquisition is dependent on both NMDA receptor and CaMKII activation but CaMKII dependent late memory consolidation does not require the activation of NMDA receptors (Figure 1). These findings support the notion that even apparently stable memory traces may undergo further molecular changes and identify NMDA-independent intrinsic activation of CaMKII as a mechanism underlying this 'lingering consolidation'. [44] This process may facilitate the preservation of LTM in the face of protein turnover or active molecular processes that underlie forgetting.

More recently, we described a novel CaMKII-dependent mechanism by which a learning-induced increase in AMPA receptor (AMPAR) levels is stabilized for ongoing consolidation of 24 h LTM after single-trial food-reward classical conditioning. [16] Six hours after training, the levels of both autophosphorylated pT305-CaMKII and GluA1 type AMPAR subunits were significantly elevated in the ganglia containing the *Lymnaea* learning circuits (buccal and cerebral ganglia). CaMKIINtide treatment significantly reduced the learning-induced elevation of both pT305-CaMKII and GluA1 levels and impaired the consolidation of associative long-term memory. Inhibition of proteasomal activity offset the deleterious effects of CaMKIINtide on both GluA1 levels and long-term memory. These findings suggest that increased levels of pT305-CaMKII play a role in AMPAR dependent memory consolidation by reducing proteasomal degradation of GluA1 receptor subunits.

There is a notable difference between MAPK and the other signalling molecules investigated in single-trial food reward conditioning in *Lymnaea*. Unlike the other factors, MAPK is activated not only in response to contingent CS + US application, but also when the CS or US were applied alone. [18] This observation, together with the fact that preventing MAPK phosphorylation after training blocks LTM [18] shows that MAPK is necessary but not sufficient for the consolidation of associative LTM after single-trial reward conditioning. PKA, NMDA receptors, CaMKII, CREB, and NOS/NO are selectively activated/upregulated by the associative training protocol [14, 15, 27, 30], so potentially each of these factors could be sufficient for memory consolidation. However, it is more likely to be the case that these and other signalling molecules make a synergistic contribution to the memory consolidation process, with neither of them alone being sufficient for LTM.

Treatment of isolated cerebral ganglia with the adenylate cyclase activator forskolin resulted in massively increased CREB1 phosphorylation [30] indicating a potential link between training-induced PKA activation and CREB phosphorylation in conditioned animals.

Whether there is any linkage between the NO, PKA and MAPK signalling pathways has yet to be determined.

A summary of the known time-windows of sensitivity of the 24 h and 42 h LTM to inhibition of the various different molecular pathways reviewed above is shown in Figure 1. For all the signalling molecules shown in Figure 1, except pT286-CaMKII [15], time-windows of requirement for 24 h LTM were only investigated in the first 6 hours after training so we cannot rule out later time windows of requirement for some or all of them for the 24 h memory trace. The requirement at 24 h post-training for the 42 h memory trace was investigated for all the signalling molecules shown on the diagram, except AMPA receptors.

The related observation that *Lym-nNOS1* mRNA levels increase at 6 h post-training [27] strongly suggests a later role for NO, perhaps in memory maintenance. The level of phosphorylated CREB1 was also found to be high at 6 h post-training [30] indicating that new transcription and translation take place after this period. The time windows of requirement for NO and PKA are both wider [12, 14] than the time window for requirement for early protein synthesis [13] indicating that their function goes beyond being upstream components of molecular pathways leading to new protein synthesis in the first hour after training.

7. PACAP and Memory Consolidation after Single-Trial Reward Conditioning

The observations concerning the role of PKA in memory consolidation in *Lymnaea* [14, 45] indicated that similar to other systems [46, 47] activation of adenylate cyclase (AC) is a key step in long-term memory (LTM) formation in *Lymnaea*. However, there was no information available on the molecules involved in the learning-induced activation of AC. A 2010 study then demonstrated both the presence and biochemical activity of a protein homologous to the vertebrate pituitary adenylate cyclase-activating polypeptide (PACAP) and its receptors in the *Lymnaea* nervous system. [19] Another study in 2010 showed that application of a PACAP receptor antagonist around the time of single-trial food-reward training with amyl acetate and sucrose blocked associative LTM. [48] This latter finding suggested that in this ‘strong’ food-reward conditioning paradigm the activation of AC by PACAP was necessary for LTM to form. Interestingly, in the ‘weak’ multitrial food-reward conditioning paradigm, lip touch paired with sucrose [4], memory formation was also dependent on PACAP. [48] Significantly, systemic application of PACAP at the beginning of multitrial tactile conditioning accelerated the formation of transcription-dependent memory. [48] In PACAP-treated animals, robust LTM formed after just 3 trials, while control animals required > 6 trials to form LTM. This memory-boosting effect of exogenously applied PACAP was blocked by the PACAP receptor antagonist PACAP6-38. These findings therefore provided the first evidence to show that in the same nervous system PACAP is both necessary and instructive for fast and robust memory formation after reward classical conditioning.

Pre-training application of the PACAP receptor antagonist resulted in a complete abolition of memory after both single-trial chemical and multi-trial tactile conditioning but not in a loss of the unconditioned feeding response. Based on this finding it is tempting to speculate that PACAP is released in response to the chemical and tactile conditioned stimuli, whereas similar to what was found in *Aplysia*, the effect of the unconditioned stimulus on AC may be mediated by different peptide or non-peptide transmitters, such as SCPs or 5-HT. [49-51]. A likely scenario is that in *Lymnaea*, the PACAP-mediated effect of the chemical or tactile CS and the non-PACAP mediated effect of sucrose US converge on AC and this convergence provides the molecular basis for coindicence detection, a fundamental requirement for associative learning. *Lymnaea* is known to differentiate learning with amyl acetate [29] from learning with touch [52] at the neuronal level within the same network (the feeding circuitry in this case), but there is no evidence for a similar differentiation at the molecular level within the same neuron. Thus, the same molecules (e.g., PACAP) can fulfil the same role (e.g., activation of AC) in different neurons leading to learning-induced changes in different pathways (e.g., activated by touch versus activated by amyl acetate).

More recently, it has been shown that the age-related decline in memory performance observed previously in *Lymnaea* [3, 53] can be reversed by systemic administration of pituitary adenylate cyclase activating polypeptide (PACAP). [54] Exogenous PACAP38 boosts memory formation in aged *Lymnaea*, where endogenous PACAP38 levels are low in the brain. [54] Treatment with insulin-like growth factor-1 (IGF-1), which in vertebrates was shown to transactivate PACAP type I (PAC1) receptors [55] also boosts memory formation in aged pond snails. [54] Due to the evolutionarily conserved nature of these polypeptides and their established role in memory and synaptic plasticity there is a very high probability that they could also act as 'memory rejuvenating' agents in humans.

8. cAMP-Dependent Molecular Cascades and Single Neuronal Plasticity Contributing to LTM After Single-Trial Reward Conditioning

An important link between cAMP-dependent molecular cascades and neuronal plasticity contributing to LTM was found in voltage-clamp experiments, which demonstrated a long-lasting cAMP-induced increase in a low-threshold persistent sodium current of the CGCs. [56] This current makes an important contribution to the CGC somal membrane potential [57, 58], which in turn becomes persistently depolarized after single-trial reward conditioning. [29]

The CGC's synaptic output also became similarly enhanced by somal injection of cAMP [56] and artificial depolarization of the soma membrane [29], lending further support to the notion that cAMP-dependent cascades might support long-lasting plastic changes in ion channel number and/or function contributing to learning-induced maintained depolarization. The mechanism by which persistent somal depolarization enhances synaptic output is based on a maintained increase in background calcium levels in axon terminals of the CGCs presynaptic to their target neurons [29], similar to what was demonstrated previously in mammalian neurons. [59].

A summary of the known molecular players whose expression or activation level changes in the different stages of the consolidation of late LTM after single-trial food-reward classical conditioning is shown in Figure 2.

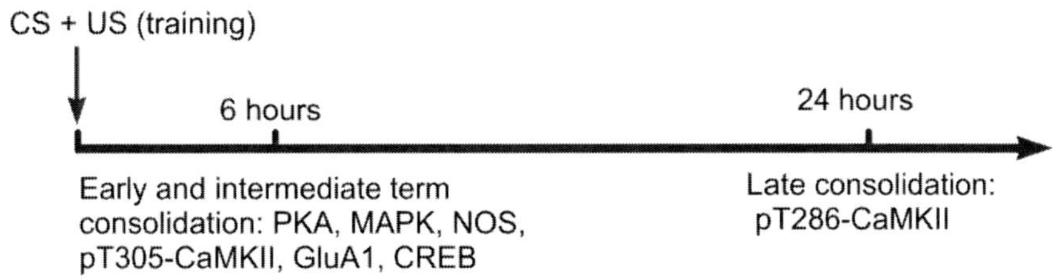

Figure 2. Involvement of specific molecular players in different stages of the consolidation of LTM after single-trial food-reward classical conditioning in *Lymnaea*. A variety of different molecular and biochemical tools were used to detect changes in the expression or activation levels of specific molecules in different stages of memory consolidation. Data sources: [12-16, 18, 30, 45, 48].

9. PKC, PKM and Memory Lapses During Early Consolidation

A detailed behavioural and electrophysiological analysis of the early (up to 4 hours post-training) stages of memory consolidation after single-trial food-reward classical conditioning revealed two periods of memory lapse, at 30 min and 2 h post-training, with no lapses at 10 min, 1 h, 3 h or 4 h. [20]. The lapse in memory expression at 30 min marks the transition from short-term memory (STM) to intermediate-term memory (ITM). The 2 h lapse indicates that ITM has an early and a late phase. No lapses in memory occurred at the transition between late ITM and LTM that first occurred 4 h after training.

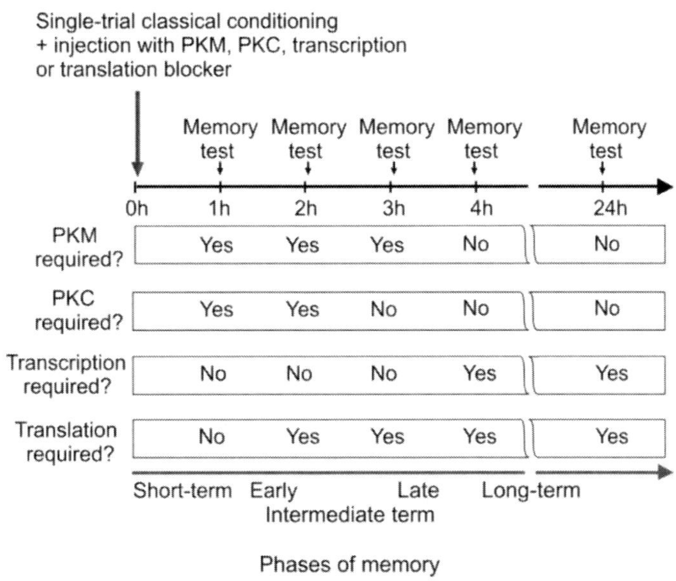

Figure 3. Differential roles for macromolecular synthesis, PKM and PKC in the early consolidation of different phases of memory after single-trial food-reward classical conditioning in *Lymnaea*. Four different groups of animals were injected with the appropriate blockers immediately after training and the conditioned response (memory) was tested at 1 h intervals for up to 4 h post-training and then again at 24 h post-training in five different groups subjected to the same post-training drug treatment. Data source: [20].

Notably, only during the two periods of lapse is ongoing memory consolidation susceptible to disruptions by external mechanical stimuli. Disruption at 30 min blocks late ITM and early LTM recall at 3 and 4 h, respectively, whereas disruption at 2 h only blocks a late form of LTM at 24 h. These shared time points of memory lapse and susceptibility were found to correspond to transitions between different phases of memory that have different molecular requirements. The consolidation of the late phase of ITM differs from that of the early one in its independence of PKC but dependence on PKM-induced processes while the consolidation of both the 4 and 24 h LTM were PKC and PKM independent (Figure 3). The occurrence of memory lapses at 30 min and 2 h after conditioning therefore indicates that during periods of molecular transitions between STM and ITM (from ~10 min to ~1 h post-training, and again between the two phases of ITM, from ~1 h to 3 h post-training) memory recall is weakened, allowing novel sensory cues to block the ongoing consolidation of intermediate-term and long-term memory, respectively.

CONCLUSION

Studies based on the single-trial food-reward classical conditioning paradigm described in this chapter have yielded valuable information on a variety of general and specific molecular mechanisms contributing to memory consolidation. Memory consolidation after single-trial classical conditioning involves the regulation of gene expression by transcription factors such as CREB and the activation of the PKA, PKC, PKM and MAPK signaling pathways as well as NMDA and AMPA receptors, CaMKII and PACAP. The key roles these molecular pathways play in memory consolidation in *Lymnaea* provide further evidence for the generality of these highly conserved mechanisms in learning, both across phylogenetic groups and across learning paradigms (non-associative or associative, single- or multi-trial, aversive or reward, operant or classical).

NO has been shown to be important for memory consolidation and this appears to involve regulation of nNOS gene expression in a specific modulatory cell type, the CGCs, key sites of plasticity after food-reward classical conditioning. [29, 57, 60] Work using *Lymnaea* as an experimental model has thus both provided important insights into the molecular mechanisms of memory consolidation at the behavioral and systems level and linked these mechanisms to learning-induced cellular and molecular changes in single identified neurons.

REFERENCES

[1] Alexander JE. Audesirk TE, Audesirk GJ. One-trial reward learning in the snail *Lymnaea stagnalis*. *J. Neurobiol.*, 1984, 15:67-72.

[2] Alexander JE. Audesirk TE, Audesirk GJ. Rapid, non-aversive conditioning in a freshwater gastropod. II. Effects of temporal relationships on learning. *Behav. Neural. Biol.*, 1982, 36:391-402.

[3] Audesirk TE, Alexander JE, Audesirk GJ, Moyer CM Rapid, nonaversive conditioning in a freshwater gastropod. I. Effects of age and motivation. *Behav. Neural. Biol.* 1982, 36:379-390.

[4] Kemenes G, Benjamin PR. Appetitive learning in snails shows characteristics of conditioning in vertebrates. *Brain Res,* 1989a, 489**:**163-166.

[5] Kemenes G, Benjamin PR. Goal-tracking behavior in the pond snail, *Lymnaea stagnalis. Behav. Neural. Biol.* 1989b, 52**:**260-270.

[6] Kemenes G, Benjamin PR. Training in a novel environment improves the appetitive learning performance of the snail, *Lymnaea stagnalis. Behav. Neural. Biol.,* 1994, 61**:**139-149.

[7] Staras K, Kemenes G, Benjamin PR. Neurophysiological correlates of unconditioned and conditioned feeding behavior in the pond snail *Lymnaea stagnalis. J. Neurophysiol.,* 1998, 79**:**3030-3040.

[8] Staras K, Kemenes G, Benjamin PR. Cellular traces of behavioral classical conditioning can be recorded at several specific sites in a simple nervous system. *J. Neurosci.,* 1999, 19**:**347-357.

[9] Benjamin PR. Distributed network organization underlying feeding behaviour in the mollusc *Lymnaea. Neural Systems & Circuits*, 2012, 2:4. doi: 10.1186/2042-1001-2-4.

[10] Whelan HA, McCrohan CR. Food-related conditioning and neuronal correlates in the freshwater snail *Lymnaea stagnalis. J. Mollusc. Stud.,* 1996, 62:483-494.

[11] Kemenes G, Staras K, Benjamin PR. *In vitro* appetitive classical conditioning of the feeding response in the pond snail *Lymnaea stagnalis. J. Neurophysiol.,* 1997, 78**:**2351-2362.

[12] Kemenes I, Kemenes G, Andrew RJ, Benjamin PR, O'Shea M. Critical time-window for NO-cGMP-dependent long-term memory formation after one-trial appetitive conditioning. *J. Neurosci.,* 2002, 22**:**1414-1425.

[13] Fulton D, Kemenes I, Andrew RJ, Benjamin PR. A single time-window for protein synthesis-dependent long-term memory formation after one-trial appetitive conditioning. *Eur. J. Neurosci.*, 2005, 21**:**1347-1358.

[14] Michel M, Kemenes I, Müller U, Kemenes G. Different phases of long-term memory require distinct temporal patterns of PKA activity after single-trial classical conditioning. *Learn Mem.*, 2008, 15:694-702.

[15] Wan H, Mackay B, Iqbal H, Naskar S, Kemenes G. Delayed intrinsic activation of an NMDA-independent CaM-kinase II in a critical time window is necessary for late consolidation of an associative memory. *J. Neurosci.,* 2010, 30:56-63.

[16] Naskar S, Wan H, Kemenes G. pT305-CaMKII stabilizes a learning-induced increase in AMPA receptors for ongoing memory consolidation after classical conditioning. *Nat. Commun.* 2014, 5:3967.

[17] Marra V, Kemenes I, Vavoulis D, Feng J, O'Shea M, Benjamin PR. Role of tonic inhibition in associative reward conditioning in Lymnaea. Front Behav. Neurosci., 2010, 4:161.

[18] Ribeiro MJ, Schofield MG, Kemenes I, O'Shea M, Kemenes, G, Benjamin PR. Activation of MAPK is necessary for long-term memory consolidation following food-reward conditioning. *Learn Mem.,* 2005, 12**:**538-545.

[19] Pirger Z, Laszlo Z, Hiripi L, Hernadi L, Toth G, Lubics A, Reglodi D, Kemenes G, Mark L. Pituitary adenylate cyclase activating polypeptide (PACAP) and its receptors are present and biochemically active in the central nervous system of the pond snail *Lymnaea stagnalis. J. Mol. Neurosci.*, 2010, 42:464-71.

[20] Marra V, O'Shea M, Benjamin PR, Kemenes I. Susceptibility of memory consolidation during lapses in recall. *Nat. Commun.*, 2013, 4:1578.
[21] Carew TJ. Molecular enhancement of memory formation. *Neuron,* 1996, 16:5-8.
[22] Grecksch G, Matthies H. Two sensitive periods for the amnesic effect of anisomycin. *Pharmacol. Biochem. Behav.,* 1980, 12:663-665.
[23] Epstein HT, Child FM, Kuzirian AM, Alkon DL. Time windows for effects of protein synthesis inhibitors on Pavlovian conditioning in *Hermissenda*: behavioral aspects. *Neurobiol. Learn Mem.,* 2003, 79:127-131.
[24] Bourtchouladze R, Abel T, Berman N, Gordon R, Lapidus K, Kandel ER. Different training procedures recruit either one or two critical periods for contextual memory consolidation, each of which requires protein synthesis and PKA. *Learn Mem.,* 1998, 5:365-374.
[25] Freeman FM, Rose SP, Scholey AB. Two time windows of anisomycin-induced amnesia for passive avoidance training in the day-old chick. *Neurobiol. Learn Mem.,* 1995, 63:291-295.
[26] Fulton D, Kemenes I, Andrew RJ, Benjamin PR. Time-window for sensitivity to cooling distinguishes the effects of hypothermia and protein synthesis inhibition on the consolidation of long-term memory. *Neurobiol. Learn Mem.*, 2008, 90:651-654.
[27] Korneev SA, Straub V, Kemenes I, Korneeva EI, Ott SR, Benjamin PR, O'Shea M. Timed and targeted differential regulation of nitric oxide synthase (NOS) and anti-NOS genes by reward conditioning leading to long-term memory formation. *J. Neurosci.,* 2005, 25:1188-1192.
[28] Ribeiro M, Straub VA, Schofield M, Picot J, Benjamin PR, O'Shea M, Korneev SA. Characterization of NO-sensitive guanylyl cyclase: expression in an identified interneuron involved in NO-cGMP-dependent memory formation. *Eur. J. Neurosci.,* 2008, 28:1157-1165.
[29] Kemenes I, Straub VA, Nikitin ES, Staras K, O'Shea M, Kemenes G, Benjamin PR. Role of delayed nonsynaptic neuronal plasticity in long-term associative memory. *Curr. Biol.,* 2006, 16:1269-1279.
[30] Ribeiro MJ, Serfozo Z, Papp A, Kemenes I, O'Shea M, Yin JC, Benjamin PR, Kemenes G. Cyclic AMP response element-binding (CREB)-like proteins in a molluscan brain: cellular localization and learning-induced phosphorylation. *Eur. J. Neurosci.,* 2003, 18:1223-1234.
[31] Sadamoto H, Sato H, Kobayashi S, Murakami J, Aonuma H, Ando H, Fujito Y, Hamano K, Awaji M, Lukowiak K, Urano A, Ito E. CREB in the pond snail *Lymnaea stagnalis*: cloning, gene expression, and function in identifiable neurons of the central nervous system. *J. Neurobiol.,* 2004, 58:455-466.
[32] Frank D, Greenberg ME. CREB: a mediator of long-term memory from mollusks to mammals. *Cell,* 1994, 79:5-8.
[33] Stevens CF. CREB and memory consolidation. *Neuron,* 1994, 13:769-770.
[34] Goda Y. Memory mechanisms. A common cascade for long-term memory. *Curr. Biol.,* 1995, 5:136-138.
[35] Yin JC, Tully T. CREB and the formation of long-term memory. *Curr. Opin. Neurobiol.,* 1996, 6:264-268.
[36] Fletcher L. Memories are made of this: the genetic basis of memory. *Mol. Med. Today,* 1997, 3:429-434.

[37] Pittenger C, Kandel E. A genetic switch for long-term memory. *C. R. Acad. Sci. III,* 1998, 321:91-96.
[38] Silva AJ, Kogan JH, Frankland PW, Kida S. CREB and memory. *Annu. Rev. Neurosci.,* 1998, 21:127-148.
[39] Alberini CM. Genes to remember. *J. Exp. Biol.,* 1997, 202:2887-2891.
[40] Lamprecht R. CREB: a message to remember. *Cell Mol. Life Sci.,* 1999, 55:554-563.
[41] Scott R, Bourtchuladze R, Gossweiler S, Dubnau J, Tully T. CREB and the discovery of cognitive enhancers. *J. Mol. Neurosci.,* 2002, 19:171-177.
[42] Tully T, Bourtchouladze R, Scott R, Tallman J. Targeting the CREB pathway for memory enhancers. *Nat. Rev. Drug Discov.,* 2003, 2:267-277.
[43] Josselyn SA, Nguyen PV. CREB, synapses and memory disorders: past progress and future challenges. *Curr. Drug Targets CNS Neurol. Disord,* 2005, 4:481-497.
[44] Dudai Y, Eisenberg M. Rites of passage of the engram: reconsolidation and the lingering consolidation hypothesis. *Neuron,* 2004, 44:93-100.
[45] Kemenes G, Kemenes I, Michel M, Papp A, Muller U. Phase-dependent molecular requirements for memory reconsolidation: differential roles for protein synthesis and protein kinase A activity. *J. Neurosci.,* 2006, 26:6298-6302.
[46] Kandel ER. The molecular biology of memory storage: a dialogue between genes and synapses. *Science,* 2001, 294:1030-1038.
[47] Margulies C, Tully T, Dubnau J. Deconstructing memory in *Drosophila. Curr Biol,* 2005, 15:R700-13.
[48] Pirger Z, László Z, Kemenes I, Tóth G, Reglodi D, Kemenes G. A homolog of the vertebrate pituitary adenylate cyclase-activating polypeptide is both necessary and instructive for the rapid formation of associative memory in an invertebrate. *J. Neurosci.,* 2010, 30:13766-13773.
[49] Trudeau LE, Castellucci VF. Contribution of polysynaptic pathways in the mediation and plasticity of *Aplysia* gill and siphon withdrawal reflex: evidence for differential modulation. *J. Neurosci.,* 1992, 12:3838-3848.
[50] Nargeot R, Baxter DA, Patterson GW, Byrne JH. Dopaminergic synapses mediate neuronal changes in an analogue of operant conditioning. *J. Neurophysiol.,* 1999, 81:1983-1987.
[51] Brembs B, Lorenzetti FD, Reyes FD, Baxter DA, Byrne JH. Operant reward learning in *Aplysia*: neuronal correlates and mechanisms. *Science,* 2002, 296:1706-1709.
[52] Jones NG, Kemenes I, Kemenes G, Benjamin PR. A persistent cellular change in a single modulatory neuron contributes to associative long-term memory. *Curr. Biol.,* 2003, 13:1064-1069.
[53] Hermann, PM, Lee A, Hulliger S, Minvielle M, Ma B, Wildering WC. Impairment of long-term associative memory in aging snails (*Lymnaea stagnalis*). *Behav. Neurosci.,* 2007, 121:1400-1414.
[54] Pirger Z, Naskar S, László Z, Kemenes G, Reglődi D, Kemenes I. Reversal of age-related learning deficiency by the vertebrate PACAP and IGF-1 in a novel invertebrate model of aging: the pond snail (*Lymnaea stagnalis*). *J. Gerontol. A. Biol. Sci. Med. Sci.,* 2014, 69:1331-1338.
[55] Delcourt, N., Thouvenot E, Chanrion B, Galéotti N, Jouin P, Bockaert J, Marin P. PACAP type I receptor transactivation is essential for IGF-1 receptor signalling and antiapoptotic activity in neurons. *Embo Journal,* 2007, 26:1542-15551.

[56] Nikitin ES, Kiss T, Staras K, O'Shea M, Benjamin PR, Kemenes G. Persistent sodium current is a target for cAMP-induced neuronal plasticity in a state-setting modulatory interneuron. *J. Neurophysiol.,* 2006, 95:453-463.

[57] Nikitin ES, Vavoulis DV, Kemenes I, Marra V, Pirger Z, Michel M, Feng J, O'Shea M, Benjamin PR, Kemenes G. Persistent sodium current is a nonsynaptic substrate for long-term associative memory. *Curr. Biol.,* 2008, 18:1221-1226.

[58] Vavoulis DV, Nikitin ES, Kemenes I, Marra V, Feng J, Benjamin PR, Kemenes G. Balanced plasticity and stability of the electrical properties of a molluscan modulatory interneuron after classical conditioning: a computational study. *Front Behav. Neurosci.,* 2010, 4:19.

[59] Awatramani GB, Price GD, Trussell LO. Modulation of transmitter release by presynaptic resting potential and background calcium levels. *Neuron,* 2005, 48:109-121.

[60] Nikitin ES, Balaban PM, Kemenes G. Nonsynaptic plasticity underlies a compartmentalized increase in synaptic efficacy after classical conditioning. *Curr. Biol.*, 2013, 23:614-619.

In: Memory Consolidation
Editors: Manabu Sakakibara and Etsuro Ito
ISBN: 978-1-63482-596-2

Chapter 9

Forgetting Another Side of Memory Consolidation As Seen by a Snail

Ken Lukowiak*
Hotchkiss Brain Institute, Cumming School of Medicine,
University of Calgary, Canada

Abstract

Following operant conditioning of aerial respiratory behavior snails form memory. Depending on the training procedure used following learning different types of memory form. The persistence of each type of memory (short, intermediate and long) is dependent on different molecular processes. Here I concentrate on long-term memory (LTM) which may persist for many days. However, as the time lengthens from when memory forms it wanes. That is, forgetting occurs. Forgetting can be prevented by various means. It is unclear, however, how forgetting occurs. Procedures used to block forgetting include cooling, placing snails under a barrier, removing the soma of a neuron that is necessary for memory formation, and finally, placing snails after the memory consolidation process is complete in an environment that prevents new memory from being made. Two main hypotheses have been put forward to explain forgetting: 1) Retrograde Interference; and 2) Decay. I attempt to differentiate between these two hypotheses in Lymnaea and conclude that the weight of evidence appears to favor the decay hypothesis.

Keywords: *Lymnaea stagnalis*, aerial respiration, operant conditioning, forgetting as an active process, retrograde interference

Introduction

Forgetting is a topic that we are all familiar with, in some cases too familiar with (e.g., Help, I forgot where I put my keys!) while at other times if only it would occur (e.g., PTSD). Forgetting has long been 'remembered' in song and verse. For example, in a song from: The Eagles, number 49 on the top 500 songs of all time was 'Hotel California'. One line in

* Calgary, AB, Canada T2N 4N1, E-mail: lukowiak@ucalgary.ca.

particular *"Some dance to remember, some dance to forget"* stands out for me in the context of this review. The writers (Don Felder, Don Henly, and Glenn Frey) in my opinion were very prescient in that, as I hope to show, forgetting is an active process that shares many similarities with learning and subsequent memory consolidation process.

Forgetting is also at the top of a lot of the general public's minds (or maybe it really should be) as evidenced by the following taken from recent summer of 2014 newspaper reports.

More and more passengers have been attempting to board Airline flights with loaded weapons in their carry-on bags, purses and pockets, according to the Transportation Security Administration. TSA told the Associate Press that officials did not keep gun confiscation statistics before 2011, but in recent years they have noticed a spike in the numbers. The most common excuse TSA hears from these passengers is, *"I forgot it was there."*

Kahlil Gibran in 1926 in his book *Sand and Foam* wrote: wrote: *Forgetfulness is a form of freedom.*

Finally, as way of introducing the topic, my first Department Chair (Keith Cooper) at the University of Calgary told me " *A clear conscience is the first sign of senility*".

So what is forgetting, how does it happen, and why is it important? My lab works on learning, memory and the effects of stress on memory formation and its persistence in a fresh water pond snail, *Lymnaea stagnalis*. One point that we have noticed is that while the learning and the initial memory is phenotypically the same, long-term memory (LTM) formation or its persistence varies dramatically depending on the training procedure used and the level of stress the snail experiences before, during or after training (Dalesman and Lukowiak, 2012a,b).

AERIAL RESPIRATORY BEHAVIOR AND OPERANT CONDITIONING IN *LYMNAEA*

The behaviour we study, aerial respiration, is a simple, easily observable, and tractable behaviour that is driven by a 3-neuron central pattern generator (CPG; Figure 1), whose necessity and sufficiency has been experimentally demonstrated (Syed, et al.,1990; Syed, et al., 1992). Since *Lymnaea* are bimodal breathers (cutaneous and acrial respiration) it is possible to modulate its aerial respiratory behaviour (e.g., teach the snail not to perform the behaviour) in a hypoxic environment where this behaviour should predominate (Lukowiak et al., 1996; 1998; 2000) without causing harm to the snail (i.e., they can still satisfy their respiratory requirements via cutaneous respiration). We examine associative learning (i.e., operant conditioning) and the subsequent three forms of memory formation: 1) short term (STM-lasting minutes), intermediate term (ITM-lasting ~ 3h) and long-term (LTM-persisting longer than 24h) following operant conditioning of aerial respiration either as a result of either 30 or 45 minute training sessions (referred to as the 'typical training procedure') or a 1-trail training (1-TT) procedure (Martens et al., 2007; Lukowiak and Dalesman, 2013). In the typical training procedure we place snails into a beaker filled with hypoxic pond-water (PW). The PW is made hypoxic by bubbling N_2 gas through it for 20 minutes. In hypoxic conditions, snails crawl to the air-water interface to open their pneumostome and perform gas exchange with the external environment. As they begin to open the pneumostome (i.e., their breathing tube which leads to a lung) we apply a negative reinforcement in the form of a gentle tactile stimulus. We call this a 'poke'. The poke is of sufficient strength to cause pneumostome closure but does not elicit the whole-animal withdrawal response. Typically with operant conditioning training, snails reduce the number of attempted pneumostome openings when placed in the hypoxic context. Memory is operationally defined as meeting the following two criteria: (1) the number of pokes in the memory-test session (MT) is significantly less than the number of attempted openings in the first training session (TS) and if two or more training sessions are used (2) the number of pokes in the memory-test session is not significantly greater than the number of pokes in the last training session (Lukowiak et al., 1996; 1998; 2000).

What does forgetting 'look' like in a typical operant conditioning experiment? Using 97 snails we determined how long LTM persisted following three 45 minute training sessions (Figure 2). Training sessions 1 and 2 (TS1 and TS2) were separated by a 1h interval and training session 3 (TS3) was given on the following day (i.e., 24h after TS2). As can be seen the number of attempted pneumostome openings (i.e., Pokes) decreased over the course of the 3 training sessions. There was a significant decrease in the number of pokes in TS2 compared to TS1. This allow us to conclude that ITM has formed. ITM in *Lymnaea* has been shown to persist for up to 3h and is dependent on new protein synthesis (Sangha et al., 2002).

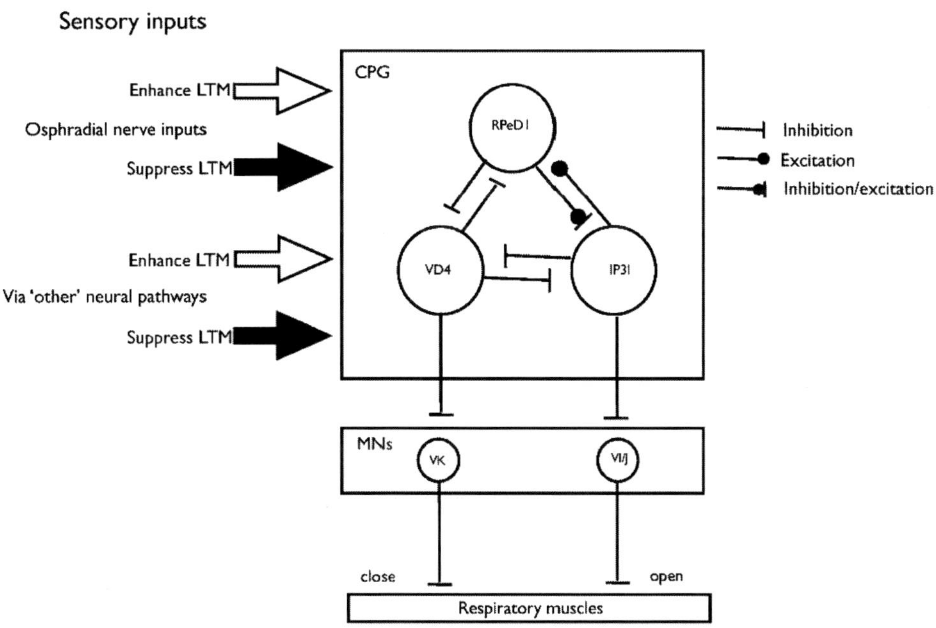

Figure 1. The CPG that drives aerial respiration in *Lymnaea*. The central pattern generator (CPG) that drives aerial respiratory behavior, associated higher order interneurons, sensory inputs and motor neuron outputs. The three CPG neurons (VD4, IP3 and RPeD1) have been shown to be both necessary and sufficient to drive aerial respiratory behavior. RPeD1 has been demonstrated to be a necessary site for long-term memory (LTM) formation, reconsolidation, extinction and forgetting. Moreover, we have recently shown that the 'state' of RPeD1 activity predicts the ability of the snail to make LTM. Rhythmogenesis is an emergent property of this network. Input from the periphery plays a major role in determining the level of spontaneous activity of neurons such as RPeD1. Neuronal activity stemming from both other parts of the CNS and periphery are shown that alter the functioning of the CPG in order to enhance or suppress the formation of memory.

There was also a significant decrease in the number of pokes in TS3 compared to TS1 and further that TS3 was not significantly greater than TS2. This allows us to conclude that the two preceding training sessions (TS1 and TS2) are sufficient to cause LTM formation, which is dependent on altered gene activity and new protein synthesis. We then randomly selected snails over the course of the next few days to determine how long the LTM persisted. As can be seen memory was seen 4 days following TS3 but not 6 days later. We thus concluded that this training procedure was sufficient to result in a memory that could persist for about 5days. The question becomes is what is the cause of the forgetting?

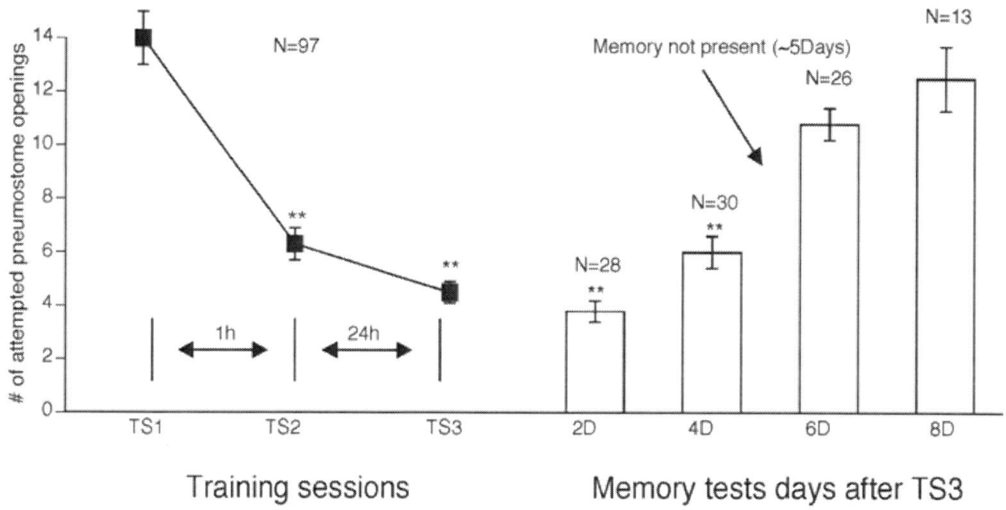

Figure 2. A forgetting curve in Lymnaea following operant conditioning of aerial respiration. Snails (n = 97) received three 30 min training sessions over the course of 1.5 days. The first two training sessions (TS1 and TS2) has a 1h interval between them. The third training session (TS3) was given 24h after TS2. Random snails (numbers indicated) were then tested for memory 2, 4, 6 and 8 days after TS3. As can be seen memory was present 4 days after training but not 6 days.

Historically, there have been two dominant theories of forgetting: 1) decay in some part of the 'memory trace'; and, 2) retrograde interference (Wixted, 2004; Sangha et al., 2005). (As an aside it is also possible that forgetting is the result of neurogenesis and/or neuronal death. However, in *Lymnaea* neither of those two possibilities are thought to play a role). Thorndike (1913; the law of disuse) hypothesized that memory decays, unless regularly used, similar to the way leg muscle atrophy occurs when a person is in a leg cast. Historically (Woodworth, 1929; Gates, 1930) it has generally been assumed that decay is a passive process; but recently, its has been reconsidered to be an active process (Hardt et al., 2013). In this view it is the loss of a substrate, for example, that is necessary to maintain memory (e.g., atypical protein kinase C isoform M-zeta (PKMz)). In support of this notion transiently inhibiting PKMz activity impairs and often abolishes fully established memories (Serrano et al., 2008), even those that are several months old (Shema et al., 2008). Consistent with this idea, over-expression of this kinase can enhance long-term memories (Shema et al., 2011).

The other hypothesis is that forgetting is produced by interference from conflicting associations, sometimes referred to as retroactive interference, RI (i.e., learning and forming memory of something new; Jenkins and Dallenbach, 1924).

The RI hypothesis views forgetting as an active process in which 'old' memories are replaced by the formation of 'new' memories. That is, new associations that resemble the naive state will be consolidated into a 'new' memory, which will permanently displace the earlier memory. Data from the 1924 involving humans (non-sense syllables; sleep vs awake and the amount of forgetting) as well as data involving an invertebrate (cockroach; Minami and Dallenbach, 1946) are great examples of type of data supporting RI as the cause of forgetting. In the study involving cockroaches forcing them to move after they received conditioning decreased retention of the conditioned response. compared to animals not forced to move. We will come back to a discussion of this below.

Both of these hypotheses in their current form view forgetting as an active process. In *Lymnaea* we have performed a number of different types of experiments in an attempt to determine how forgetting occurs. However, whatever the actual cause of forgetting is, it needs to be remembered that a positive result of forgetting is that it may allow an animal to react more appropriately to current conditions, rather than continuing to exhibit a previously learned, possibly maladaptive behaviour based on previous experience. As will become apparent the initial data from *Lymnaea* more readily supported the notion that forgetting is due to RI, but more recent data (Lukowiak et al. 2014b) are more supportive of the active-decay hypothesis.

We define learning as the acquisition of a new or different behaviour resulting from training and define memory as the ability to retain that new behaviour. Thus, forgetting is the loss of the learned response. We also do not attribute forgetting in *Lymnaea* to a 'retrieval problem'. That is, the memory is still present but the snail can't access it. We believe this to be correct since in *Lymnaea* we are dealing with a non-declarative memory, defined as changes in skilled behaviour and the ability to respond appropriately to stimuli through practice as the result of conditioning (Milner et al., 1998). Retrieval failure has more to do with declarative memories.

Non-declarative memory is stored within the same neuronal circuit that mediates the behaviour (Milner et al., 1998), in our case the respiratory CPG (Scheibenstock et al., 2002). Thus, if the behaviour can be performed, the memory, if present, is accessible.

Do we know that the memory formed following operant conditioning is located within the respiratory CPG? The answer question is yes. This is based on both accumulated correlational data and data that can be considered causative. These data are quite strong. They show that in both isolated ganglionic preparations (i.e., only the central nervous system-CNS; Spencer et al., 1999) and semi-intact preparations (i.e., pneumostome and foot still connected to the CNS; Spencer et al., 2002: McComb et al., 2005) the electrophysiological activity of one of the CPG neurons, RPeD1, the neuron that initiates rhythmic activity in the CNS, is significantly altered in preparations from operantly conditioned animals compared to naive or other control animals. In all cases the activity of RPeD1 was significantly less, the neuron was less excitable, and its ability to elicit rhythmic electrical activity in the CNS or pneumostome closure activity was significantly diminished. In addition, in naive, untrained snails RPeD1 activity is significantly different between smart and average snails (Braun et al., 2011; Braun and Lukowiak, 2012). Finally, a unique advantage possessed by our model system is that it is possible to surgically remove the soma of RPeD1, one of the CPG neurones, without altering aerial respiratory behaviour (Scheibenstock et al., 1992). However, even though the remaining neurites have the ability for new protein synthesis (i.e., mRNA is present and there are ribosomes to make the protein), there is no nucleus.

Thus, there cannot be altered gene activity (necessary for LTM formation) as there are no genes present. In this situation snails can learn and form ITM (only dependent on new protein synthesis) but cannot form LTM. We have extended these findings by showing that following soma ablation of RPeD1 reconsolidation and extinction do not occur (Sangha et al., 2003 a,b; Martens et al., 2007). As discussed below removal of RPeD1's soma after memory has formed also blocks forgetting (Sangha et al., 2005). Thus, we are confident when we say that we have shown that this neuron, RPeD1 is a necessary site for LTM formation as well as forgetting.

We have designed experiments in order to determine how forgetting occurs. That is, does the RI hypothesis or the decay hypothesis explain our experimental findings.

Submersion Experiments

If the RI hypothesis of forgetting is correct it should be possible to block forgetting by preventing a new learned interfering event. Since we operantly condition snails not to perform aerial respiration preventing aerial respiratory behaviour should, if the hypothesis is correct, prevent forgetting and thus extend memory persistence. To block the 'interfering event' we placed snails following operant conditioning training and memory consolidation under a barrier in the aquarium which prevents them from performing aerial respiration (Sangha et al., 2003e; Figure 3). Snails can only perform aerial respiration when they are at the surface of the water. It is a stereotyped behaviour that has been fully described elsewhere (Lukowiak et al., 1996), and it enables the lung to directly connect with the atmosphere via the pneumostome, the respiratory orifice. As long as the pond water is not hypoxic snails can be kept under a barrier for months.

We first showed that preventing aerial respiration in a eumoxic environment by the use of the barrier did not alter the normal homeostatic respiratory behaviour when they were allowed to come to the surface. Preventing aerial respiration for 3 or 7 days did not alter subsequent aerial respiratory behaviour in the hypoxic training/testing apparatus.

We then asked whether blocking the ability to perform aerial respiration following memory formation would extend the persistence of memory (i.e., block forgetting; Figure 3). Snails received two 45 min training sessions separated by a 1h interval. This training procedure results in an LTM that persists for 2 days. Immediately following the last training session half the snails were placed below the barrier in a eumoxic aquarium and prevented from performing aerial respiration for 7 days whilst the other half were maintained above the barrier. The half above the barrier performed aerial respiration over the course of the 7 days. We then tested all snails for memory, 7 days after the last training session. In the snails maintained below the barrier LTM was still present 7 days after the last training session. On the other hand, in the snails maintained in the same aquarium but above the barrier they did not exhibit LTM 7 days later. Because the only difference between the two groups was the ability to perform aerial respiration, these data are fully consistent with the hypothesis that RI is a major factor in causing forgetting. If forgetting is simply the result of the memory decaying with time due to a necessary 'memory-substrate (i.e., PKMz), submerging the snails should not alter this decay.

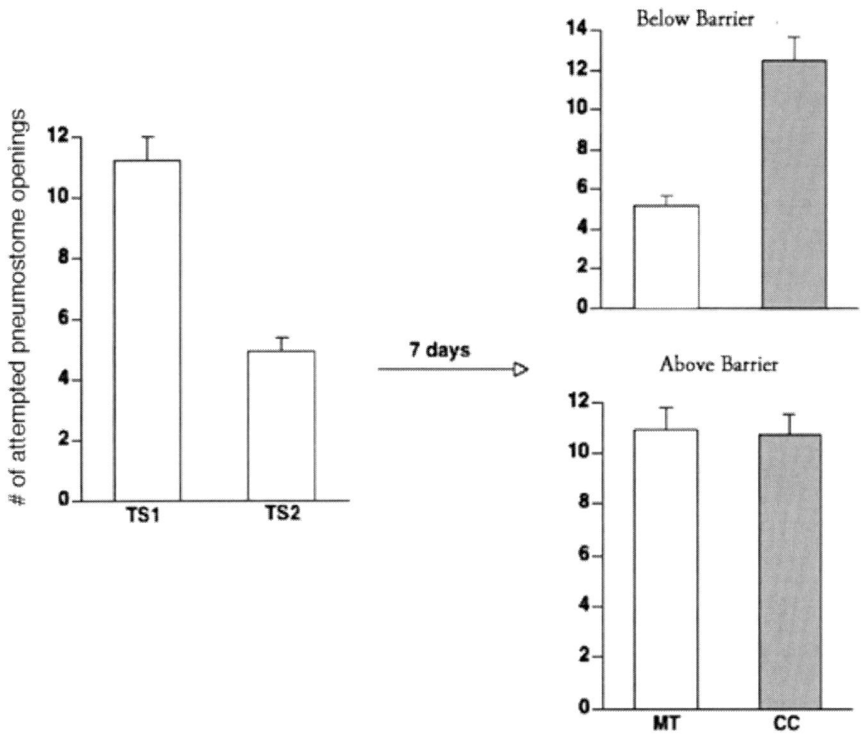

Figure 3. Preventing snails from performing aerial respiration following the operant conditioning procedure prevents forgetting. Submersion impedes forgetting. Snails (n 28) received the operant-conditioning procedure shown in Figure 1. Immediately (within 30 s) after Training Session 2 (TS2), the snails were randomly divided into two equal groups. One cohort was placed below a barrier and thus used only cutaneous respiration, whereas the other cohort was placed above the barrier in a eumoxic aquarium. Seven days later, both cohorts were blindly tested for long-term memory (LTM; memory test [MT]). The snails that were housed below the barrier and thus prevented from performing aerial respiratory behavior exhibited LTM, whereas the snails that were able to perform aerial respiratory behavior did not exhibit LTM. Both cohorts also received a change of context (CC) challenge. In the snails maintained below the barrier, there were a significantly greater number of attempted pneumostome openings in CC than in MT; whereas in the cohort maintained above the barrier, there was no significant difference between CC and MT. Thus, preventing snails from performing aerial respiratory behavior impeded forgetting.

The data presented up to this point show that submersion extended the duration of memory (i.e., retard forgetting). The data are consistent with the hypothesis that unreinforced behaviour is necessary for forgetting to occur. It is not clear how these data support the decay hypothesis.

COOLING BLOCKS FORGETTING

We found (Sangha et al., 2003c) that we could block memory formation after operant conditioning training by subjecting snails to a 'cold block' (1h @ 4° C) Cooling the snails for 1h immediately after training (i.e., within 10 min) was sufficient to block LTM. Cooling snails for a similar period 10 or 15 min after cessation of training, failed to block LTM formation. That is, once the memory consolidation period was initiated (within 10 min following training) the cold block did not prevent LTM formation. The advantages of using the cooling procedure are that it is non-toxic, easily reversible, and can be applied and removed for discrete amounts of time.

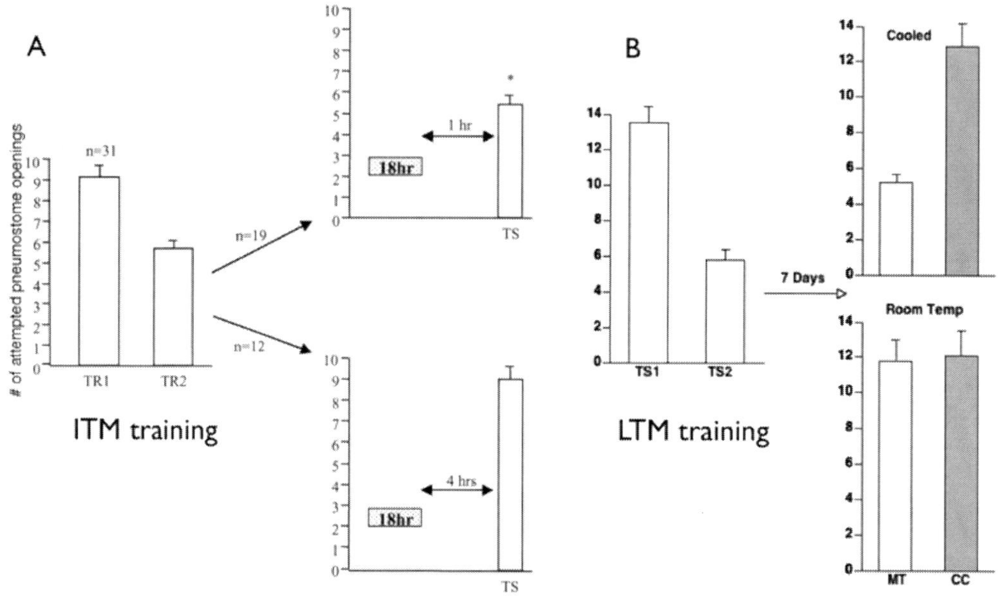

Figure 4. Cooling extends both ITM and LTM following operant conditioning. Cooling extends intermediate-term memory (A). All animals (n = 31) first received two 30-min training sessions (TR1, TR2) separated by 30 min and 1 h later, were then put in 4 C water for 18 h. Snails were tested either 1 or 4 h after this treatment. All animals demonstrated learning: TR2 was significantly lower (p < :01) than TR1. The test for savings at 1 h (n = 19) was significantly lower (p < :01) than the test for savings at 4 h (n = 12). One hour after TR2 animals (n . 19) were put into 4 C water for 18 h (indicated by box). One hour before the 30-min memory-test session (TS), animals were placed in room temperature water. Animals demonstrated memory: TS was significantly lower than TR1 and not significantly different than TR2. In the lower panel one hour after TR2 animals (n = 12) were put into 4 C water for 18 h (indicated by box). Animals were then put in room temperature water until time of testing 4 h later. Since ITM only lasts 3 h, memory was not demonstrated: TS session was not significantly different than TR1 and was significantly higher than TR2. B) Cooling to 4 °C prevents forgetting and thus extends LTM. After receiving operant conditioning, 1 hr after Training Session 2 (TS2), the snails (n = 30) were randomly divided into two equal cohorts. One cohort was placed into 4 °C pond water for 7 days, whereas the other cohort was maintained at room temperature for the same period of time. The snails were then blindly tested for long-term memory (LTM) 7 days later (memory test [MT]). All snails were tested for LTM at room temperature. LTM was present in the snails maintained at 4 °C for the 7 days; whereas in the snails maintained for 7 days at room temperature, LTM was not observed. In both cohorts, 1 hr following MT the snails were challenged with a change of context (CC) test session. In the snails maintained at 4 °C there were significantly more attempted pneumostome openings in CC than in MT; whereas, in the snails maintained at room temperature, CC and MT were not statistically different.

Cooling had been used as a means to halt or slow down metabolic activity, including protein synthesis (Sommerville, 1999; Thieringer, Jones, & Inouye, 1998); and was thus used to disrupt memory formation in both vertebrate and invertebrate preparations (Cartford et al., 1997; Morrison & van der Kooy, 1997; Sekiguchi et al., 1997; Yamada et al., 1992). Since cooling blocked LTM we hypothesized that if forgetting was due to RI (i.e., learning anew and then consolidating that new learning into memory resulting in the behavioral phenotype similar to the naive state) then it should be possible to disrupt the forgetting process (i.e., block the new consolidation process for the 'interfering events') by using cooling. We first found (Figure 4) that if we examined breathing behavior in the hypoxic environment that subjecting snails to an 8 day period @ 4 °C did not alter the breathing behavior when tested 1h after being returned to room temperature. That is, the amount of time spent breathing in the pre-cooling session was not significantly different from that in the post-cooling session. We next determined if we could extend the duration of LTM (Figure 4). One hour after the training procedure that results in LTM that persists for only 2 days at room temperature the snails were put in 4 °C water for 8 days. One hour before the memory test snails were put in room temperature water. When tested these snails exhibited memory. We then maintained these snails for a further 4 days at room temperature. When tested again for memory they did not now exhibit memory. What is interesting is that within the same individual animals, we see that the memory can be extended up to 7 days by cooling and then subsequently forgotten when the snails are returned to room temperature.

These data, while consistent with the RI hypothesis could also be thought to be consistent with the decay process. That is, cooling could slow down the decay of a substance such as PKMz and thus forgetting would be slowed down. However, when we add in the data obtained with the submersion experiments we feel the data are in further support of the RI hypothesis.

LOW CALCIUM POND WATER AS AN EXTERNAL ENVIRONMENT

We then used another strategy for blocking forgetting that did not entail submersion or cooling. This strategy made use of the finding that a low calcium pond water environment (Dalesman and Lukowiak, 2010) blocks LTM formation. Dalesman et al., (2011) found that placing snails in a low calcium pond water environment (20 mg/l) for as short as 1h was sufficient to block LTM formation. They further showed that if snails were maintained for 1 week in the low calcium environment then placing snails in a normal calcium environment (80 mg/l) for 1h allowed snails to form LTM. They suggested that the detection of sufficient environmental calcium (i.e., 80 mg/l) rapidly reversed the genomic changes in neurons (e.g., RPeD1) induced by the low environmental calcium (20 mg/l), such that the necessary molecular events that cause the formation of LTM can ensue.

If forgetting in *Lymnaea* is an active process (i.e., RI) involving the acquisition of a newer memory that interferes with the older memory then blocking new LTM formation using low calcium pond water, should prevent forgetting. Using this strategy (Figure 5) Knezevic et al. (2011) predicted that placing snails in the low Ca^{2+} environment after they had been operantly conditioned in standard Ca^{2+} would prevent forgetting as no new LTM could be formed and destroy the old memory. They found that snails maintained in the low Ca^{2+} condition following training in the standard Ca^{2+} condition did not exhibit forgetting. They also showed that the prevention of forgetting was due to the prevention of new memory formation and not to some 'side-effect' of placing snails in a low Ca^{2+} environment. To do this they performed a number of experiments in which snails were maintained in the low calcium environment for different time periods. For example, snails were held immediately following training in the standard Ca^{2+} environment for 72h, then placed in the low Ca^{2+} environment 24h before the memory test in the standard Ca^{2+} environment. These snails exhibited forgetting. Likewise, the snails were maintained in low Ca^{2+} for 72h and then transferred to standard Ca^{2+} for 24h before memory was tested in the standard Ca^{2+} also showed forgetting. That is, 24h was sufficient time for the snails to learn and form new memory that interfered with the older memory and thus caused forgetting. These data are more easily explained by the RI hypothesis. It is unclear how we could explain the low calcium experiments with a decay process.

Thus, if snails form LTM and then are: 1) cooled, 2) prevented from performing aerial respiration; or 3) placed into a low calcium environment forgetting is blocked. In each of these situations we have blocked forgetting by 1) reducing the ability to form a new memory (cooling) albeit with snails performing reduced aerial respiration (because of the increased oxygen in the cooler pond water); 2) prevented aerial respiration from occurring (barrier experiments); 3) prevented any further LTM forming by placing snails in a low calcium environment, even though the number of pneumostome openings does not change. All these data are consistent with the hypothesis that forgetting is an active process dependent on altered gene activity and new protein synthesis in which forgetting is new learning and memory caused by the unreinforced pneumostome openings that occur when the snail performs aerial respiration and has the capacity to form new LTM. This new memory (i.e., RI) replaces the old memory and whose phenotype bears resemblance to the naive state.

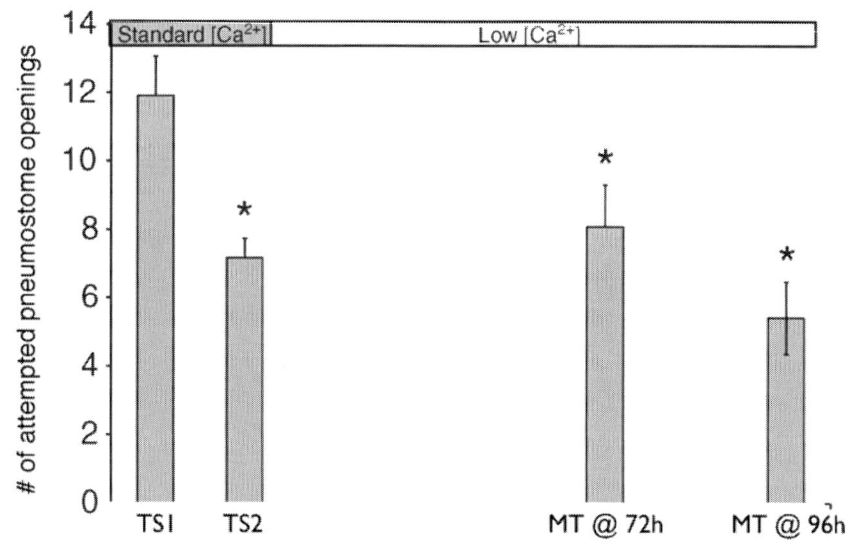

Figure 5. A low calcium environment prevents forgetting. Snails maintained in the low Ca2+ condition following training in the standard Ca2+ condition do not exhibit forgetting. The number of attempted pneumostome openings (means ± s.e.) in 30 min during training sessions (TR1 and TR2) and memory test sessions (Test) 72 and 96 h later. Column shading indicates conditions during training/testing (grey, standard Ca2+). All training and testing sessions were performed in standard Ca2+. Shading of the horizontal bar above the columns indicates conditions whilst snails were held in their eumoxic aquaria (grey, standard Ca2+; white, low Ca2+). *Number of pneumostome openings differs significantly from that in TR1.

ABLATION OF RPeD1 SOMATA AFTER LTM FORMATION

LTM formation in *Lymnaea*, as in other organisms (McGaugh, 2000), depends on *de novo* protein and RNA synthesis (i.e., altered gene activity; Sangha et al., 2003). Thus, procedures such as cooling or the application of specific transcription and translation blockers immediately after operant conditioning (i.e., during the consolidation period), prevent LTM formation. Moreover, the molecular processes necessary for LTM formation, memory reconsolidation (i.e., restabilization of the memory after it has been made active), and extinction of LTM have been directly shown to occur within one of these neurones, RPeD1 (Sangha, Scheibenstock, & Lukowiak, 2003; Sangha, Scheibenstock, McComb, & Lukowiak, 2003d; Sangha, Scheibenstock, Morrow, & Lukowiak, 2003b; Scheibenstock et al., 2002).

A signature advantage of the *Lymnaea* preparation is the capability of removing (i.e., ablating) the somata of a single neuron, leaving behind a functional and intact neurite where local protein synthesis can occur (Scheibenstock et al., 2002; Spencer et al., 2000). Successful RPeD1 soma ablation prior to operant conditioning does not block aerial respiratory behavior, associative learning, ITM formation but does prevent LTM formation from occurring. If our hypothesis is correct that RI causes forgetting and couple that with the fact that RPeD1 soma ablation prevents new LTM from forming; it would be logically correct to predict that removal of RPeD1's somata after they have formed LTM that they would be incompetent to form a new long-term memory (i.e., unreinforced aerial respiratory behavior) and thus RI could not occur and forgetting would not be observed.

As can be seen (Figure 6) one hour after TS2, to allow for consolidation (Sangha, Scheibenstock, & Lukowiak, 2003a), the somata of either RPeD1 or LPeD1 was ablated (downward arrow). All the snails were observed to be capable of performing aerial respiration. It was found in the control group (a neuron of a similar size, LPeD1 but not involved with controlling aerial respiratory behavior, had its soma ablated) that LTM was not present on Day 7. LTM was present in the RPeD1-ablated group. To further ensure that the RPeD1-ablated animals were not adversely affected by the ablation procedure, causing a decrease in pneumostome openings not related to memory, a change of context test was used. The RPeD1-ablated snails recognized this as a different context than the one in which they were trained and thus showed an increase in pneumostome openings when exposed to this context. They exhibited learning, ITM but not LTM.

However, they could not form a new memory because the soma of RPeD1 was not present; thus they did not form an LTM for the carrot context. Notice however, that in the LPeD1 soma ablated cohort the snails learned formed ITM and LTM in the new context. These results in carrot replicate previous findings that LPeD1-ablated animals are capable of forming new long-term memories, whereas RPeD1-ablated animals are not (Scheibenstock et al., 2002). Thus, we were able to impair the forgetting of a memory that typically lasts 2 days by ablating the soma of RPeD1, thereby extending the memory to at least 15 days.

Our finding that snails whose RPeD1 somata were ablated (i.e., genes are not present as the nucleus has been removed) after learning and memory consolidation did not forget shows more directly that blocking the establishment of new LTM of a conflicting association impedes forgetting. That is, forgetting is an active process. To our knowledge, this was the first demonstration of the necessity of molecular processes occurring in a single neuron that are required for forgetting. However, those data do not really help to differentiate between the two competing hypotheses of what causes forgetting.

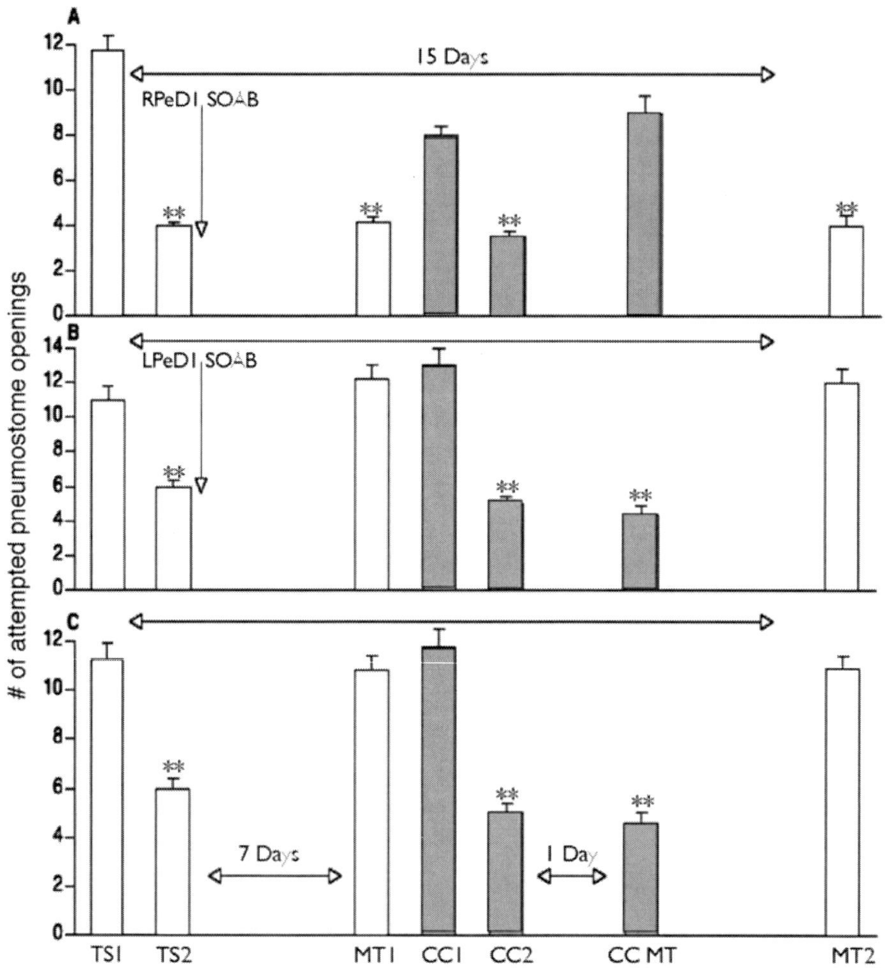

Figure 6. RPeD1 soma ablation following operant conditioning and memory consolidation prevents forgetting. Right Pedal Dorsal 1 (RPeD1) soma ablation prevents forgetting. A: After receiving operant conditioning, 1 hr after Training Session 2 (TS2), the snails' RPeD1 soma was ablated (n 10; downward arrow). We tested for long-term memory (LTM) 7 days later (Memory Test 1 [MT1]) and found LTM to be present. To demonstrate that these snails were still capable of performing aerial respiration, they were challenged with Change of Context Session 1 (CC1) and then Change of Context Session 2 (CC2). The number of attempted openings in CC1 was significantly larger than in MT1. These RPeD1 soma-ablated snails were able to learn in the new context as evidenced by the reduction in pneumostome openings between CC1 and CC2; however, they did not show LTM for the carrot context memory test (CC MT) 1 day later. Finally, these snails were once again challenged with a memory test (MT2) 15 days later, and LTM for the original context was still present. B: The same procedure as in Panel A was used, but the snails' Left Pedal Dorsal 1 (LPeD1) soma was ablated (n 9; downward arrow). Snails did not exhibit LTM at MT1 or MT2. However, both learning (CC1 and CC2) and LTM (CC MT) were observed for the carrot context. C: The same training procedure as in Panels A and B were used, but no ablation procedure was performed (n 13). Results were the same as those in Panel B.

Can we perform an experiment to better differentiate between the RI and the decay hypotheses? On looking back on our experiments we were struck with one result that appeared to better support the RI theory rather than the decay hypothesis. This was the submersion experiment. Snails prevented from performing aerial respiration had their memory extended (Figure 2). In some ways this experiment complemented the experiment performed in Miami and Dallenbach in 1946, on cockroaches which many investigators have referenced in support of the RI hypothesis. In that study it was demonstrated that after intervals of inactivity in which the cockroaches were immobilized in small boxes filled with tissue paper, memory retention and relearning was far superior to those insects that received corresponding intervals of normal rest. In addition, this same study illustrated that forced activity following learning led to savings scores that were much poorer than after corresponding intervals of normal rest. These data are in agreement with the much earlier study on humans which many believe demonstrated that RI caused forgetting. In that study, Jenkins and Dallenbach (1924) found in human subjects that after periods of sleep, retention of nonsense syllables was superior than after corresponding periods of normal waking activity in the same subjects. Together, these studies demonstrate that it is not the passage of time that results in memory decay; rather it is a result of interference from new events (Minami and Dallenbach, 1946).

So, we went back and performed the submersion experiments again but this time we injected into the snails 5-AZA, a DNA methyltransferase inhibitor (DNMT), which blocks DNA methylation. It is thought that DNA methylation plays a prominent role in LTM formation (Zovic et al., 2013). We first demonstrated (Lukowiak et al., 2014b) that 5-AZA injection did not alter (i.e., neither blocked nor extended the duration of LTM) memory formation when snails were given two 0.5h training sessions with a 1h interval between the training sessions. We then trained snails in a similar manner but now injected half of the snails with saline and the other half with 5-AZA. All snails were placed under the barrier (Figure 7). We found that snails injected with saline had extended LTM, in agreement with our previous findings (Sangha et al., 2003; 2005). However, snails injected with 5-AZA did not exhibit LTM (Lukowiak et al., 2014a). We concluded in that study that the persistent memory (i.e., delaying forgetting) induced by submersion in *Lymnaea* is blocked by the DNMT inhibitor, 5-AZA, given prior to training. Thus, it appears that DNMT's play an important role in enhanced memory persistence. It is possible that the DNMT's play a role in maintaining the substrate necessary for LTM support and thus interference with this substrate by 5AZA injection causes forgetting in this example. Thus, the data from these the submersion experiments while initially favoring the RI hypothesis now appear to favor the decay hypothesis.

However, there are 'problems' with both the hypotheses put forward to explain forgetting in *Lymnaea*. Dalesman and Lukowiak (2012a) found that different memories formed at the same time with the aerial respiratory operant conditioning training procedure are forgotten at different rates. Typically the number of attempted pneumostome openings is assessed in the training and memory test sessions.

However, in the Dalesman and Lukowiak 2012 study two other behaviors were also assessed: shell tilt behavior and reduction in pneumostome area. Using a training procedure (tactile stimulation of the pneumostome as it attempts to open) consisting of two 0.5h training sessions separated by a 1h interval they found that there was LTM for the # of attempted openings 24h after the last training session, consistent with all previous findings in that strain of snail. However, they also found that the memory for the alteration of pneumostome area persisted for 72h, and that there was no LTM for shell tilt behavior. The shell-tilt behavior did however, exhibit short-term memory (STM). Thus, following 'typical' operant conditioning of aerial respiratory training 3 different behaviors are altered. One, shell-tilting, is only altered for less than 1h (i.e., STM), whilst the two other exhibit LTM. The question arises why do the 3 behaviors have different rates of forgetting?

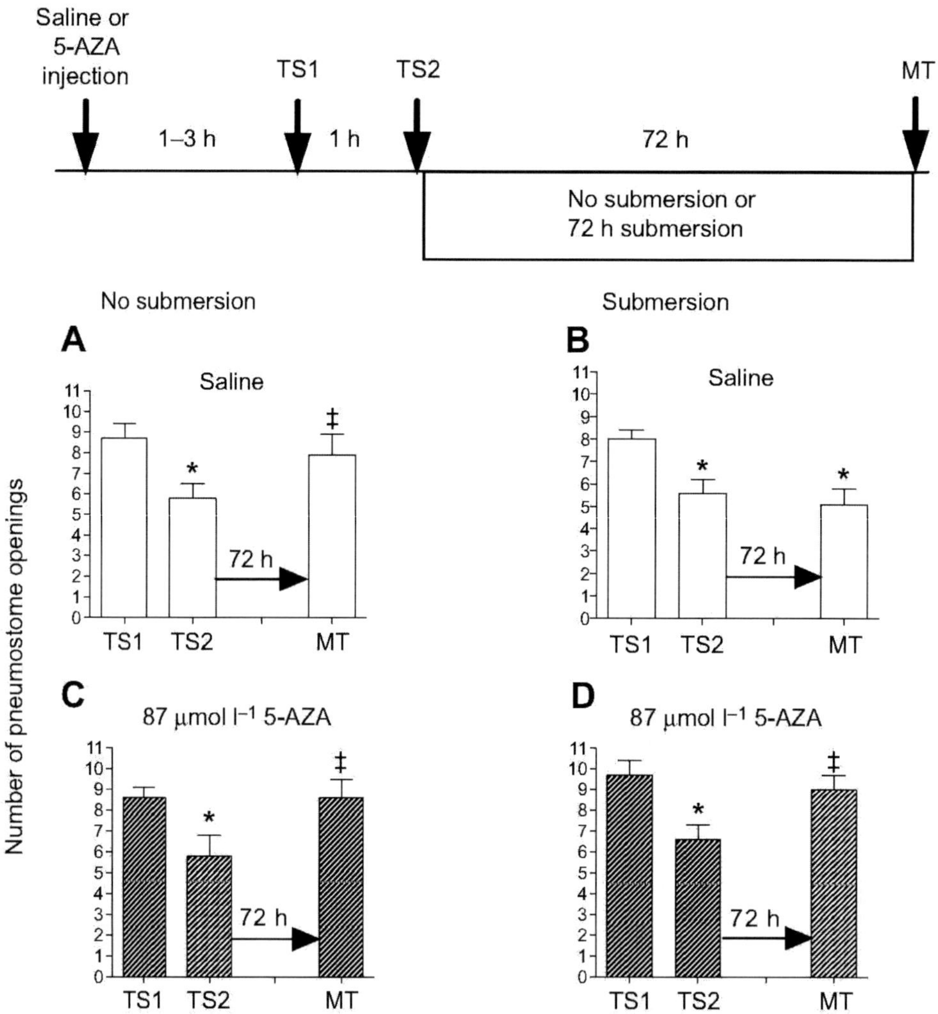

Figure 7. 5-AZA enhances forgetting of snails kept under a barrier. As was shown in Figure 3 placing snails under a barrier extends the duration of memory. Here we asked whether there was an epigenetic basis to this effect by using the methylation blocker 5-AZA. Submersion of *Lymnaea* extends memory that is prevented by 5-AZA treatment. (A–D) Mean ± s.e.m. number of pneumostome openings for the following treatments: (A) saline injection 1–3 h prior to TS1, no submersion (N=18); (B) saline injection 1–3 h prior to TS1, 72 h submersion (N=30); (C) 87 μmol l−1 5-AZA injection 1–3 h prior to TS1, no submersion (N=17); (D) 87 μmol l−1 5-AZA injection 1–3 h prior to TS1, 72 h submersion (N=23). Injection of 5-AZA prevented the memory enhancing effects of submersion on memory measured 72 h later. Timeline for treatment is shown at the top.

One possible explanation for the very short memory of shell-tilt, consistent with the RI hypothesis is that the snail is constantly moving the shell relative to its body, thus it may rapidly forget the memory relating to shell angle. That is, the snail is forming a new memory that moving the shell does not result in being poked. In contrast to the very short time period over which shell tilting is maintained, the reduction in pneumostome opening area is still apparent 72h later. However, memory for the number of attempted pneumostome openings only persists for 24h. As both these behaviors (i.e., pneumostome area and attempted pneumostome openings) will be performed an equal number of times in the eumoxic aquaria between training and testing (unlike the shell tilting behavior) and therefore undergo a similar number of unreinforced events, the RI hypothesis cannot be satisfactorily used to explain the difference in memory retention times.

However, it is likely that reducing the number of attempted pneumostome opening occurrences carries a greater cost (i.e., snail more likely to be hypoxic) than reducing the size of the opening. Thus, because the snail 'does not want to be hypoxic' forgetting this learned behavior sooner, has a more positive value. This in essence is 'adaptive forgetting' (Kraemer and Golding, 1997). The 'cost' associated with behavioral plasticity may therefore limit the extent or duration of the changed behavioral phenotype. Thus, the reduction in the number of opening attempts carries a greater cost to the snail and will therefore be forgotten faster in the absence of reinforcement. Obviously, this 'cost-hypothesis' will have to be experimentally tested to determine if it is a valid hypothesis. The 'cost hypothesis' might be viewed as being more in line with the decay hypothesis. We will have to experimentally verify this, perhaps using the 5AZA strategy.

Some final notes on forgetting in *Lymnaea*. The majority of studies occurring worldwide examining learning and memory in *Lymnaea* (see chapters in Menzel and Benjamin's 2013 book) make use of a strain of snails originally collected in polders near Utrecht in The Netherlands in the 1950s. This strain of snails, which we call 'Dutch' snails, has therefore been extensively inbred over the course of 60 years. Remarkably, these snails have not 'forgotten' the 'smell' (i.e., kairomone- interspecies communication due to chemicals released by another species) of a crayfish predator even though they have not experienced a crayfish or its 'smell' for over 250 generations. Encountering either crayfish or its kairmone alters many behaviors in *Lymnaea*, including their ability to form LTM (Orr et al., 2007; Orr and Lukowiak, 2008). Detecting the presence of the crayfish predator results in enhanced LTM formation that is:1) more easily formed; and 2) more resistant to forgetting and extinction. Enhanced memory formation is also found in freshly collected specimens of *Lymnaea* from other European countries when they detect the crayfish kairomone (The UK and Germany). However, *Lymnaea stagnalis* collected in Alberta do not respond to crayfish kairomone (Orr et al., 2009a,b). Crayfish are not natively present in Alberta. Alberta *Lymnaea* do however, respond to the kairomone from tiger salamanders, which prey on them. Thus, the question becomes did Alberta *Lymnaea* forget that crayfish are a predator, or did they never have that memory?

Finally, different strains of *Lymnaea* possess different abilities to form LTM. Some strains, which we have referred to as 'smart' form memory far easier and their LTM persists far longer than memories Dutch snails form. It is not, however, the case that all freshly collected snails have such superior memory forming capabilities. Some freshly collected snails' ability to form memory is no different from what the 'Dutch' snails can do (Dalesman et al., 2011b). Moreover, laboratory reared individuals from the 'smart' strains continue to show enhanced memory forming capabilities. We are attempting to determine why some strains of *Lymnaea* more easily form LTM and why forgetting occurs more slowly in these snails. One might have thought that if the 'smart' snails form LTM more easily, then the unreinforced openings that occur should have more easily disrupted the memory. That is, RI should have occurred faster in these snails. We also do not know the 'cost' of having a better ability to form memory (i.e., less forgetting) in snails. We are presently attempting to determine this. It may be that our 'smarter' snails cannot cope as well with stress as the 'average' snails do.

We do not know why unreinforced behavior is not more easily learned and remembered in the 'smart' snails to displace the memory resulting from the reinforced behavior. In like manner, if forgetting is due to a decay process we have to determine why the decay is slower in the smart snails. What those findings point out is that while much of our data are consistent with the either of the two hypotheses, the idea that forgetting is an active process is not in question. The question is whether forgetting is due to RI, decay or some combination of both. In that scenario it could be that retrograde interference drives the decay process such that with 'more' RI the rate of decay increases. We need to perform many more experiments using new strategies before we will be really able to come to a causal mechanism of forgetting, even in this relatively simple model system.

Memory formation is a dynamic process that can be modulated, for example, by stressful events occurring during, before or after the learning and memory consolidation processes (Lukowiak et al., 2008; 2013). The effect that stressors have on memory formation varies greatly depending on the type of stress experienced, the level of stress applied and the timing of the stressful event relative to the learning process, and can either enhance or block memory formation. What has not often been acknowledged or rigorously tested is the stressors may also be altering the forgetting process. It is possible that the stressors that 'block' memory formation are actually enhancing the 'forgetting' processes. It will be interesting to attempt to design experiments that address this question in the future.

REFERENCES

Braun MH, Lukowiak K. (2011) Intermediate and long-term memory are different at the neuronal level in *Lymnaea stagnalis* (L.). *Neurobiol. Learn. Mem.*96:403-416.

Braun MH, Lukowiak K, Karnik V, Lukowiak K. (2012) Differences in neuronal activity explain differences in memory forming abilities of different populations of *Lymnaea stagnalis*. *Neurobiol. Learn Mem.* 97:173182.

Cartford, M. C., Gohl, E. B., Singson, M., & Lavond, D. G. (1997). The effects of reversible inactivation of the red nucleus on learning- related and auditory-evoked unit activity in the pontine nuclei of classically conditioned rabbits. *Learning & Memory*, 3, 519–531.

Dalesman, S. and Lukowiak, K. (2010). Effect of acute exposure to low environmental calcium alters respiration and locomotion of *Lymnaea stagnalis* (L.). *J. Exp. Biol.* 213, 1471-1476.

Dalesman, S., Braun, M. H. and Lukowiak, K. (2011a). Low environmental calcium blocks long term memory formation in a pulmonate snail. *Neurobiol. Learn. Mem.* 95, 393-403.

Dalesman S, Rundle SD, Lukowiak K (2011b) Microgeographic variability in long-term memory formation in the pond snail, *Lymnaea stagnalis. Anim. Behav.* 82:311–319.

Dalesman S, Lukowiak K. (2012a) How stress alters memory in "smart" snails. *PLoS ONE*;7:e32334.

Dalesman, S., and Lukowiak, K. (2012b) Alternate behavioral measurements following a single operant training regime demonstrate differences in memory retention. *Animal Cognition* 15: 483-494.

Gates, A. I. (1930). *Psychology for students of education.* New York: MacMillan.

Hardt, O., Nader, K., and Nadel, L. (2013). Decay happens: the role of active forgetting in memory. *Trends Cog Neurosci* 17: 111 - 121.

Jenkins, J. G., & Dallenbach, K. M. (1924). Obliviscence during sleep and waking. *American Journal of Psychology*, 35, 605–612.

Knezevic B, Dalesman S, Karnik V, Byzitter J, Lukowiak K. (2011) Low external environmental calcium levels prevent forgetting in *Lymnaea. J. Exp. Biol.* 214:2118-2124.

Kraemer PJ and Golding JM (1997) Adaptive forgetting in animals. Psych Bull Rev 4:480–491.

Lukowiak, K., Ringseis, E., Spencer, G., Wildering, W., & Syed, N. (1996). Operant conditioning of aerial respiratory behaviour in *Lymnaea* stagnalis. *Journal of Experimental Biology*, 199, 683–691.

Lukowiak K, Cotter R, Westly J, Ringseis E, Spencer G, Syed N (1998) Long-term memory of an operantly conditioned respiratory behaviour pattern in *Lymnaea stagnalis. J. Exp. Biol.* 201(6):877–882.

Lukowiak, K., Adatia, N., Krygier, D., & Syed, N. (2000). Operant conditioning in *Lymnaea*: Evidence for intermediate- and long-term memory. *Learning & Memory*, 7, 140–150.

Lukowiak K, Martens K, Rosenegger D, Browning K, de Caigny P, Orr M. (2008)The perception of stress alters adaptive behaviours in *Lymnaea stagnalis. J Exp Biol.* 211:17471756.

Lukowiak, K and Dalesman, S (2013). Operant Conditioning of Respiration in *Lymnaea*: The Environmental Context. *Invertebrate Learning and Memory,* Chapter 23 (eds Manzell and Benjamin) Elsevier.

Lukowiak, K., Sunada, H., Teskey, M., Lukowiak, K. And Dalesman, S. (2013). Environmentally relevant stressors alter memory formation in the pond snail *Lymnaea. J. Exp. Biol.* (In Press).

Lukowiak, K., Sunada, H., Teskey, M., Lukowiak, K.S., and Dalesman, S (2014a) Environmentally relevant stressors alter memory formation in the pond snail *Lymnaea. J. Exp. Biol.* 217: 79.

Lukowiak, K., Heckler, B., Bennett, T., Schriner, E., Wyrick, K., Jewett, C., Todd, R., and Sorg, B. (2014b). Enhanced memory persistence is blocked by DNA methyltransferase inhibitor in the snail *Lymnaea stagnalis. J. Exp. Biol.* 217: 2920 - 2929.

Martens, K., Amarell, M., Parvez, K., Hittel, K., De Caigny, P., Ito, E. And Lukowiak, K. (2007). One-trial conditioning of aerial respiratory behaviour in *Lymnaea stagnalis. Neurobiol. Learn. Mem.* 88, 232-242.

McComb, C., Rosenegger, D., Varshney, N., Kwok, H-Y., and Lukowiak, K. (2005) Operant conditioning of an *in vitro* CNS-pneumostome preparation of *Lymnaea. Neurobiology of Learning and Memory* 84: 9 - 24.

McGaugh, J. (2000, January 14). Memory: A century of consolidation. *Science,* 287, 248–251.

Menzel R. and Benjamin P. Eds 2013 *Invertebrate Learning and Memory.* Vol 22 Handbook of Behavioral Neuroscience Academic Press London UK.

Milner, B., Squire, L. R., & Kandel, E. R. (1998). Cognitive neuroscience and the study of memory. *Neuron, 20,* 445–468.

Minami, H. and Dallenbach, K. (1946). The effect of activity upon learning and retention in the cockroach, Periplaneta americana. *Amer J. Psychol.* 59, 1-58.

Morrison, G. E., & van der Kooy, D. (1997). Cooling before associative conditioning blocks memory retrieval, but cooling after conditioning blocks memory retention in Caenorhabditis elegans. *Behavioral Neuroscience*, 111(3), 564–578.

Orr MV, El-Bekai M, Lui M, Watson K, Lukowiak K. Predator detection in *Lymnaea stagnalis. J. Exp. Biol.* 2007;210:41504158.

Orr MV, Lukowiak K. (2008) Electrophysiological and behavioral evidence demonstrating that predator detection alters adaptive behaviors in the snail *Lymnaea. J. Neurosci.*;28:27262734.

Orr M, Hittel K, Lukowiak K (2009a) Different strokes for different folks: geographically isolated strains of *Lymnaea stagnalis* only respond to sympatric predators and have different memory forming capabilities. *J. Exp. Biol.* 212:2237–2247

Orr M, Hittel K, Lukowiak KS, Han J, Lukowiak K (2009b) Differences in LTM-forming capability between geographically different strains of Alberta *Lymnaea stagnalis* are maintained whether they are trained in the lab or in the wild. *J. Exp. Biol.* 212(23):3911–3918.

Sangha, S., McComb, C., & Lukowiak, K. (2003e). Forgetting and the extension of memory in *Lymnaea. Journal of Experimental Biology, 206,* 71–77.

Sangha, S., Morrow, R., Smyth, K., Cooke, R., & Lukowiak, K. (2003c). Cooling blocks ITM and LTM formation and preserves memory. *Neurobiology of Learning and Memory, 80,* 130–139.

Sangha, S., Scheibenstock, A., & Lukowiak, K. (2003a). Reconsolidation of a long-term memory in *Lymnaea* requires new protein and RNA synthesis and the soma of Right Pedal Dorsal 1. *Journal of Neuroscience, 23,* 8034–8040.

Sangha, S., Scheibenstock, A., McComb, C., & Lukowiak, K. (2003d). Intermediate and long-term memories of associative learning are differentially affected by transcription versus translation blockers in *Lymnaea. Journal of Experimental Biology, 206,* 1605–1613.

Sangha, S., Scheibenstock, A., Morrow, R., & Lukowiak, K. (2003b). Extinction requires new RNA and protein synthesis and the soma of the cell Right Pedal Dorsal 1 in *Lymnaea stagnalis. Journal of Neuro- science, 23,* 9842–9851.

Sangha, S., Scheibenstock, A., Martens, K., Varshney, N., Cooke, R. and Lukowiak, K. (2005). Impairing forgetting by preventing new learning and memory. *Behav. Neurosci.* 119, 787-796.

Scheibenstock, A., Krygier, D., Haque, Z., Syed, N., & Lukowiak, K. (2002). The soma of RPeD1 must be present for long-term memory formation of associative learning in *Lymnaea. Journal of Neurophysiology, 88,* 1584–1591.

Sekiguchi, T., Yamada, A., & Suzuki, H. (1997). Reactivation- dependent changes in memory states in the terrestrial slug *Limax flavus. Learning & Memory,* 4, 356–364.

Serrano, P. et al. (2008) PKMz maintains spatial, instrumental, and classically conditioned long-term memories. *PLoS Biol.* 6, e318.

Shema, R. et al. (2009) Boundary conditions for the maintenance of memory by PKM in neocortex. *Learn. Mem.* 16, 122–128.

Shema, R. et al. (2011) Enhancement of consolidated long-term memory by overexpression of protein kinase Mzeta in the neocortex. *Science* 331, 1207–1210.

Sommerville, J. (1999). Activities of cold-shock domain proteins in translation control. *BioEssays*, 21, 319–325.

Spencer, G. E., Syed, N. I., & Lukowiak, K. (1999). Neural changes after operant conditioning of the aerial respiratory behavior in *Lymnaea stagnalis*. *Journal of Neuroscience, 19,* 1836–1843.

Spencer, G. E., Syed, N. I., van Kesteren, E., Lukowiak, K., Geraerts, W. P., & van Minnen, J. (2000). Synthesis and functional integration of a neurotransmitter receptor in isolated invertebrate axons. *Journal of Neurobiology, 44,* 72–81.

Spencer, G. E., Kazmi, M. H., Syed, N. I., & Lukowiak, K. (2002). Changes in the activity of a CPG neuron after the reinforcement of an operantly conditioned behavior in *Lymnaea. Journal of Neurophysiology, 88,* 1915–1923.

Syed, N. I., Bulloch, A. G. M., & Lukowiak, K. (1990). In vitro reconstruction of the respiratory central pattern generator of the mollusk *Lymnaea. Science, 250,* 282–285.

Syed, N. I., Ridgway, R., Lukowiak, K., & Bulloch, A. G. M. (1992). Transplantation and functional integration of an identified respiratory interneuron in *Lymnaea stagnalis. Neuron, 8,* 767–774.

Thieringer, H. A., Jones, P. G., & Inouye, M. (1998). Cooling and adaptation. *BioEssays,* 20, 49–57.

Thorndike, E.L. (1913) *Educational Psychology,* Teachers College, Columbia University.

Wixted, J. T. (2004). The psychology and neuroscience of forgetting. *Annual Review of Psychology, 55,* 235–269.

Woodworth, R. S. (1929). *Psychology.* New York: Holt and Co., p. 93.

Yamada, A., Sekiguchi, T., Suzuki, H., & Mizukami, A. (1992). Behavioral analysis of internal memory states using cooling- induced retrograde amnesia in *Limax flavus. Journal of Neuroscience*, 12(3), 729–735.

Zovkic, I. B., Guzman-Karlsson, M. C. and Sweatt, J. D. (2013a). Epigenetic regulation of memory formation and maintenance. *Learn. Mem.* 20, 61-74.

In: Memory Consolidation
Editors: Manabu Sakakibara and Etsuro Ito

ISBN: 978-1-63482-596-2

Chapter 10

ROLE OF DOPAMINE IN MEMORY CONSOLIDATION

***Attilio Iemolo*[1], *Maria De Risi*[1,2] *and Elvira De Leonibus*[1,2,*]**

[1]Institute of Genetics and Biophysics "Adriano Buzzati-Traverso", Naples, Italy

[2]Telethon Institute of Genetics and Medicine (TIGEM), Pozzuoli, Naples, Italy

ABSTRACT

Dopamine (3,4-dihydroxyphenethylamine, DA) is a neuroendocrine transmitter belonging to the catecholamine and phenethylamine families. DA in the brain plays a critical role in modulating a variety of actions on neuroendocrine and behavioral systems. Pathological change in DA transmission is a key feature of many neurological and psychiatric disorders, such as schizophrenia, Parkinson's disease, Tourette's syndrome, and many others. Theories on the behavioral function of DA in the brain have advanced from the original proposal that DA could mediate the hedonic impact of rewards and motor behavior, to the hypothesis that this system may mediate incentive salience to reward-related stimuli. More recently it has been proposed that DA mediates reinforcement learning. Even if the theoretical debate around DA function continues, the central role of this neurotransmitter as a mediator of memory formation is being widely recognized.

Memory has been distinguished, according to a temporal gradient, as: *short-term memory,* which lasts from seconds to minutes, and *long term memory,* which lasts from hours to years. The two forms of memory rely on different neural substrates and are affected by different pharmacological manipulations. Long-term memory has been further distinguished into different categories depending on the kind of learning mechanism activated. The multiple memory system theory, in fact, posits that long-term memory can be distinguished into explicit and implicit memories. *Explicit memory* derives from relational and/or action-outcome learning mechanisms that lead to the formation of mental representation of events, contexts, or instrumental actions. Explicit memory is thought to be dependent upon the activity of regions located in the cortico-

[*] Correspondence and requests for materials should be addressed to: Elvira De Leonibus (deleonibus@tigem.it) Institute of Genetics and Biophysics "Adriano Buzzati-Traverso", Via Pietro Castellino 111, 80131 Naples, Italy, Phone: +39.081.6132267.

limbic systems (i.e., hippocampus). *Implicit memory* is formed through stimulus-response learning and leads to the incremental acquisition of skills and habits, and it is dependent on the activity of sensorimotor systems (i.e., the dorsal striatum).

DA modulates the activity of both of these neuronal pathways, and therefore it mediates the formation of different forms of memory.

In this chapter we will first describe the characteristics and general functions of the dopaminergic system, including an overview of the pharmacological agents available to modulate it. Then we will discuss the involvement of DA in learning and memory within this theoretical framework. In particular, we will review pre-clinical behavioral pharmacological experiments using DAergic drugs to manipulate DA activity, supporting the role for DA in modulating different forms of memory during the consolidation phase.

1. Introduction

Memory is a complex process consisting of several steps during which new acquired information becomes consolidated and stored. According to the human information processing theory (Atkinson and Shiffrin, 1971) the formation of long-term memory (LTM) requires the subject to be exposed to a new stimulus, which is processed first by the sensorial system then by the working memory (WM) system and finally stored in the short-term memory (STM) for seconds to minutes. Information transfer to LTM depends on an offline process called memory consolidation. Upon necessity, information stored in LTM, is retrieved and re-handled by the WM system. After modification, the information eventually undergoes re-consolidation, the process by which new information is integrated into previously existing memory traces (see figure 3, bottom, for a schematic representation of the entire process). WM has been distinguished from STM, although both, unlike LTM, have limited duration. WM is generally defined as a process that actively maintains and manipulates information in the mind for a short time interval, while STM is considered to be more like a passive storage space. This differentiation suggests that WM comprises STM, *plus* executive processes for manipulation of information. Whether or not they should be considered different processes remains controversial (Baddeley, 1992). In this chapter we will not make a distinction between the two, generally referring to WM also to include STM. Conventionally, WM was thought to rely on cortical regions, such as the prefrontal cortex (PFC) and the perirhinal cortex and acts under little or no control by hippocampus (HP) (Murphy et al., 1996; Aggleton and Brown, 2005; Ainge et al., 2006; Brown et al., 2008; for review see Brozoski et al., 1979). Accordingly, memory deficits in patients with HP lesions mainly result in altered long-term storage of information, leaving WM unharmed.

WM and LTM have also been distinguished from a neurobiological point of view. Both require the activation of glutamate ionotropic receptors, such as N-methyl-D-aspartate (NMDA) and α-amino-3-hydroxy-5-methyl-4-isoxazolepropionic acid (AMPA) receptors (Mele et al., 1996; Aultman and Moghaddam, 2001; Dalley et al., 2005; Ferretti et al., 2005; Bethus et al., 2010; Wang et al., 2010). When NMDA receptors are stimulated, however, calcium-sensitive enzymes, including calcium-calmodulin-dependent-kinases (CaMK), protein-kinase-C (PKC) and protein-kinase-A (PKA), are activated in response to elevated intracellular calcium. This leads to LTM formation, regulated by the HP (Abel et al., 1997; Tinsley et al., 2009; Barsegyan et al., 2010). The activation of these protein kinases ultimately leads to the synthesis of new proteins that are required for the formation of new synaptic contacts, a process requiring hours. Protein kinase activation was previously understood to inhibit WM within the PFC (for review see Dash et al., 2007); nevertheless, this hypothesis has been recently challenged by experimental findings indicating that protein kinases activation is also necessary for WM maintenance (Silva et al., 1996; Michel et al., 2008; Wang et al., 2008). In WM, despite the uncertainty, active protein kinases appear to induce transient modulation of activity of substrate proteins, rather than to promote *de novo* protein synthesis (for review see Dash et al., 2007).

A distinction between different forms of memory has been done not only on the basis of memory duration (WM *vs* LTM), but also on the basis of the nature of learning mechanism leading to LTM formation. Explicit memory, such as allocentric spatial memory and/or instrumental goal-directed learning, is formed based on stimulus-stimulus association and/or action-outcome associations. It is mediated by the activation of the HP-medial striatum (including the nucleus accumbens, NAc) pathway (DeCoteau and Kesner, 2000; De Leonibus et al., 2005; De Leonibus et al., 2011; Pennartz et al., 2011). In contrast, implicit memory, such as procedural memory or skill memory, is acquired through stimulus-response learning mechanisms and is believed to rely on the activation of the motor cortex-dorsolateral striatum pathway (De Leonibus et al., 2003; De Leonibus et al., 2005; Pennartz et al., 2011). Finally, emotional memory, which is acquired through Pavlovian conditioning, has been shown to rely principally on the integrity of the amygdala (Cheng et al., 2003). Although it is increasingly evident that the reported distinctions between different forms of memory is a simplistic view that needs to be integrated into a wider complexity of the brain function, it has been extremely useful to dissect the main neuronal and molecular components associated with specific brain dysfunctions, such as amnesia and frontal dysfunctions. While cerebral glutamate transmission is necessary for synaptic fleeting of information, there are many other molecules, both neurotransmitters and hormones, which influence memory function in several brain regions (Roozendaal, 2002). One of such, the catecholamine DA, has been shown to modulate each step of information processing, and although initially it was thought DA to modulate performance, accumulating evidence suggests that a correct balance between DA release and DA receptor activation is necessary for memory formation in different regions of the brain. In particular, DA has been shown to regulate the process of consolidation, during which memories go from their initial labile states to subsequent, more stable ones (Roozendaal, 2002).

In this chapter we will first describe the characteristics and general functions of the dopaminergic system, which is composed by neurons that use DA as neurotransmitters and their receptors, including an overview of the pharmacological agents available to modulate it. Then we will review pre-clinical experimental evidence on the role of DA in modulating memory, focusing on memory consolidation.

2. The Dopaminergic System

DA is a neurotransmitter found predominantly in the brain, but also in the peripheral nervous system. Dysfunctional DA has been associated with various neuropsychiatric disorders, such as Parkinson's disease (PD), drug abuse, schizophrenia, attention deficit hyperactivity disorder (ADHD) and mania. DA dysfunction is also involved in other medical conditions such as hyperprolactinemia (Colao et al., 2006), restless legs syndrome (for review see Zintzaras et al., 2010), hypertensive crisis (for review see Varon, 2008), erectile dysfunction (for review see Carson, 2007), nausea and vomiting (for review see Axelrod, 1997). DA plays a role in this wide range of disorders due to its involvement in the function of many organs, including in neuronal activity in different regions of the brain.

In the brain, as illustrated in figure 1, DA controls different neuronal circuitries. Four major dopaminergic pathways have been identified within the mammalian brain: 1) the nigrostriatal, which begins at the substantia nigra (SN) A9 neurons and contains neurons that supply the striatum, both the dorsal and the ventral striatum, which comprises the NAc and the olfactory tubercle. The NAc shell is one of the NAc subregions, which has been originally distinguished from the NAc core and the rostrale pole on the basis of anatomical and some functional differences. Neurons from the SN also project to the PFC, the HP and the amygdala; 2) the mesolimbic and 3) the mesocortical pathways, mainly originates in the A10 neurons from the ventral tegmental area (VTA) and sends more dense innervation, as compared to the nigrostriatal pathway, to the shell of the NAc, the HP, the amygdala and also the ventral PFC and other cortical regions, such as parahippocampal cortexes respectively (Beckstead, 1978; Swanson, 1982); 4) the tuberoinfundibular dopaminergic (TIDA) pathway, that originates from both the A12 (arcuate nucleus) and A14 (mediobasal hypothalamus) and projects to the median eminence where they mediate prolactin secretion (Lyons et al., 2012).

The network of neurons forming these pathways is critically involved in various vital central nervous system functions and processes, including voluntary movement, food intake, regulation of emotions, sleep, attention, learning and memory. In the periphery system, DA plays important physiological roles in the regulation of olfaction, retinal processes, hormonal regulation, cardiovascular functions, sympathetic regulation, immune system and renal functions (Snyder et al., 1970; O'Malley et al., 1992; Santambrogio et al., 1993).

Below we will briefly describe each single aspect of DA synthesis and metabolism, from its biosynthesis, to its degradation and post-synaptic actions. Dysregulation occurring in any of these steps has been shown to lead to cognitive deficits, in the domains of attention, impulse control, and memory.

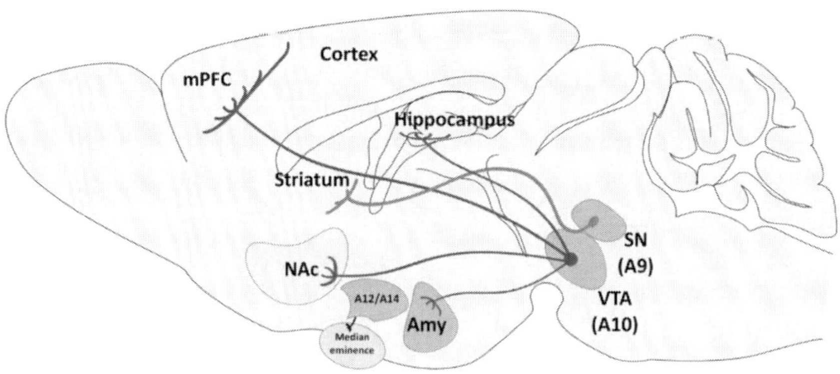

Figure 1. Cross-section of a mouse brain showing how DA neurons in the VTA (A10), SN (A9), nucleus arcuate (A12) and mediobasal hypothalamus (A14) are connected with other brain areas.

2.1. Biosynthesis

DA can be produced from either tyrosine or phenylalanine, two amino acids found plentifully in dietary proteins (see figure 2A). Phenylalanine is converted to tyrosine by the enzyme phenylalanine hydroxylase in the liver, or by tyrosine hydroxylase (TH) in the dopaminergic neuron (for review see Fisher et al., 1972; Elsworth and Roth, 1997). When tyrosine crosses the blood-brain barrier through a low-affinity amino acid transport system, it is subsequently taken up by high- and low-affinity amino acid transporters, moving from brain extracellular fluid to dopaminergic neurons (for review see Elsworth and Roth, 1997). Once tyrosine enters the neuron, it is converted to dihydroxyphenylalanine (L-DOPA) by the cytosolic enzyme TH. Finally, aromatic L-amino acid decarboxylase (AADC, DOPA decarboxylase) catalyzes the cytosolic conversion of L-DOPA to DA. In dopaminergic neurons, no further enzymatic modification occurs. However, in some neurons, DA is the substrate for dopamine-β-hydroxylase (DBH). DBH catalyzes the oxidative hydroxylation of DA to norepinephrine, which in turn can be transformed into epinephrine (adrenaline) by phenylethanolamine-N-methyltransferase (for review see Opmeer et al., 2010). A common metabolic pathway between DA and noradrenaline gives already an idea on how much these two neurotransmitters cooperatively act on regulating many behavioral functions, included memory consolidation. In this chapter, however, we are not going to discuss the well-established interaction of DA with other neurotransmitters, such as acetylcholine, noradrenaline, serotonin in mediating memory consolidation.

2.2. Storage in Synaptic Vesicles and Release

The reserpine-sensitive vesicular monoamine transporter (VMAT2) is responsible for actively taking DA from the cytosol and transporting it to specialized storage granules (vesicles) in preparation for release into the synapse after an action potential has occurred (for review see Elsworth and Roth, 1997). The mechanism that favors storage of DA, and other catecholamines within the vesicles, consists of an ATP-dependent process involving a proton pump (see figure 2B). Reserpine is a specific, irreversible inhibitor of the vesicular amine pump, which prohibits vesicles from collecting amines (for review see Kuhar, 1999).

DA can also be synthesized and released from dendrites; here, it appears to be stored both in classical vesicles and in the smooth endoplasmic reticulum (Nirenberg et al., 1996). When an action potential occurs, changes in the membrane proteins conformation allow calcium ions to influx. This response is a key component of the stimulus responsible for the fusion between DA-containing vesicles and neuronal membrane (exocytosis). During exocytosis (see figure 2B), vesicles release DA into the synaptic cleft (Grace and Bunney, 1984).

2.3. Reuptake and Metabolism

Dopaminergic terminals have transporters that have crucial roles in terminating DA action and in maintaining DA homeostasis. One such protein, the plasmalemmal dopamine transporter (DAT), provides DA re-uptake by transporting DA from the synaptic cleft back into the neuron with the help of energy derived from the electrochemical gradients of sodium and chloride (see figure 2B). Additionally, some evidence suggests that extracellular DA may be taken up and metabolized by glial and non-DA neurons (for review see Opmeer et al., 2010).

Two key metabolic enzymes, methyltransferase (COMT) and monoamine oxidase (MAO), are responsible for the metabolism of DA. Their function in DA metabolism is secondary to DA re-uptake.

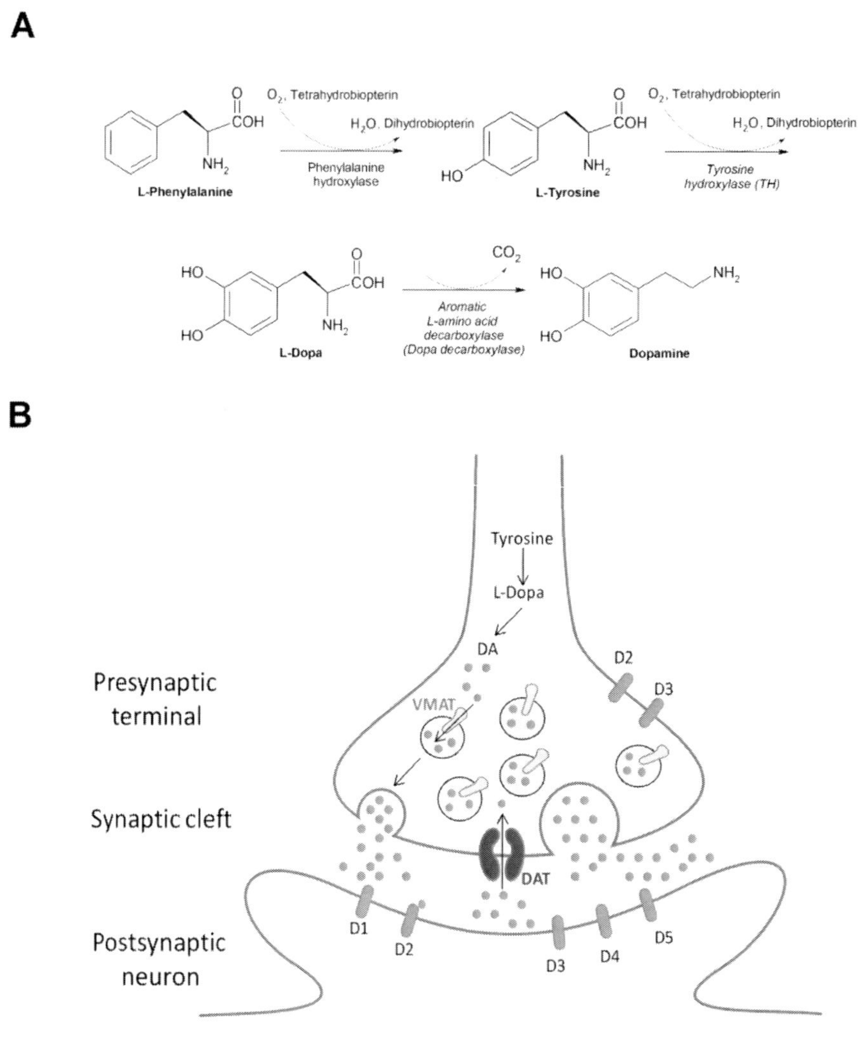

Figure 2. Dopamine biosynthesis (A). Both L-phenylalanine and L-tyrosine are the precursors of DA. Tyrosine hydroxylase (TH) is the enzyme that catalyzes the reaction in which L-tyrosine is converted to levodopa (L-DOPA); L-DOPA is subsequently decarboxylated to DA by the enzime dopa decarboxylase. Schematic representation of a DA neuron synapse (B). DA transporters VMAT and DAT are localized at presynaptic level, where they are crucial for the reuptake and termination of DA signal and the maintenance of presynaptic DA storage.

COMT converts DA into metanephrine (3-MT) by catalyzing the methylation of the 3-OH group. Other catechol-containing structures, such as norepinephrine, epinephrine and the

DA precursor L-DOPA, are COMT substrates. COMT has a soluble form (S-COMT) and a membrane-bound form (MB-COMT), both of which are encoded by the same gene (Lundstrom et al., 1991). S-COMT is the predominant form in all tissue types, except for the brain, in which MB-COMT and S-COMT levels are found at 70% and 30%, respectively. In the brain, S-COMT is abundandly found in the glia, indicating its minor influence in DA metabolism (Naudon et al., 1992). Conversely, MB-COMT is abundantly localized in intracellular compartments (cell body, axons and dendrites) in postsynaptic terminals of neurons and in perisynaptic locations (Schott et al., 2010). Interestingly, the C-terminal catalytic domain of MB-COMT is found in the extracellular space. Thus, it likely plays an important role in direct inactivation of extra-synaptic or diffused DA (Chen et al., 2011).

MAO A and B are mitochondrially-bound enzymes differing by substrate affinity: MAO A preferentially oxidizes serotonin and norepinephrine, whereas MAO B preferentially oxidizes phenylethylamine (Greenawalt and Schnaitman, 1970). MAO A and MAO B also differ in terms of their anatomical localizations. MAO A is localized in catecholaminergic neurons, particularly within the locus coeruleus, NAc, hypothalamus and mammillary complex, whereas MAO B is found in serotonergic neurons, as well as in histaminergic cells and astrocytes (for review see Bortolato et al., 2008). While both forms can oxidize DA, MAO B is the major metabolic enzyme of DA in humans and primates, whereas MAO A is the primary player in rodents (Fornai et al., 1999). The process of metabolism of DA by MAOs occurs, for the most part, in the presynaptic terminal after reuptake by the DAT.

2.4. Dopamine Receptors

The effect of DA on cognition and behavior is mediated by its action on different types of DA receptors. In a pioneering study, DA receptors were shown to exist in two distinct populations, D1 and D2 subtypes, positively and negatively coupled to adenylyl cyclase (AC) activation (Spano et al., 1978). The two receptors subtypes were also distinguished on the basis of their pharmacological properties (for review see Kebabian and Calne, 1979). Later on, after using genetic cloning approaches, several groups reported that DA could activate multiple receptor subtypes (Bunzow et al., 1988; Dearry et al., 1990; Monsma et al., 1990; Sokoloff et al., 1990; Zhou et al., 1990; Sunahara et al., 1991; Tiberi et al., 1991; Van Tol et al., 1991).

On the basis of their distinct structural, pharmacological, and biochemical properties, these receptors are now classified as either D1-class DA receptors, made up of D1 and D5 subtypes, or D2-class DA receptors, divided into D2, D3 and D4 subtypes (Andersen et al., 1990; Tiberi et al., 1991). It is commonly accepted that D1-class DA receptors (both D1 and D5) activate the $G\alpha_{s/olf}$ family of G proteins to stimulate cAMP (cyclic adenosine monophosphate) production by AC and are found exclusively post-synaptically on DA receptive cells, such as the GABAergic medium spiny neurons (MSNs) in the striatum (Monsma et al., 1990; Sokoloff et al., 2006; for review see Rondou et al., 2010). The D2-class DA receptors (D2, D3, and D4) couple to the $G\alpha_{i/o}$ family of G proteins and induce inhibition of AC.

In contrast to the D1-class DA receptors, D2 and D3 DA receptors are expressed both post-synaptically, on DA target cells, and pre-synaptically, on dopaminergic neurons

(Sokoloff et al., 2006; for review see Rondou et al., 2010). Alternative splicing of an 87-base pair exon between introns 4 and 5 of the D2 DA receptor leads to the generation of two major D2 DA receptor variants that have been named D2S (D2-short) and D2L (D2-long) (Giros et al., 1989; Monsma et al., 1989). D2L is longer than D2S by 29 amino acid residues in the third intracellular loop. These D2 DA receptor variants have distinct anatomical, physiological, signaling, and pharmacological properties. D2S expression has been mostly found pre-synaptically, and its activity mostly involved in autoreceptor functions, whereas D2L seems to be a predominantly postsynaptic isoform in both its location and function (Usiello et al., 2000; De Mei et al., 2009).

2.4.1. Distribution of Dopamine Receptors

DA receptors are widely distributed in the brain and in the periphery. In the brain, D1 DA receptors are highly expressed in the nigrostriatal, mesolimbic, and mesocortical areas, such as the caudate-putamen (striatum), NAc, SN, olfactory bulb, amygdala, and frontal cortex, as well as at lower levels, in the HP, cerebellum, thalamic areas, and hypothalamic areas. D5 DA receptors are expressed at low levels in multiple brain regions, including pyramidal neurons of the PFC, the premotor cortex, the cingulated cortex, the entorhinal cortex, SN, hypothalamus, the HP, and the dentate gyrus. A very low level of expression has also been observed in the medium spiny neurons (MSNs) of the caudate nucleus and NAc (Bergson et al., 1995). The highest levels of D2 DA receptors are found in the striatum, the NAc, and the olfactory tubercle. D2 receptors are also expressed at significant levels in the SN, VTA, hypothalamus, cortical areas, septum, amygdala, and HP (Bouthenet et al., 1991; for review see Seeman, 2006).

The D3 DA receptor has a more limited pattern of distribution. Its highest level of expression is observed in the limbic areas, such as in the shell of the NAc, the olfactory tubercle, and the islands of Calleja (Bouthenet et al., 1991; Levesque et al., 1992; Sokoloff et al., 2006). At significantly lower levels, the D3 DA receptor is also detectable in the striatum, the SN pars compacta, the VTA, the HP, the septal area, and in various cortical areas.

The D4 DA receptor has the lowest level of expression in the brain, with documented expression in the frontal cortex, amygdala, HP, hypothalamus, globus pallidus, SN pars reticulata, and thalamus.

To summarize, the highest density of all DA receptors, except D4, are found in the NAc and in the olfactory tubercle. D1 and D2 receptors are highly expressed also in the striatum, the VTA, and the amygdala. Low to moderate low expression of all DA receptors subtypes is found in the SN, HP, frontal cortex and the hypothalamus.

D1, D2, and D4 DA receptors have also been observed in the retina, and prominent levels of expression of D2 DA receptors have been detected in the pituitary gland (Van Tol et al., 1991; Cohen et al., 1992; O'Malley et al., 1992; for review see Rondou et al., 2010). In the periphery, all subtypes of DA receptors have been observed in varying proportions in the kidney, adrenal glands, sympathetic ganglia, gastrointestinal tract, blood vessels, and heart (Amenta et al., 2002).

2.4.2. Targeting Dopaminergic Receptors Using Specific Agonist and Antagonist Compounds

Over the last decades, a considerable number of pharmacologically active compounds has been designed to interact with dopaminergic transmission. Such drugs can either increase (agonists) or impede (antagonists) the actions of DA. Many of these compounds have significantly contributed to the understanding of structural, genetic and pharmacological properties of DA receptors, the plasticity of intracellular DA signaling and the possible clinical applications of pharmaceutical drugs to several central nervous system disorders.

Since Ehringer and Hornykiewicz carried out a pioneering study in 1960, leading to their discovery of reduced DA levels in the striatum of a patient suffering from PD, several molecules capable of activating DA receptors have been developed to symptomatically treat this debilitating disease. This led to the discovery of the DA precursor L-DOPA (Cotzias et al., 1967; Cotzias et al., 1969) a few years later. Afterwards, several DA receptor agonists were introduced for the clinical intervention in PD (for review see Bonuccelli and Pavese, 2007; Millan, 2010); however, L-DOPA remains the most effective treatment for PD. The most studied indirect agonists of DA are psychostimulants, such as cocaine, which elevates the extracellular concentration of DA by blocking DAT (Fienberg et al., 1998), and amphetamine. Amphetamine increases and prolongs DA action on its receptors through different mechanisms. It is a substrate for the DAT that competitively inhibits DA uptake; second, it facilitates the movement of DA out of vesicles and into the cytoplasm; and third, it is a substrate for VMAT, being taken up by VMAT into the vesicles (Fleckenstein et al., 2007). Indeed, although with different potency, all drugs of abuse have been shown to increase the amount of DA in the synaptic cleft in the brain, mainly within the mesocorticolimbic pathway. It must be said, however, that most of these drugs also increase the release of norepinephrine, serotonin, acetylcholine etc., and therefore their effects are not specific to the DA system. More recently, highly selective direct DA agonists, together with DA itself, have been developed, such as apomorphine, bromocriptine, quinpirole, SKF 38393, fenoldopam etc. These drugs can be highly selective for D1-like or D2-like receptors, such as SKF 38393 for D1 receptors and quinpirole for D2 receptors.

The generation of DA antagonists has been mainly linked to the treatment of schizophrenia, as the disorder is generally associated with DA system hyperactivity. The first generation of antipsycotic drugs ("typical antipsychotics"), including chlorpromazine, haloperidol, fluphenazine and other compounds, commonly block D2 receptors with high affinity, but result in strong extrapyramidal side effects such as tardive dyskinesia and neuroleptic malignant syndrome (for reviews see Lindenmayer and Khan, 2004; Miyamoto et al., 2005; Lieberman et al., 2008). Second generation ("*atypical*") antipsychotic drugs, such as risperidone, olanzapine, quetiapine, aripiprazole, block D2 receptors. These drugs have been modeled after the prototype drug, clozapine (for review see Divac et al., 2014). Clozapine-like drugs induce transient D2 receptor antagonism, which is sufficient to obtain antipsychotic effects with minor side effects. From a clinical point of view, atypical antipsychotics are very useful. Nevertheless, their usefulness to dissect the specific role of DA in memory processing is limited by their broad aspecific action on neurotransmitter systems other than DA, like serotonin and noradrenaline. As reported in table 1 drugs that selectively act on one of the two receptors classes, such as (-)-Sulpiride on D2 receptors, and SCH 23390 on D1 receptors are useful for research use.

Table 1. The table shows values of KD (dissociation constant that is defined as the molar concentration of a drug that occupies 50% of the receptors) expressed in mg/kg of some DA antagonists and agonists evaluated in different regions of rodents' brain.

	D1 Receptors	D2 Receptors
Antagonist		
SCH 23390	0.0000014 (Billard et al., 1984)	
SCH 39166	0.0053 (Alburges et al., 1992)	
Sulpiride		0.011 (Jastrow et al., 1984)
Haloperidol	0.00021 (Seeman and Van Tol, 1995)	0.000034 (Seeman et al., 2003)
Agonist		
SKF 38393	0.025 (Dubois et al., 1986)	
Apomorphine	0.000027 (Murphy and Vodyanoy, 1984)	0.000085 (Murphy and Vodyanoy, 1984)
7-OH DPAT		0.0000089 (Gonzalez and Sibley, 1995)
Quinpirole		0.0000050 (Levant et al., 1992)

It must be said, however, that the interpretation of the results obtained by studying D2-like acting drugs is complicated by the contemporary action that the drugs have on pre- and post-synaptically localized receptors; indeed, pre-synaptic receptors provide an important negative feedback mechanism that modulates neuronal firing rate, synthesis and neurotransmitter release in response to changes in extracellular neurotransmitter levels (for review see Sibley, 1999). There are no drugs that can distinguish between pre- and post-synaptic D2 receptors, or that selectively act on the D2S isoform. However, a concentration of DA agonists lower than that needed to activate postsynaptic receptors can activate D2-class autoreceptors. Therefore, DA agonists can produce a biphasic effect depending on whether it is acting on DA release or on post-synaptic receptors (for review see Beaulieu and Gainetdinov, 2011). As a consequence, any effect induced by D2 drugs, although highly selective for this receptors class, might involve different activation of D1 receptors by DA.

Nevertheless, sophisticated experimental research appropriately designed has taken advantage of the discovery of highly selective drugs acting on specific receptors subtypes to address the biological mechanisms underlying learning and memory. Most of this research was performed in animal models. Such models exhibit behavioral outcomes assumed to represent specific aspects of cognitive function, and these are measured following the administration of different doses of these drugs. Thus, the use of these animal models has allowed us to identify selective molecular mechanisms that can be targeted by potential cognitive enhancers.

2.4.3. Intracellular Dopamine Signaling

All DA receptors contain the structure common to G protein-coupled receptors (GPCRs), which consists of seven transmembrane domains with a high degree of similarity amongst primary amino acid sequences. These proteins are capable of activating heterotrimeric G proteins, which induce intracellular signaling mechanisms (Probst et al., 1992; Sidhu, 1998). They may also activate G protein-independent signaling events (Luttrell and Lefkowitz, 2002).

A subset of 16 heterotrimeric G protein subtypes has been functionally classified into four broad classes, $G\alpha_s$, $G\alpha_i$, $G\alpha_q$, and $G\alpha_{12}$. In general, G proteins are associated with three protein subunits, α, β, and γ. The classification of G proteins is based on the nature of the α-subunit sequence and their functional characteristics (for review see Pierce et al., 2002).

The α-subunit is bound to GDP (guanosine diphosphate) and to a tightly associated βγ-complex, forming an inactive trimeric protein complex. A series of mechanisms following agonist binding result in GDP release, GTP (guanosine triphosphate) binding to the α-subunit, and the dissociation of α-subunit from the βγ-complex. Both the α-subunit and the βγ-complex can then transduce the signal to activate a relatively small number of effector systems inside the cell. When GTP hydrolysis occurs, the GDP-bound α-subunit and the βγ-subunit complex reassociate into the heterotrimeric inactive G protein complex (for review see Pierce et al., 2002). In Gαs/olf-coupled receptors, such regulation might be carried out by PKA, PKC, MAP kinases, or several other kinases (for review see Hamm and Gilchrist, 1996). Several substrates of PKA, such as CREB (cAMP response element-binding protein), ionotropic glutamate receptors (AMPA and NMDA), and certain ion channels, have been shown to be affected by DA receptor stimulation (for review see Greengard, 2001). In particular, CREB is a transcription factor that has been described as an important factor in LTM formation (Yin and Tully 1996). Another PKA substrate is the dopamine- and cAMP-regulated phosphoprotein (DARPP-32). DARPP-32 is a multifunctional phosphoprotein that is predominantly expressed in MSNs. In these cells, DARPP-32 acts as an integrator in the modulation of cell signaling in response to multiple neurotransmitters, including DA (for review see Svenningsson et al., 2004). *In vivo* studies of DARPP-32 function in striatal slice preparations have shown that this regulator acts as an amplifier of PKA signaling. Mice lacking DARPP-32 display deficits in their responses to dopaminergic drugs, such as cocaine.

Results obtained from heterologous cell culture systems suggest that both D1- and D2-class DA receptors regulate the MAP (Mitogen-activated protein) kinases extracellular signal-regulated kinases 1 and 2 (ERK1 and ERK2) (Beom et al., 2004; Kim et al., 2006). Pharmacological and genetic characterizations of the DA receptor types that are involved in the activation of striatal ERK have revealed that D1 DA receptors are essential for the activation of such kinases in MSNs (Liu et al., 2006a). In contrast, D2-class DA receptors, in particular D3 DA receptors, have been shown to mediate the inhibition of ERK-mediated signaling in the striatum (Zhang et al., 2004b). It has been shown that ERK 1/2 signaling is involved in protein synthesis in LTM within the PFC (Nagai et al., 2007).

In addition to their effects on PKA, DA receptors may also act on other proteins that are directly activated by cAMP, Epac1 and Epac2 (exchange **protein** directly activated by cAMP), highly expressed in the striatum (for review see Gloerich and Bos, 2010).

DA receptors may also couple to Gαq to regulate phospholipase C (PLC). Activation of PLC leads to the production of inositol trisphosphate (IP3) and diacylglycerol (DAG). Induced IP3 production results in the activation of PKC by DAG and increased mobilization of intracellular calcium, which occurs in response to IP3. Although several lines of evidence indicate that DA regulates PKC activity and intracellular calcium signaling, the exact mechanisms involved in these effects are yet to be determined (Lee et al., 2004; Sahu et al., 2009).

The molecular determinants that regulate the alternative coupling of D1-class receptors to Gαq are still not completely understood. Various possible mechanisms have been postulated. For example, proteins interacting with DA receptors may regulate receptor coupling to different G proteins *in vivo* (Bergson et al., 2003) or activation of DA receptors may enhance the signaling of other GPCRs through the PLC/IP3 pathway (Dai et al., 2008). Alternatively, it has also been shown that D1/D2 DA receptor heterodimers regulate DAG and IP3 signaling in transfected cells (Lee et al., 2004).

There is also evidence that calcium-dependent D1 DA receptor signaling mechanisms regulate Brain Derived Neurotropic Factor (BDNF) production in the NAc of adult rats (Hasbi et al., 2009) and for a potential contribution of D1/D2 DA receptor heterodimers in the regulation of DA-responsive behaviors in adult rats (Perreault et al., 2010). Indeed, it is possible that D1/D2 DA receptor heterodimers occur *in vivo* and regulate calcium-dependent cell signaling in some neuronal populations.

D2-class receptors also modulate intracellular calcium levels by acting on ion channels or by triggering the release of intracellular calcium stores (Vallar et al., 1990; Tang et al., 1994).

In addition to regulating neuronal ion channels through G protein- mediated signaling, DA receptors have been shown to interact directly with several proteins, including ion channels and ionotropic receptors in ionotropic glutamate and GABA receptors. The interaction between D5 DA receptors and GABA-A reduces GABA-A receptor-mediated whole-cell currents (Liu et al., 2000). The interaction between D2 DA receptors and NR2B NMDA receptor subunits disrupts the association between NMDA receptors and CaMKII, inhibiting the phosphorylation of this receptor and the resulting NMDA receptor-mediated currents (Liu et al., 2006b). Finally, the different interactions between D1 DA receptors and NMDA receptors have several functional consequences for NMDA receptor functions and appear to be important in regulating WM in rats (Nai et al., 2010). Some evidence also supports that the interaction with NMDA receptors affects D1 DA receptor cell surface expression and trafficking, resulting in increased insertion of D1 DA receptors into the plasma membrane (Fiorentini et al., 2003; Pei et al., 2004).

In conclusion, based on the assumption that PKA-pathway activation is necessary for CREB phosphorylation and consequent protein synthesis necessary for LTM, an opposite role of D1 and D2 receptors on LTM has been originally postulated. Indeed, D1 activates AC, while D2 inhibits AC. Nevertheless, more recent findings reported here suggest that the scenario is more complex at the system level. Indeed, D2 receptors activation can positively modulate LTM by acting on different pathways such as PLC causing calcium mobilization that activates calcium-dependent proteins that can contribute to CREB phosphorylation (for review see Neve et al., 2004).

3. Role of Dopamine Transmission in Memory

In this section we will review most of the experimental work done in the last 50 years. We will briefly first summarize major findings revealing on the role of DA in mediating WM. Then, we will focus on the effects of dopaminergic agents on memory consolidation in rodents. As noted in this chapter most of the available drugs act on D1-like and D2-like receptor classes, with little capacity to distinguish between subclasses (D1 vs D5 receptors and/or D3 *vs.* D2, respectively). For this purpose, newer drugs are being developed. Therefore, we will mainly summarize the effects of DA drugs on memory consolidation that are able to selectively distinguish between these two receptor classes, but not between the subreceptors classes belonging to them. The higher relevance of D1 and D2 receptors in modulating DA-memory dependent effects is also confirmed by the results in mutant mice, specifically lacking the different receptors subclasses, showing no major learning and memory impairment in D3, D4 and D5 knock-out mice in the different behavioral tasks reviewed in this chapter (El-Ghundi et al., 1999; Holmes et al., 2001; Glickstein et al., 2002; for review see Holmes et al., 2004).

3.1. A Behavioral Pharmacological Approach to the Study of Memory Consolidation

In 1917 Lashley (Lashley, 1917) first demonstrated that pharmacological agents could improve maze learning in rats. In studies investigating the role of DA drugs on WM, agonists or antagonists were injected before testing the subject to the behavioral assay. In addition, permanent selective lesions of DA neurons, through the injection of toxins such as 6-hydroxydopamine (6-OHDA), were also useful tools for establishing the necessary role of DA in task performance, including WM-dependent tasks. Nevertheless, the study on memory consolidation was dramatically impacted by subsequent findings: a lab operating under the supervision of McGaugh demonstrated that post-training injection of drugs could improve memory retention for up to one to two weeks in different behavioral tasks (McGaugh, 1966; for reviews see McGaugh, 1973; McGaugh and Roozendaal, 2009). Introducing such post-training drug injections represented a powerful approach for the experimental investigation of memory formation. When an animal is given a chemical agent before being trained in a behavioral task (pre-training) or prior to memory retention testing (pre-testing), alterations in behavioral performance may also include changes in general performance, such as sensorimotor function, attention, motivation, or anxiety. This result may skew how we interpret experimental findings, such as alterations in memory processing. Furthermore, pre-training and pre-testing drug injections do not allow exclusion of state-dependency effects. State-dependence is the phenomenon by which memory retrieval is most efficient when an individual is in the same internal status (i.e., state of consciousness) as he/she was when the memory was formed.

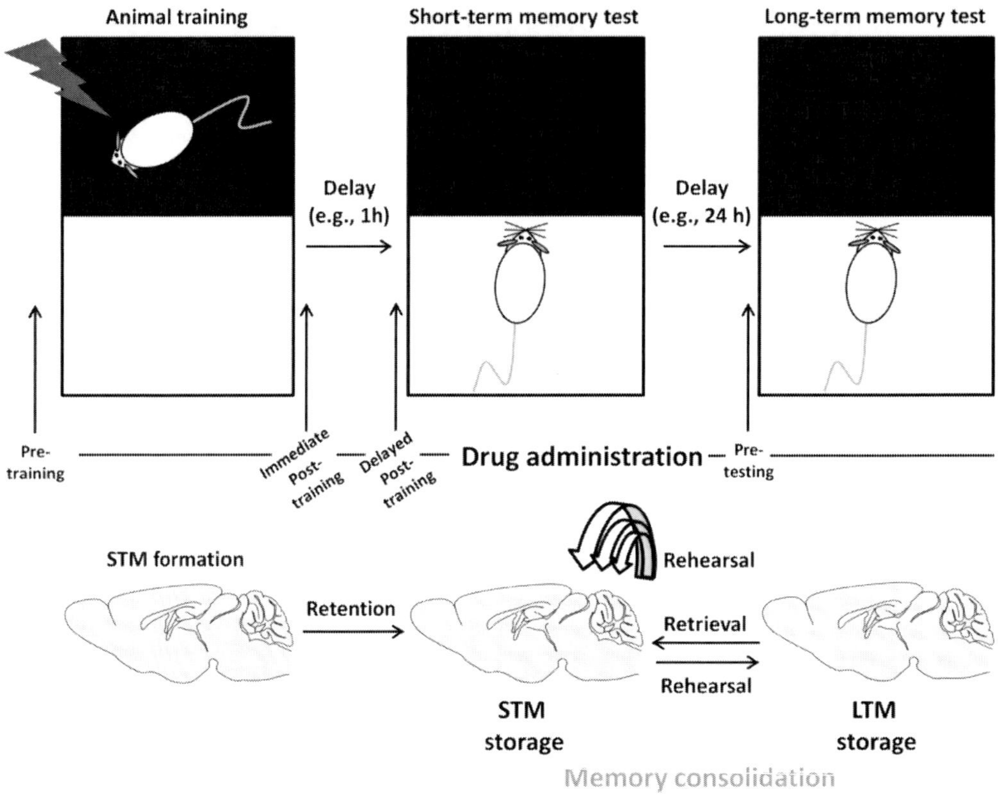

Figure 3. Interfacing the passive avoidance (PA) paradigm (top figure), a common behavioral test for the study of learning and memory processes, with the different stages of memory. Learning, short-term memory (STM) storage and long-term memory (LTM) storage that occur in the brain (bottom figure) during animal training and testing after a short or long period of time, respectively. The figure also shows how pharmacological manipulations, given at different time points, can affect memory during a particular processing phase. In the passive avoidance test animals are initially trained in a two-compartment apparatus separated by a sliding door. The animals are placed in the white compartment facing the door. After that, the door is opened and the time an animal takes to enter the dark compartment is recorded. When the animal steps into the dark section with all four paws, the door is closed and a mild footshock is given (top left). The animal is then removed and returned to its home cage. At the time of the testing trial (after a short or a long delay), the animal is returned to the white compartment and latency to enter the dark compartment is recorded, but the animal is not shocked (top middle and right figure). Drug administration before training will affect both memory acquisition (i.e., learning) and memory consolidation. Pre-training drugs injection allows the study of the effect of the drug on learning and STM/WM, immediate post-training injection and testing at long delays allows the study of modulatory actions that specifically influence early memory consolidation, 120-min delayed post-training injection allows to control for pro-active long-lasting effect of the drug on memory retrieval. Finally, pre-testing drug injection is used to investigate how drugs affect the capability to retrieve information stored in LTM.

Drug administration immediately after training (i.e., post-training), on the other hand, enables the investigation of memory consolidation without affecting other aspects of behavior (for reviews see McGaugh, 1989; McGaugh and Roozendaal, 2009; Roesler and McGaugh, 2010) as long as the drug does not have long-term proactive effects (see figure 3). Indeed, even if the drug's half-life is brief, its behavioral aspecific effects might be long-lasting, so as to disturb test performance even days after administration. For example, if the drug induces side effects, even if not directly related to action on the central nervous system (such as intestinal cramps, etc.), such adverse effects may affect other fundamental biological processes such as sleep or rest, even hours after the drug is eliminated. This may result in altered performance during the test phase. To exclude the long-lasting pro-active effects that the drugs have on performance, a delay-dependency of the memory effects must be proved. To encourage such effects, the drug is injected 2 hours after training. The time frame is decided based on original findings, which demonstrated that during this time interval, the consolidation process is completed, and that any further pharmacological manipulation no longer affects the task performance when the subject is tested 24 hours (or more) later. Therefore, if the drug, when injected immediately after training, affects task performance 24 hours later as a consequence of its pro-active long-lasting effect, a similar or more drastic impairment may be found when the subject is injected 2 hours after training. In contrast, if it affects memory consolidation, the 2 hour delay injection should not affect performance since memory consolidation process is already completed by then. This concept has been more recently modified in view of the finding supporting that memory consolidation is a phasic process, characterized by waves of protein synthesis and synaptic rearrangement (Bekinschtein et al., 2007). Accordingly, administration of the same drug in different brain regions may affect LTM formation at different time points after learning (from minutes to hours). This post-training approach to the study of memory consolidation is extremely useful with tasks that are learned in one single training trial, and therefore requiring one-single administration of the drug injected either systemically or in a specific region of the brain. Training duration is also relevant, given that during a training session lasting many hours, by the time the drug is injected, the consolidation process may already be underway from the beginning, which thus minimizes the efficacy of the post-training treatment. In rodent studies the posttraining approach has been extensively used to test the effects of hundreds of drugs on memory consolidation, and it has been widely established that systemic and focal brain injections of the protein synthesis inhibitor anysomicin affects performance at long delays (hours), but not at short delay (minutes). This posed on a firm biological substrate the distinction between WM and LTM on the only base of the retention interval.

3.2. Role of Dopamine in Working Memory

DA is critically involved in a number of physiological processes, such as motor control, rewarding properties of natural stimuli and drug of abuse, attention and impulse control (for review see Beaulieu and Gainetdinov, 2011). Dopaminergic neurons fire and DA is released into the axon terminal of several brain regions when the subject is confronted with novel stimuli, which are biologically relevant and/or unexpected (Baudonnat et al., 2013). Based on these observations it has been suggested that DA might be important for learning new associations. Anatomical distribution of DA projections provides a logical basis for proposing a role for DA in WM, which is thought to mainly rely on the activation of the PFC. In 1979 Brozoski *et al.* (Brozoski et al., 1979) revealed that DA depletion in the PFC of monkeys caused severe impairment during a WM test, the delayed response task.

In rodents WM can be tested using the *8-arm radial maze test*. The apparatus consists of eight equidistantly spaced arms, each radiating from a small circular central platform. At the end of each arm there is a small cavity filled with food, the content of which is not visible from the central platform. In brief, WM is assessed when the animal enters each arm a single time. Re-entry into the same arm would result in a WM error (see figure 4). See reference (Dudchenko, 2004) for a comprehensive review of WM tasks. Using this behavioral assay, WM deficits have been detected as a result of DA mesencephalic lesion in rats (Simon, 1981; Bubser and Schmidt, 1990). Interestingly, the administration of the DA precursor L-DOPA rescued the WM defects induced by the lesion, which further supported the direct role of DA in the identified WM defect (Brozoski et al., 1979; Stam et al., 1989). These results have been confirmed also in human studies (Cools et al., 2004).

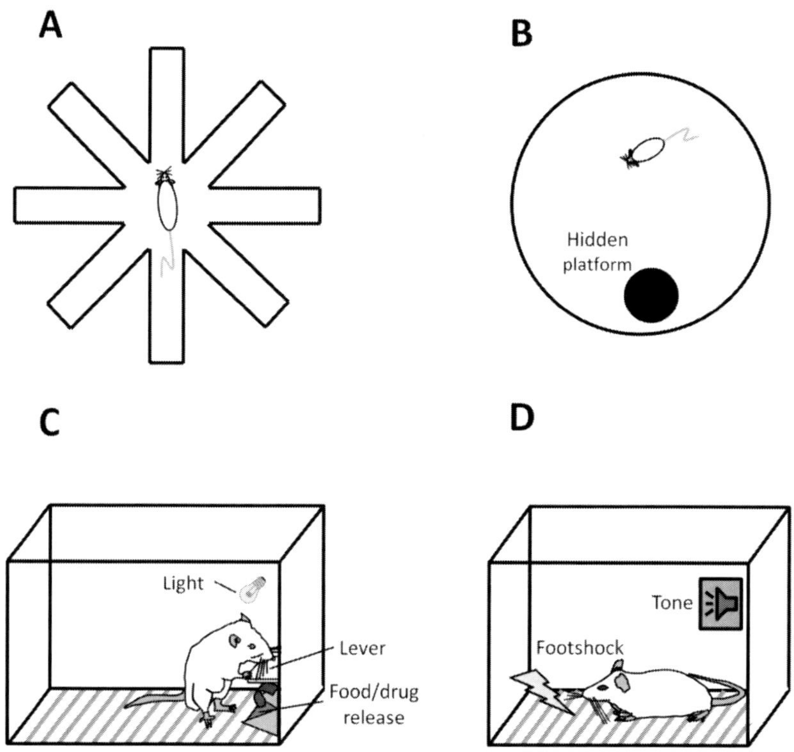

Figure 4. Different behavioral tasks are used to detect changes in memory functions. From top left: A) 8-arm radial maze, B) Morris water maze, C) Skinner box, D) fear conditioning. Each of this apparatus can be used to test different forms of memory depending on the training protocol, and therefore to probe the role of different brain regions in the task. (A) The radial maze is classical used to test WM, as described in the text. But it can also be modified to study reference spatial LTM, and or procedural memory depending on the training protocol. The hidden version of the water maze test (B) is used to test spatial memory (both LTM and WM) while the cued or visible version is used to study procedural memory. The Skinner box (C) is an experimental chamber that provides the subject a lever it can learn to repeatedly press. Pressing the lever will deliver a "reinforcement" (such as food or drug), which will enhance the probability of lever pressing. This task can be used to study action, action-outcome association learning, as well as stimulus-response learning depending on the protocol used. Similarly to the PA task described in figure 3, fear memory can be measured using the fear conditioning chamber (D), which differs from the PA because the animal is located directly into the shock-compartment where after a couple of minutes it will receive the footshock. Memory is measured as the time the animal spends freezing once replaced in the same context. Memory in this task is used to study Pavlovian association to the context or to a single cue (either visual or auditory), depending on the protocol.

The receptor basis for DA role in WM has been extensively studied with selective pharmacological agents. Blockade of D1-like receptors in PFC disrupts WM (Sawaguchi and Goldman-Rakic, 1991; Arnsten et al., 1994) and the administration of the D1 agonist SFK 38393 in animals is sufficient to rescue the WM defect associated to DA depletion (Arnsten et al., 1994; Mizoguchi et al., 2000). The role of D2 receptors in WM is more controversial given that D2-like receptor antagonists have been reported to impair WM and delay-dependent PFC neuronal activity in some studies (Didriksen, 1995; Murphy et al., 1996; Arnsten and Goldman-Rakic, 1998), but not in others (Bushnell and Levin, 1993; Sawaguchi and Goldman-Rakic, 1994; Williams and Goldman-Rakic, 1995; Verma and Moghaddam, 1996; Aultman and Moghaddam, 2001). Recent evidence suggests that D4 receptors, which are detected mainly in the corticolimbic areas, with particularly high levels in PFC (for reviews see Tarazi and Baldessarini, 1999; Oak et al., 2000), are involved in mediating WM in rodents (Zhang et al., 2004a).

In general, based on the current literature, both receptors classes, D1 and D2, are involved in WM; an optimal level of DA receptors simulation is required for optimal WM performance, which means that too low or too high DA levels in the PFC are disruptive for this cognitive function. Optimal levels are also influenced by the basal level of DA (as well as WM capacity) in the subjects, which is regulated by genetic variation in DA-related genes expression. Indeed, polymorphism in the genes important for the synthesis and metabolism of DA, has been identified both in normal subjects and subjects affected by different neuropsychiatric disorders. This has been shown to regulate WM.

3.3. Effects of Post-Training Administration of Dopaminergic Drugs on Memory in Different Behavioral Tasks

We will review the findings on the behavioral effects of DA drugs on consolidation of different forms of memory. We will consider the different behavioral tasks as if they could ideally make a clear distinction between emotional memory (with the passive avoidance [PA], and the fear conditioning tasks), spatial memory (with the hidden version of the water maze, the winshift version of the radial maze, and the odor-event associative task) and procedural memory (with the visible version of the water maze, and the win-stay version of the radial maze). It must be considered, however, that each of these tasks requires multiple learning mechanisms that can be manipulated by varying the behavioral procedure. This allows dissecting which of these mechanisms are being affected by the injected DA drugs. For example, only by contrasting the effects of the same drug on the visual and hidden version of the water maze (see below for details on the two tasks) it is possible to establish whether it is affecting spatial memory, or procedural memory to perform it. Here, we will first summarize evidence on the effects of systemic administration of DA drugs, and then we will report the effects obtained with focal brain injections.

Emotional memory: emotional memory is formed when a neutral stimulus, after being paired with a negative or positive stimulus, acquires the same emotional characteristics of the latter, a process known as Pavlovian associative learning. Traumatic emotional memory may be the cause of different anxiety-linked pathologies such as post-traumatic stress disorder and or panic disorders. Although much attention is given to emotional trauma, emotional memory is also related to positive rewards. Based on findings showing that DA is released upon the expectation of appetitive rewards, it was intuitive to expect that DA might contribute to the formation of emotional positive memory; based on the same assumption, one might have expected an opposite effect of DA on the formation of traumatic memory. Nevertheless, accumulating experimental findings have suggested that DA is released not only in response to natural positive rewards (Di Chiara and Imperato, 1988; for review see Wise and Rompre, 1989), but also in response to stressful, aversive events (for reviews see Salamone et al., 1997; Berridge and Robinson, 1998), such as a predator odor for animals or mild footshocks (for review see Trainor, 2011). The vast majority of pre-clinical studies addressing the role of DA in memory uses the PA task (one-trial inhibitory avoidance) for mice and rats (see figure 3 for a description of the task), and/or the step-down version of the task where they learn to avoid step-down (instead of step through) on an electrified grid floor, which gives a mild footshock during training. These two tasks require the formation of Pavlovian association (dark compartment-footshock). More debated is whether or not the mice also require instrumental learning as they "learn" to avoid the electrified compartment. Using these tasks, a group of researchers, led by Castellano, provided strong experimental evidence that psychostimulants, such as cocaine and amphetamine, affected aversive memory. In particular they showed that post-training systemic administration of 2.5, 5, or 10 mg/kg of cocaine improved performance in the PA task when the mice were tested 24 hr later. This effect was dose-dependent, as lower doses of cocaine (1 or 2.5 mg/kg) were not effective. In the same study the authors showed that the administration of the highest dose of cocaine did not affect performance 24 hr later if injected 120 min after training. They also showed that the animal's behavior was not modified if no shock was administered during training, and that the cocaine injection did not affect shock sensitivity (Puglisi-Allegra et al., 1994). This rules out that the performance improvement observed 24 hr later was due to any not-memory-related effect of the drug. Similar results were obtained using the DA and noradrenaline re-uptake inhibitor nomifensine (2.5, 5 and 10 mg/kg but not with 1 mg/kg) (Puglisi-Allegra et al., 1994; Cestari and Castellano, 1996), and D-amphetamine in mice (0.5 mg/kg) (Manago et al., 2013) and rats (2 and 4 mg/kg but not 0.5 and 1 mg/kg) (Wiig et al., 2009). In the latter, similar effects were obtained with L-amphetamine and L-methanfetamine (Wiig et al., 2009). All together these studies suggest that increasing amounts of catecholamines improve emotional LTM.

It is extremely interesting that more or less of the same dose of these indirect agonists were able to induce impairment, rather than improvement, in a different mouse inbred strain, the DBA/2 mice. In this strain, differently from C57BL/6 inbred mice, CD1 outbred mice, and Long Evans rats previously described, systemic administration of amphetamine or cocaine impairs memory consolidation 24 hr later (Puglisi-Allegra et al., 1994; Cabib and Castellano, 1997). This finding was one of the first sets of experimental evidence showing that cathecolamines can differently modulate memory consolidation depending on the genetic background. The same group of authors, in the last twenty years, has shown that these two lines of inbred mouse strains (C57BL/6 and DBA/2 mice) have different reactivity to drug of abuse (Cabib et al., 2000), and different coping strategies (Cabib and Puglisi-Allegra, 1994) associated with different reactivity of the dopaminergic system to environmental stimuli (Ventura et al., 2001). This created the basis for future studies using genetic engineer approaches, such as the generation of knock-out or knock-in mice, to demonstrate the role of genes in controlling memory processes.

None of these drugs are specific for DA, as they also affect serotonin and noradrenaline. Although noradrenaline has been also demonstrated to mediate emotional memory consolidation (Soeter and Kindt, 2012), a set of behavioral studies consistently showed that DA contributes to the memory enhancing effects of psychostimulants, and that similar effects can be obtained by administering direct agonists. For instance the post-training systemic administration of the D1 agonist SKF 23390 (8-10 mg/kg) improves memory consolidation 24 hr later (Castellano et al., 1994; Mele et al., 1997; Ouyang et al., 2012) in C57BL/6 and NMRI mice, with the exception of one study in which no effect was found in CD1 outbred mice using the same dose range and even higher dose (3, 10 and 30 mg/kg) (Gasbarri et al., 1993). Similarly, systemic administration of the highly selective D2 receptor agonist quinpirole (or LY 171555) dose-dependently improves memory consolidation in mice (0.5 and 1 mg/kg but not 0.025 mg/kg) (Castellano et al., 1991; Gasbarri et al., 1993; Mele et al., 1996). Consistently, there are findings showing that post-training administration of both D1 and D2 receptor antagonists SCH 23390 (0.05-0.1mg/kg, but not 0.025 mg/kg) and sulpiride (12.5-25 mg/kg, but not 6mg/kg), dose-dependently impairs performance 24 hr later (Castellano et al., 1991). Moreover, low doses of the same drugs that are not-effective when administered alone, block the memory-improving effects of cocaine or nomifensine in C57BL/6 mice (Castellano et al., 1991; Puglisi-Allegra et al., 1994). Similarly, the LTM defect induced by cocaine or amphetamine in DBA/2 mice is also rescued by coadministration of the same D1 and D2 receptors antagonists (Castellano et al., 1994; PuglisiAllegra et al., 1994). All together these data suggest that DA receptors stimulation is necessary for PA memory consolidation, and that the memory enhancing effects of dopaminergic drugs are dependent on the genetic background of the subject.

This set of experiments has been followed by site-specific administration studies in which the drug is directly administered post-training within specific brain regions. This allows to specifically dissect the neuronal circuit involved in the behavior under investigation, and the specific role of DA in that brain region.

The amygdala is known to have a pivotal role in the formation and maintenance of emotional LTM, in particular for traumatic memory. Accordingly, intra-amygdala administration of DA in rats improves memory in PA when tested 48 hr later, and this improvement is blocked by concurrent injection of both D1 or D2 receptor antagonists, SCH 23390 and sulpiride, respectively (Lalumiere et al., 2004; LaLumiere et al., 2005). The memory enhancing effects induced by DA within the amygdala can be also deterred by concurrent blockade of DA receptors within the NAc shell (LaLumiere et al., 2005). Accordingly, the amygdala and, specifically, the basolateral nucleus (BLA) projects to the NAc, both the core and shell in their medial part (LaLumiere et al., 2005). Thus, within the NAc, the projections from the BLA are integrated with inputs from the HP and the PFC. This brain circuit is a limbic circuit that has been demonstrated to control the selection of complex and appropriate behavioral output on the basis of learned associations (Mogenson et al., 1980; Kelley et al., 1982; Everitt et al., 1991; Sargolini et al., 1999). Consistently with this view, inhibiting both D1 and D2 receptors within the shell and the core of the NAc blocks memory consolidation in the PA (Manago et al., 2009). Intra-HP immediate post-training administration of SCH 23390 in rats does not generally affect memory consolidation; however if the drug is injected before training, performance is impaired 24 hr later, but not 30 min later (Bernabeu et al., 1997; O'Carroll et al., 2006; Rossato et al., 2013); furthermore, delayed (3-6 hr) post-training injection of the D1 agonist SKF 38393 (Bernabeu et al., 1997; Bevilaqua et al., 1997) has been shown to improve LTM in the PA task. HP post-training focal administration of the D2 receptor antagonists or agonists does not affect memory consolidation of Pavlovian aversive memory (Liao et al., 2013; Rossato et al., 2013).

Concerning the PFC, a little available experimental evidence suggests that D1 receptors activation might be involved in memory consolidation, not only in its very early stage (0 min post-training), but also in later stages (up to 180 min after training) (Izquierdo et al., 2007), which is consistent with recent findings showing that the anterior cingulate cortex is crucial for remote fear memory (Restivo et al., 2009). Recent findings also show that intra-ventral pallidum (VP) administration of the D1 receptor agonist SKF 38393 improves PA memory consolidation, and that the effect lasts for more than two weeks (Peczely et al., 2014). This is consistent with findings showing that the VP, which is an output station from the NAc, modulates the activity of DA neurons.

Spatial memory: spatial memory records spatial information, such as the position or location of objects and places. Studies in rodents have shown that hippocampal neurotransmission modulates learning and long-term retention in some spatial memory tasks, and, in particular, ones that require finding the path using allocentric spatial memory, such as the hidden platform version of the water maze, the reference memory and the win-shift versions of the radial arm maze (for reviews see Morris et al., 1982; Sutherland and McNaughton, 2000; Clark et al., 2005). The *hidden version of the water maze task* (see figure 4), for example, is a classic test used for studying spatial memory in rodents (Morris, 1984). The subject (mouse/rat) is initially released into a circular pool filled with water surrounded by visual distant cues that serve as reference points. The subject swims around the pool until, by chance, it discovers a hidden platform, just below the surface of the water, onto which it can climb. Usually, the subject first swims around the edge of the pool before deciding to move toward the center where it may finally stumble upon the hidden platform. However, as time passes the amount of time needed to locate the platform decreases; trained animals swim almost immediately to the platform after being released into the water (for review see Stewart et al., 2011). Using this task, it has been shown that post-training systemic administration of sulpiride (10 mg/kg, but not 30 or 100 mg/kg) improves performance 48 hr later (Setlow and McGaugh, 2000), which is in contradiction with what occurs in the PA. Nevertheless, this finding is also in contradiction with other findings using different spatial memory tasks also depending on the HP integrity. Indeed, post-training systemic administration of moderate doses of amphetamine or quinpirole (2 mg/kg, but not 1 or 4 mg/kg) improves performance 24 hr later in the win-shift version of the radial arm maze; systemic injection of SKF 38393 has no effect on the same task (Packard and White, 1991). The findings with quinpirole are paralleled by those obtained with HP site-specific administration of the same drug in the same task; furthermore, immediate post-training (but not 120 min delayed) site-specific administration of sulpiride or SCH 23390 into the NAc impairs performance in different spatial tasks (Setlow and McGaugh, 1998; Mele et al., 2004). This is not surprising considering that the NAc is also involved in spatial memory consolidation (Sargolini et al., 1999; Ferretti et al., 2005) and is the main recipient of HP outputs. Nevertheless, when site-specifically injected within the HP the D1 receptor agonist, SKF 38393, induces memory improvement as well (5ug/side, but not 2 or 10 ug/side). This finding, on the one hand, is consistent with the finding showing that post-training intra-HP injection of SCH 23390 (30ng/side) impairs memory consolidation 24 hr later, in the contextual fear conditioning (see figure 4 for a description of the task), which strongly relies on the activation of the HP (Liao et al., 2013). On the other hand it contradicts the recent evidence showing that 15 min post-training administration of SCH 23390 (5ug/side) does not affect performance 6 hour later in the delayed not matching version of the water maze (O'Carroll et al., 2006). Similar results were obtained in another HP-dependent task, requiring to the animals to learn six paired-associates in an "event arena" involving a repeated association between specific flavors of food and locations in space (Bethus et al., 2010); intra-HP injection of SCH 23390 has been shown to affect memory at long delay (6 or 24 hr), but not at short delay (20 min), if administered around 15 min before learning.

Procedural memory: procedural memory is a category that includes many different types of memory, although in rodents' studies, it generally refers to memory relying on the acquisition of a habit or stimulus-response association. Behavioral tasks considered to require procedural memory are the visual (cued) version of the water maze, or the win-stay version of the radial maze, or the active avoidance test. These tasks rely on the integrity of the sensorimotor corticalstriatal pathway (Kelley et al., 2003). Post-training systemic administration of haloperidol has been shown to impair active avoidance memory when tested 7 days later, and this effect is reversed by co-administration of otherwise ineffective doses of apomorphine (Gozzani and Izquierdo, 1976; Grecksch and Matties, 1981). Post-training systemic administration of amphetamine and quinpirole, but not of SKF 38393, improves performance in this win-stay version of the radial maze (Packard and White, 1991); these effects were replicated by intra-striatum (but not intra-HP) injection of the same drugs. Additionally, as well as for the spatial version of the task, SKF 38393 was effective when site-specifically (5ug/side) injected within the striatum (Packard and White, 1991).

CONCLUSION

All together these findings suggest that DA modulates early stages of consolidation (immediate post-training) of emotional, spatial and procedural memory by acting on both D1 and D2 receptors. Furthermore, they suggest that the effects of dopaminergic drugs on memory consolidation, as well as for WM, depend on the genetic background of the subjects, which may explain discrepancies between the effects of dopaminergic drugs on the different species (rats *vs.* mice) used in pre-clinical studies. There is at least one experimental evidence showing that each of the brain regions of the mesocorticolimbic system (PFC, NAc, CPu, VP, HP, and BLA) is involved in memory consolidation. These brain regions have a different role in the different kinds of memory we considered in this chapter; although a direct comparison cannot be made on the basis of the few studies published in literature, we can conclude that if the brain region is involved in memory consolidation, DA regulates the process. This is supported by the fact that post training site-specific administration of DA drugs modulates: (i) emotional LTM within the BLA, (ii) spatial LTM within the HP, (iii) procedural LTM dependent on the caudate-putamen, (iv) the NAc role on emotional and spatial LTM. This modulation generally implies that memory consolidation is improved by the administration of DA agonists and impaired by DA receptors antagonists. Contrarily to the results obtained in WM tasks, the U-inverted shape dose-response curve is not always evident with post-training site-specific administration (Packard and White, 1991; Cabib and Castellano, 1997; LaLumiere et al., 2003; Wiig et al., 2009). This is likely due to the fact that with post-training injection of the drugs, the subjects are off-drug both during training and testing; therefore, non-mnemonic confounding effects of their injection are avoided.

Specific roles of the different receptors subtypes seem to be allocated to certain brain regions and to be dependent on the task performed. In particular, systemic administration of both D1 and D2 receptor antagonists blocks emotional memory, and these effects are paralleled when the same drugs are injected in the BLA-NAc pathway, which is consistent with a crucial role of these two brain regions in aversive emotional memory. In contrast, within the other regions of the corticolimbic system, D2 receptors are not involved in emotional aversive memory. A necessary role of D1 receptors activation is evidenced within a different time framework. These data demonstrate that D1 receptor-mediated effects do not affect learning or early memory, but do affect memory at later times. DA has no effect on the previously formed memories and is not required at the time of retrieval (O'Carroll et al., 2006). This suggests that although D1 receptors activation during learning is not necessary for WM, is necessary for memory consolidation. The time-frame of the effects of the D1 receptor antagonists on HP dependent-memory consolidation supports the Synaptic Tagging and Capture model developed by Frey and Morris (Frey and Morris, 1997). This model has been generated in relation to the role of DA in long-term potentiation, which is a form of synaptic plasticity supporting LTM formation. In brief, this hypothesis suggests that activating the DA by a reward, novelty of any biological relevant stimulus, should enhance HP-dependent episodic/spatial memory of information present at the time of reward. According to this hypothesis reward-related activation of DA is associated with memory enhancement only after long retention intervals, compatible with the effects of intra-HP injections of specific DA antagonists. This time-dependent effect is supposed to be due to DA associated upregulation of plasticity proteins, necessary for LTM formation (but not for WM). Although data consistent with such a mechanism responsible for memory consolidation have been reported (Moncada and Viola, 2007), and this model is extensively presented in the mentioned reviews (Frey and Morris, 1997, 1998), it still requires further investigation to explain experimental data in support of a role of DA in encoding short-term memory (Frey and Morris, 1997).

Although the time-limited action of D1 agents within the HP is also confirmed by spatial memory tasks, the lack of effect that D2 agents have on these tasks is not replicated.

Accordingly, post-training administration of D2 receptor antagonists or agonists modulates memory consolidation in spatial tasks, thus suggesting that DA, through these two receptors classes, is modulating two different processes within the HP during spatial memory formation. As previously suggested (Bethus et al., 2010) the possibility that there are actions of DA on memory consolidation at or around the time of encoding (pre-training) and, separately, at later time points (post-training), are not mutually exclusive. The necessary role of post-training D2 receptors activation for LTM suggests its role in off-line processes, which has been suggested to strength the formed association, such as sleep-dependent memory consolidation (for review see O'Neill et al., 2010).

The role of D1 in memory consolidation is consistent with the intracellular pathway associated with D1 receptors activation, and its interaction with NMDA receptor-dependent plasticity, which leads to PKA-dependent CREB-phosphorylation and associated gene modification. Nevertheless, D2 has an opposite effect on AC, and therefore is expected it might be detrimental for memory formation. Recent findings, however, suggest that D1, in excitatory circuits, and D2, in inhibitory circuits, act cooperatively to drive synaptic plasticity in the PFC (Xu and Yao, 2010). This circuit mechanism may allow DA to gate associative synaptic modifications globally, regulating the time frame at individual local synapses. This mechanism can play a crucial role in the different steps of LTM formation.

DA cells can be silent, fire single spikes that occur somewhat irregularly (tonic mode), or fire brief bursts of spikes. This firing is modulated by GABAergic inputs from the VP, in which cells have high spontaneous activity (Yim and Mogenson, 1983). As a result of this inhibition, up to 50% of DA neurons are in the silent mode at baseline. Activation of the NAc can occur through input from the HP, the BLA or the PFC, which are integrated within this brain region (Mogenson et al., 1980). The HP brings in information about the context or the novelty of the incoming stimulus (mismatch with previous experience); it is interesting, in this regards, that signal transmission from the HP to the subpallidal output neurons and in turn to the pedunculopontine nucleus is modulated by a D2, but not D1, receptor-mediated mechanism in the NAc (Yang and Mogenson, 1987). The BLA likely signals the emotional relevance of the events likely integrating noradrenergic and serotoninergic inputs that reach the nucleus. The PFC modulates executive functions such as attention-related and behavioral motor programs to cope with the incoming stimulus. The integration of these signals within the NAc, leads to its activation, which results in the inhibition of the VP ultimately resulting in DA release in the terminal regions where, through a fine interaction between the activity of D1 and D2 receptors, the formation and the strengthen of LTM is allowed (Floresco et al., 2001; Floresco et al., 2003).

Based on these findings it might be suggested that DA mainly through its action on D1 receptors would help to make LTM highly efficient, by initially selecting relevant information for LTM storage during encoding. During memory consolidation, through D2 receptors activation, by favoring off-line reactivation mechanisms while permitting the decay of irrelevant memory traces, might strengthen the selected association.

Relevance to Human Studies

Unfortunately, although the role of DA in WM has been broadly investigated in humans, little experimental evidence indicating the role of DA in memory consolidation is available. Methylphenidate inhibits the reuptake of the catecholamines DA and noradrenaline by blocking the transporters, and thus enhances catecholamine availability (Volkow et al., 1998; Hannestad et al., 2010). There are also observations of methylphenidate-induced cognition enhancement in healthy adults (Elliott et al., 1997; Mehta et al., 2000); more recently, findings have shown that methylphenidate improved delayed recall of word lists. Since no improvement was observed at immediate recall, enhanced delayed recall is likely caused by better consolidation (Linssen et al., 2012).

Furthermore, the pharmacologically similar compound amphetamine has been shown to improve consolidation and retention of word lists, while leaving acquisition unaffected (Advokat, 2010). Experiments from Soetens and Zeeuws (Soetens et al., 1993; Soetens et al., 1995; Zeeuws and Soetens, 2007) have shown that amphetamine facilitates verbal memory recall if it is active in the brain during the consolidation phase (Soetens et al., 1995).

These studies in humans confirm the role of DA in memory consolidation. More interestingly, a recent study has set up a cognitive paradigm used to test the role of reward in memory consolidation in humans. It was found that the expectation of receiving a reward (e.g., monetary gain) improved long-term episodic memory for novel stimuli. This reward-related memory enhancement was associated with a co-activation of SN/VTA, striatum, and HP, as detected by functional Magnetic Resonance Imaging (Wittmann et al., 2005; Adcock et al., 2006).

The results have been interpreted as dependent on DA-mediated influence on memory consolidation. This opens the opportunity for future studies in humans aimed at selectively manipulating the DA system at different time points before and after learning, to establish his specific role in memory consolidation.

ACKNOWLEDGMENTS

The authors thank Alexis Burton for editorial assistance.

REFERENCES

Abel T, Nguyen PV, Barad M, Deuel TA, Kandel ER, Bourtchouladze R (1997) Genetic demonstration of a role for PKA in the late phase of LTP and in hippocampus-based long-term memory. *Cell* 88:615-626.

Adcock RA, Thangavel A, Whitfield-Gabrieli S, Knutson B, Gabrieli JD (2006) Rewardmotivated learning: mesolimbic activation precedes memory formation. *Neuron* 50:507517.

Advokat C (2010) What are the cognitive effects of stimulant medications? Emphasis on adults with attention-deficit/hyperactivity disorder (ADHD). *Neuroscience and biobehavioral reviews* 34:1256-1266.

Aggleton JP, Brown MW (2005) Contrasting hippocampal and perirhinal cortex function using immediate early gene imaging. *The Quarterly journal of experimental psychology* 58:218-233.

Ainge JA, Heron-Maxwell C, Theofilas P, Wright P, de Hoz L, Wood ER (2006) The role of the hippocampus in object recognition in rats: examination of the influence of task parameters and lesion size. *Behavioural brain research* 167:183-195.

Alburges ME, Hunt ME, McQuade RD, Wamsley JK (1992) D1-receptor antagonists: comparison of [3H]SCH39166 to [3H]SCH23390. *Journal of chemical neuroanatomy* 5:357-366.

Amenta F, Ricci A, Tayebati SK, Zaccheo D (2002) The peripheral dopaminergic system: morphological analysis, functional and clinical applications. Italian journal of anatomy and embryology = Archivio italiano di anatomia ed embriologia 107:145-167.

Andersen PH, Gingrich JA, Bates MD, Dearry A, Falardeau P, Senogles SE, Caron MG (1990) Dopamine receptor subtypes: beyond the D1/D2 classification. *Trends in pharmacological sciences* 11:231-236.

Arnsten AF, Goldman-Rakic PS (1998) Noise stress impairs prefrontal cortical cognitive function in monkeys: evidence for a hyperdopaminergic mechanism. *Archives of general psychiatry* 55:362-368.

Arnsten AF, Cai JX, Murphy BL, Goldman-Rakic PS (1994) Dopamine D1 receptor mechanisms in the cognitive performance of young adult and aged monkeys. *Psychopharmacology* 116:143-151.

Atkinson RC, Shiffrin RM (1971) The control of short-term memory. *Scientific American* 225:82-90.

Aultman JM, Moghaddam B (2001) Distinct contributions of glutamate and dopamine receptors to temporal aspects of rodent working memory using a clinically relevant task. *Psychopharmacology* 153:353-364.

Axelrod RS (1997) Antiemetic therapy. *Comprehensive therapy* 23:539-545.

Baddeley A (1992) Working memory. *Science* 255:556-559.

Barsegyan A, Mackenzie SM, Kurose BD, McGaugh JL, Roozendaal B (2010) Glucocorticoids in the prefrontal cortex enhance memory consolidation and impair working memory by a common neural mechanism. *Proceedings of the National Academy of Sciences of the United States of America* 107:16655-16660.

Baudonnat M, Huber A, David V, Walton ME (2013) Heads for learning, tails for memory: reward, reinforcement and a role of dopamine in determining behavioral relevance across multiple timescales. *Frontiers in neuroscience* 7:175.

Beaulieu JM, Gainetdinov RR (2011) The physiology, signaling, and pharmacology of dopamine receptors. *Pharmacological reviews* 63:182-217.

Beckstead RM (1978) Afferent connections of the entorhinal area in the rat as demonstrated by retrograde cell-labeling with horseradish peroxidase. Brain research 152:249-264.

Bekinschtein P, Cammarota M, Igaz LM, Bevilaqua LR, Izquierdo I, Medina JH (2007) Persistence of long-term memory storage requires a late protein synthesis- and BDNF-dependent phase in the hippocampus. *Neuron* 53:261-277.

Beom S, Cheong D, Torres G, Caron MG, Kim KM (2004) Comparative studies of molecular mechanisms of dopamine D2 and D3 receptors for the activation of extracellular signalregulated kinase. *The Journal of biological chemistry* 279:28304-28314.

Bergson C, Levenson R, Goldman-Rakic PS, Lidow MS (2003) Dopamine receptor-interacting proteins: the Ca(2+) connection in dopamine signaling. *Trends in pharmacological sciences* 24:486-492.

Bergson C, Mrzljak L, Smiley JF, Pappy M, Levenson R, Goldman-Rakic PS (1995) Regional, cellular, and subcellular variations in the distribution of D1 and D5 dopamine receptors in primate brain. The *Journal of neuroscience : the official journal of the Society for Neuroscience* 15:7821-7836.

Bernabeu R, Bevilaqua L, Ardenghi P, Bromberg E, Schmitz P, Bianchin M, Izquierdo I, Medina JH (1997) Involvement of hippocampal cAMP/cAMP-dependent protein kinase signaling pathways in a late memory consolidation phase of aversively motivated learning in rats. *Proceedings of the National Academy of Sciences of the United States of America* 94:7041-7046.

Berridge KC, Robinson TE (1998) What is the role of dopamine in reward: hedonic impact, reward learning, or incentive salience? *Brain research Brain research reviews* 28:309369.

Bethus I, Tse D, Morris RG (2010) Dopamine and memory: modulation of the persistence of memory for novel hippocampal NMDA receptor-dependent paired associates. *The Journal of neuroscience : the official journal of the Society for Neuroscience* 30:16101618.

Bevilaqua L, Ardenghi P, Schroder N, Bromberg E, Schmitz PK, Schaeffer E, Quevedo J, Bianchin M, Walz R, Medina JH, Izquierdo I (1997) Drugs acting upon the cyclic adenosine monophosphate/protein kinase A signalling pathway modulate memory consolidation when given late after training into rat hippocampus but not amygdala. *Behavioural pharmacology* 8:331-338.

Billard W, Ruperto V, Crosby G, Iorio LC, Barnett A (1984) Characterization of the binding of 3H-SCH 23390, a selective D-1 receptor antagonist ligand, in rat striatum. *Life sciences* 35:1885-1893.

Bonuccelli U, Pavese N (2007) Role of dopamine agonists in Parkinson's disease: an update. *Expert review of neurotherapeutics* 7:1391-1399.

Bortolato M, Chen K, Shih JC (2008) Monoamine oxidase inactivation: from pathophysiology to therapeutics. *Advanced drug delivery reviews* 60:1527-1533.

Bouthenet ML, Souil E, Martres MP, Sokoloff P, Giros B, Schwartz JC (1991) Localization of dopamine D3 receptor mRNA in the rat brain using in situ hybridization histochemistry: comparison with dopamine D2 receptor mRNA. *Brain research* 564:203-219.

Brown MW, Warburton EC, Aggleton JP (2008) Recognition memory: material, processes, and substrates. *Hippocampus* 20:1228-1244.

Brozoski TJ, Brown RM, Rosvold HE, Goldman PS (1979) Cognitive deficit caused by regional depletion of dopamine in prefrontal cortex of rhesus monkey. *Science* 205:929-932.

Bubser M, Schmidt WJ (1990) 6-Hydroxydopamine lesion of the rat prefrontal cortex increases locomotor activity, impairs acquisition of delayed alternation tasks, but does not affect uninterrupted tasks in the radial maze. *Behavioural brain research* 37:157-168.

Bunzow JR, Van Tol HH, Grandy DK, Albert P, Salon J, Christie M, Machida CA, Neve KA, Civelli O (1988) Cloning and expression of a rat D2 dopamine receptor cDNA. *Nature* 336:783-787.

Bushnell PJ, Levin ED (1993) Effects of dopaminergic drugs on working and reference memory in rats. *Pharmacology, biochemistry, and behavior* 45:765-776.

Cabib S, Puglisi-Allegra S (1994) Opposite responses of mesolimbic dopamine system to controllable and uncontrollable aversive experiences. *The Journal of neuroscience : the official journal of the Society for Neuroscience* 14:3333-3340.

Cabib S, Castellano C (1997) Impairments produced by amphetamine and stress on memory storage are reduced following a chronic stressful experience. *Psychopharmacology* 129:161-167.

Cabib S, Orsini C, Le Moal M, Piazza PV (2000) Abolition and reversal of strain differences in behavioral responses to drugs of abuse after a brief experience. *Science* 289:463-465.

Carson CC, 3rd (2007) Central nervous system-acting agents and the treatment of erectile and sexual dysfunction. *Current urology reports* 8:472-476.

Castellano C, Cestari V, Cabib S, Puglisi-Allegra S (1991) Post-training dopamine receptor agonists and antagonists affect memory storage in mice irrespective of their selectivity for D1 or D2 receptors. *Behavioral and neural biology* 56:283-291.

Castellano C, Cestari V, Cabib S, Puglisi-Allegra S (1994) The effects of morphine on memory consolidation in mice involve both D1 and D2 dopamine receptors. *Behavioral and neural biology* 61:156-161.

Cestari V, Castellano C (1996) Caffeine and cocaine interaction on memory consolidation in mice. *Archives internationales de pharmacodynamie et de therapie* 331:94-104.

Chen J, Song J, Yuan P, Tian Q, Ji Y, Ren-Patterson R, Liu G, Sei Y, Weinberger DR (2011) Orientation and cellular distribution of membrane-bound catechol-O-methyltransferase in cortical neurons: implications for drug development. *The Journal of biological chemistry* 286:34752-34760.

Cheng DT, Knight DC, Smith CN, Stein EA, Helmstetter FJ (2003) Functional MRI of human amygdala activity during Pavlovian fear conditioning: stimulus processing versus response expression. *Behavioral neuroscience* 117:3-10.

Clark RE, Broadbent NJ, Squire LR (2005) Hippocampus and remote spatial memory in rats. *Hippocampus* 15:260-272.

Cohen AI, Todd RD, Harmon S, O'Malley KL (1992) Photoreceptors of mouse retinas possess D4 receptors coupled to adenylate cyclase. *Proceedings of the National Academy of Sciences of the United States of America* 89:12093-12097.

Colao A, Di Sarno A, Guerra E, De Leo M, Mentone A, Lombardi G (2006) Drug insight: Cabergoline and bromocriptine in the treatment of hyperprolactinemia in men and women. *Nature clinical practice Endocrinology & metabolism* 2:200-210.

Cools R, Clark L, Robbins TW (2004) Differential responses in human striatum and prefrontal cortex to changes in object and rule relevance. *The Journal of neuroscience : the official journal of the Society for Neuroscienc*e 24:1129-1135.

Cotzias GC, Van Woert MH, Schiffer LM (1967) Aromatic amino acids and modification of parkinsonism. *The New England journal of medicine* 276:374-379.

Cotzias GC, Papavasiliou PS, Gellene R (1969) Modification of Parkinsonism--chronic treatment with L-dopa. *The New England journal of medicine* 280:337-345.

Dai R, Ali MK, Lezcano N, Bergson C (2008) A crucial role for cAMP and protein kinase A in D1 dopamine receptor regulated intracellular calcium transients. *Neuro-Signals* 16:112123.

Dalley JW, Laane K, Theobald DE, Armstrong HC, Corlett PR, Chudasama Y, Robbins TW (2005) Time-limited modulation of appetitive Pavlovian memory by D1 and NMDA receptors in the nucleus accumbens. *Proceedings of the National Academy of Sciences of the United States of America* 102:6189-6194.

Dash PK, Moore AN, Kobori N, Runyan JD (2007) Molecular activity underlying working memory. *Learning & memory* (Cold Spring Harbor, NY 14:554-563.

De Leonibus E, Oliverio A, Mele A (2005) A study on the role of the dorsal striatum and the nucleus accumbens in allocentric and egocentric spatial memory consolidation. *Learning & memory* (Cold Spring Harbor, NY) 12:491-503.

De Leonibus E, Lafenetre P, Oliverio A, Mele A (2003) Pharmacological evidence of the role of dorsal striatum in spatial memory consolidation in mice. *Behavioral neuroscience* 117:685-694.

De Leonibus E, Costantini VJ, Massaro A, Mandolesi G, Vanni V, Luvisetto S, Pavone F, Oliverio A, Mele A (2011) Cognitive and neural determinants of response strategy in the dual-solution plus-maze task. *Learning & memory* (Cold Spring Harbor, NY) 18:241244.

De Mei C, Ramos M, Iitaka C, Borrelli E (2009) Getting specialized: presynaptic and postsynaptic dopamine D2 receptors. *Current opinion in pharmacology* 9:53-58.

Dearry A, Gingrich JA, Falardeau P, Fremeau RT, Jr., Bates MD, Caron MG (1990) Molecular cloning and expression of the gene for a human D1 dopamine receptor. *Nature* 347:7276.

DeCoteau WE, Kesner RP (2000) A double dissociation between the rat hippocampus and medial caudoputamen in processing two forms of knowledge. *Behavioral neuroscience* 114:1096-1108.

Di Chiara G, Imperato A (1988) Drugs abused by humans preferentially increase synaptic dopamine concentrations in the mesolimbic system of freely moving rats. *Proceedings of the National Academy of Sciences of the United States of America* 85:5274-5278.

Didriksen M (1995) Effects of antipsychotics on cognitive behaviour in rats using the delayed non-match to position paradigm. *European journal of pharmacology* 281:241-250.

Divac N, Prostran M, Jakovcevski I, Cerovac N (2014) Second-Generation Antipsychotics and Extrapyramidal Adverse Effects. *BioMed research international* 2014:656370.

Dubois A, Savasta M, Curet O, Scatton B (1986) Autoradiographic distribution of the D1 agonist [3H]SKF 38393, in the rat brain and spinal cord. Comparison with the distribution of D2 dopamine receptors. *Neuroscience* 19:125-137.

Dudchenko PA (2004) An overview of the tasks used to test working memory in rodents. *Neuroscience and biobehavioral reviews* 28:699-709.

El-Ghundi M, Fletcher PJ, Drago J, Sibley DR, O'Dowd BF, George SR (1999) Spatial learning deficit in dopamine D(1) receptor knockout mice. *European journal of pharmacology* 383:95-106.

Elliott R, Sahakian BJ, Matthews K, Bannerjea A, Rimmer J, Robbins TW (1997) Effects of methylphenidate on spatial working memory and planning in healthy young adults. *Psychopharmacology* 131:196-206.

Elsworth JD, Roth RH (1997) Dopamine synthesis, uptake, metabolism, and receptors: relevance to gene therapy of Parkinson's disease. *Experimental neurology* 144:4-9.

Everitt BJ, Morris KA, O'Brien A, Robbins TW (1991) The basolateral amygdala-ventral striatal system and conditioned place preference: further evidence of limbic-striatal interactions underlying reward-related processes. *Neuroscience* 42:1-18.

Ferretti V, Florian C, Costantini VJ, Roullet P, Rinaldi A, De Leonibus E, Oliverio A, Mele A (2005) Co-activation of glutamate and dopamine receptors within the nucleus accumbens is required for spatial memory consolidation in mice. *Psychopharmacology* 179:108-116.

Fienberg AA et al. (1998) DARPP-32: regulator of the efficacy of dopaminergic neurotransmission. *Science* 281:838-842.

Fiorentini C, Gardoni F, Spano P, Di Luca M, Missale C (2003) Regulation of dopamine D1 receptor trafficking and desensitization by oligomerization with glutamate N-methyl-Daspartate receptors. *The Journal of biological chemistry* 278:20196-20202.

Fisher DB, Kirkwood R, Kaufman S (1972) Rat liver phenylalanine hydroxylase, an iron enzyme. *The Journal of biological chemistry* 247:5161-5167.

Floresco SB, Todd CL, Grace AA (2001) Glutamatergic afferents from the hippocampus to the nucleus accumbens regulate activity of ventral tegmental area dopamine neurons. *The Journal of neuroscience : the official journal of the Society for Neuroscience* 21:49154922.

Floresco SB, West AR, Ash B, Moore H, Grace AA (2003) Afferent modulation of dopamine neuron firing differentially regulates tonic and phasic dopamine transmission. *Nature neuroscience* 6:968-973.

Fornai F, Chen K, Giorgi FS, Gesi M, Alessandri MG, Shih JC (1999) Striatal dopamine metabolism in monoamine oxidase B-deficient mice: a brain dialysis study. *Journal of neurochemistry* 73:2434-2440.

Frey U, Morris RG (1997) Synaptic tagging and long-term potentiation. *Nature* 385:533-536.

Frey U, Morris RG (1998) Weak before strong: dissociating synaptic tagging and plasticityfactor accounts of late-LTP. *Neuropharmacology* 37:545-552.

Gasbarri A, Introini-Collison IB, Packard MG, Pacitti C, McGaugh JL (1993) Interaction of cholinergic-dopaminergic systems in the regulation of memory storage in aversively motivated learning tasks. *Brain research* 627:72-78.

Giros B, Sokoloff P, Martres MP, Riou JF, Emorine LJ, Schwartz JC (1989) Alternative splicing directs the expression of two D2 dopamine receptor isoforms. *Nature* 342:923-926.

Glickstein SB, Hof PR, Schmauss C (2002) Mice lacking dopamine D2 and D3 receptors have spatial working memory deficits. *The Journal of neuroscience : the official journal of the Society for Neuroscience* 22:5619-5629.

Gloerich M, Bos JL (2010) Epac: defining a new mechanism for cAMP action. *Annual review of pharmacology and toxicology* 50:355-375.

Gonzalez AM, Sibley DR (1995) [3H]7-OH-DPAT is capable of labeling dopamine D2 as well as D3 receptors. *European journal of pharmacology* 272:R1-3.

Gozzani JL, Izquierdo I (1976) Possible peripheral adrenergic and central dopaminergic influences in memory consolidation. *Psychopharmacology* 49:109-111.

Grace AA, Bunney BS (1984) The control of firing pattern in nigral dopamine neurons: burst firing. *The Journal of neuroscience : the official journal of the Society for Neuroscience* 4:2877-2890.

Grecksch G, Matties H (1981) The role of dopaminergic mechanisms in the rat hippocampus for the consolidation in a brightness discrimination. *Psychopharmacology* 75:165-168.

Greenawalt JW, Schnaitman C (1970) An appraisal of the use of monoamine oxidase as an enzyme marker for the outer membrane of rat liver mitochondria. *The Journal of cell biology* 46:173-179.

Greengard P (2001) The neurobiology of slow synaptic transmission. *Science* 294:1024-1030.

Hamm HE, Gilchrist A (1996) Heterotrimeric G proteins. *Current opinion in cell biology* 8:189196.

Hannestad J, Gallezot JD, Planeta-Wilson B, Lin SF, Williams WA, van Dyck CH, Malison RT, Carson RE, Ding YS (2010) Clinically relevant doses of methylphenidate significantly occupy norepinephrine transporters in humans in vivo. *Biological psychiatry* 68:854-860.

Hasbi A, Fan T, Alijaniaram M, Nguyen T, Perreault ML, O'Dowd BF, George SR (2009) Calcium signaling cascade links dopamine D1-D2 receptor heteromer to striatal BDNF production and neuronal growth. *Proceedings of the National Academy of Sciences of the United States of America* 106:21377-21382.

Holmes A, Lachowicz JE, Sibley DR (2004) Phenotypic analysis of dopamine receptor knockout mice; recent insights into the functional specificity of dopamine receptor subtypes. *Neuropharmacology* 47:1117-1134.

Holmes A, Hollon TR, Gleason TC, Liu Z, Dreiling J, Sibley DR, Crawley JN (2001) Behavioral characterization of dopamine D5 receptor null mutant mice. *Behavioral neuroscience* 115:1129-1144.

Izquierdo LA, Barros DM, da Costa JC, Furini C, Zinn C, Cammarota M, Bevilaqua LR, Izquierdo I (2007) A link between role of two prefrontal areas in immediate memory and in long-term memory consolidation. *Neurobiology of learning and memory* 88:160-166.

Jastrow TR, Richfield E, Gnegy ME (1984) Quantitative autoradiography of [3H]sulpiride binding sites in rat brain. *Neuroscience letters* 51:47-53.

Kebabian JW, Calne DB (1979) Multiple receptors for dopamine. *Nature* 277:93-96.

Kelley AE, Domesick VB, Nauta WJ (1982) The amygdalostriatal projection in the rat--an anatomical study by anterograde and retrograde tracing methods. *Neuroscience* 7:615630.

Kelley AE, Andrzejewski ME, Baldwin AE, Hernandez PJ, Pratt WE (2003) Glutamatemediated plasticity in corticostriatal networks: role in adaptive motor learning. *Annals of the New York Academy of Sciences* 1003:159-168.

Kim SY, Choi KC, Chang MS, Kim MH, Kim SY, Na YS, Lee JE, Jin BK, Lee BH, Baik JH (2006) The dopamine D2 receptor regulates the development of dopaminergic neurons via extracellular signal-regulated kinase and Nurr1 activation. *The Journal of neuroscience : the official journal of the Society for Neuroscience* 26:4567-4576.

Kuhar MJC, P.R.; Lambert, P.D. (1999) Storage and Release of Catecholamines. In: Siegel GJ, Agranoff BW, Albers RW, et al., editors. Basic Neurochemistry: Molecular, Cellular and Medical Aspects. 6th edition. Philadelphia: Lippincott-Raven.

LaLumiere RT, Buen TV, McGaugh JL (2003) Post-training intra-basolateral amygdala infusions of norepinephrine enhance consolidation of memory for contextual fear conditioning. *The Journal of neuroscience : the official journal of the Society for Neuroscience* 23:6754-6758.

Lalumiere RT, Nguyen LT, McGaugh JL (2004) Post-training intrabasolateral amygdala infusions of dopamine modulate consolidation of inhibitory avoidance memory: involvement of noradrenergic and cholinergic systems. *The European journal of neuroscience* 20:2804-2810.

LaLumiere RT, Nawar EM, McGaugh JL (2005) Modulation of memory consolidation by the basolateral amygdala or nucleus accumbens shell requires concurrent dopamine receptor activation in both brain regions. *Learning & memory* (Cold Spring Harbor, NY) 12:296301.

Lashley KS (1917) The Effects of Strychnine and Caffeine Upon the Rate of Learning. *Psychobiology* Vol 1(2):141-169.

Lee SP, So CH, Rashid AJ, Varghese G, Cheng R, Lanca AJ, O'Dowd BF, George SR (2004) Dopamine D1 and D2 receptor Co-activation generates a novel phospholipase Cmediated calcium signal. *The Journal of biological chemistry* 279:35671-35678.

Levant B, Grigoriadis DE, DeSouza EB (1992) Characterization of [3H]quinpirole binding to D2-like dopamine receptors in rat brain. *The Journal of pharmacology and experimental therapeutics* 262:929-935.

Levesque D, Diaz J, Pilon C, Martres MP, Giros B, Souil E, Schott D, Morgat JL, Schwartz JC, Sokoloff P (1992) Identification, characterization, and localization of the dopamine D3 receptor in rat brain using 7-[3H]hydroxy-N,N-di-n-propyl-2-aminotetralin. *Proceedings of the National Academy of Sciences of the United States of America* 89:8155-8159.

Liao Y, Shi YW, Liu QL, Zhao H (2013) Glucocorticoid-induced enhancement of contextual fear memory consolidation in rats: Involvement of D1 receptor activity of hippocampal area CA1. *Brain research* 1524:26-33.

Lieberman JA, Bymaster FP, Meltzer HY, Deutch AY, Duncan GE, Marx CE, Aprille JR, Dwyer DS, Li XM, Mahadik SP, Duman RS, Porter JH, Modica-Napolitano JS, Newton SS, Csernansky JG (2008) Antipsychotic drugs: comparison in animal models of efficacy, neurotransmitter regulation, and neuroprotection. *Pharmacological reviews* 60:358-403.

Lindenmayer JP, Khan A (2004) Pharmacological treatment strategies for schizophrenia. *Expert review of neurotherapeutics* 4:705-723.

Linssen AM, Vuurman EF, Sambeth A, Riedel WJ (2012) Methylphenidate produces selective enhancement of declarative memory consolidation in healthy volunteers. *Psychopharmacology* 221:611-619.

Liu F, Wan Q, Pristupa ZB, Yu XM, Wang YT, Niznik HB (2000) Direct protein-protein coupling enables cross-talk between dopamine D5 and gamma-aminobutyric acid A receptors. *Nature* 403:274-280.

Liu NY, Zhang L, Wang XN, Zhang L (2006a) Dopamine receptors oppositely regulate cocaineinduced transcription factor CREB activation. Nan fang yi ke da xue xue bao = *Journal of Southern Medical University* 26:715-718.

Liu XY, Chu XP, Mao LM, Wang M, Lan HX, Li MH, Zhang GC, Parelkar NK, Fibuch EE, Haines M, Neve KA, Liu F, Xiong ZG, Wang JQ (2006b) Modulation of D2R-NR2B interactions in response to cocaine. *Neuron* 52:897-909.

Lundstrom K, Salminen M, Jalanko A, Savolainen R, Ulmanen I (1991) Cloning and characterization of human placental catechol-O-methyltransferase cDNA. *DNA and cell biology* 10:181-189.

Luttrell LM, Lefkowitz RJ (2002) The role of beta-arrestins in the termination and transduction of G-protein-coupled receptor signals. *Journal of cell science* 115:455-465.

Lyons DJ, Hellysaz A, Broberger C (2012) Prolactin regulates tuberoinfundibular dopamine neuron discharge pattern: novel feedback control mechanisms in the lactotrophic axis. *The Journal of neuroscience : the official journal of the Society for Neuroscience* 32:8074-8083.

Manago F, Castellano C, Oliverio A, Mele A, De Leonibus E (2009) Role of dopamine receptors subtypes, D1-like and D2-like, within the nucleus accumbens subregions, core and shell, on memory consolidation in the one-trial inhibitory avoidance task. *Learning & memory* (Cold Spring Harbor, NY) 16:46-52.

Manago F, Lopez S, Oliverio A, Amalric M, Mele A, De Leonibus E (2013) Interaction between the mGlu receptors 5 antagonist, MPEP, and amphetamine on memory and motor functions in mice. *Psychopharmacology* 226:541-550.

McGaugh JL (1966) Time-dependent processes in memory storage. *Science* 153:1351-1358.

McGaugh JL (1973) Drug facilitation of learning and memory. *Annual review of pharmacology* 13:229-241.

McGaugh JL (1989) Dissociating learning and performance: drug and hormone enhancement of memory storage. *Brain research bulletin* 23:339-345.

McGaugh JL, Roozendaal B (2009) Drug enhancement of memory consolidation: historical perspective and neurobiological implications. *Psychopharmacology* 202:3-14.

Mehta MA, Owen AM, Sahakian BJ, Mavaddat N, Pickard JD, Robbins TW (2000) Methylphenidate enhances working memory by modulating discrete frontal and parietal lobe regions in the human brain. *The Journal of neuroscience : the official journal of the Society for Neuroscience* 20:RC65.

Mele A, Fontana D, Pert A (1997) Alterations in striatal dopamine overflow during rotational behavior induced by amphetamine, phencyclidine, and MK-801. *Synapse* (New York, NY) 26:218-224.

Mele A, Castellano C, Felici A, Cabib S, Caccia S, Oliverio A (1996) Dopamine-N-methyl-Daspartate interactions in the modulation of locomotor activity and memory consolidation in mice. *European journal of pharmacology* 308:1-12.

Mele A, Avena M, Roullet P, De Leonibus E, Mandillo S, Sargolini F, Coccurello R, Oliverio A (2004) Nucleus accumbens dopamine receptors in the consolidation of spatial memory. *Behavioural pharmacology* 15:423-431.

Michel M, Kemenes I, Muller U, Kemenes G (2008) Different phases of long-term memory require distinct temporal patterns of PKA activity after single-trial classical conditioning. *Learning & memory* 15:694-702.

Millan MJ (2010) From the cell to the clinic: a comparative review of the partial D(2)/D(3)receptor agonist and alpha2-adrenoceptor antagonist, piribedil, in the treatment of Parkinson's disease. *Pharmacology & therapeutics* 128:229-273.

Miyamoto S, Duncan GE, Marx CE, Lieberman JA (2005) Treatments for schizophrenia: a critical review of pharmacology and mechanisms of action of antipsychotic drugs. *Molecular psychiatry* 10:79-104.

Mizoguchi K, Yuzurihara M, Ishige A, Sasaki H, Chui DH, Tabira T (2000) Chronic stress induces impairment of spatial working memory because of prefrontal dopaminergic dysfunction. *The Journal of neuroscience : the official journal of the Society for Neuroscience* 20:1568-1574.

Mogenson GJ, Jones DL, Yim CY (1980) From motivation to action: functional interface between the limbic system and the motor system. *Progress in neurobiology* 14:69-97.

Moncada D, Viola H (2007) Induction of long-term memory by exposure to novelty requires protein synthesis: evidence for a behavioral tagging. *The Journal of neuroscience : the official journal of the Society for Neuroscience* 27:7476-7481.

Monsma FJ, Jr., McVittie LD, Gerfen CR, Mahan LC, Sibley DR (1989) Multiple D2 dopamine receptors produced by alternative RNA splicing. *Nature* 342:926-929.

Monsma FJ, Jr., Mahan LC, McVittie LD, Gerfen CR, Sibley DR (1990) Molecular cloning and expression of a D1 dopamine receptor linked to adenylyl cyclase activation. *Proceedings of the National Academy of Sciences of the United States of America* 87:6723-6727.

Morris R (1984) Developments of a water-maze procedure for studying spatial learning in the rat. *Journal of neuroscience methods* 11:47-60.

Morris RG, Garrud P, Rawlins JN, O'Keefe J (1982) Place navigation impaired in rats with hippocampal lesions. *Nature* 297:681-683.

Murphy BL, Arnsten AF, Goldman-Rakic PS, Roth RH (1996) Increased dopamine turnover in the prefrontal cortex impairs spatial working memory performance in rats and monkeys. *Proceedings of the National Academy of Sciences of the United States of America* 93:1325-1329.

Murphy RB, Vodyanoy V (1984) Functional reconstitution of rat striatal dopamine agonist receptors into artificial lipid bimolecular membranes. *Biophysical journal* 45:22-23.

Nai Q, Li S, Wang SH, Liu J, Lee FJ, Frankland PW, Liu F (2010) Uncoupling the D1-Nmethyl-D-aspartate (NMDA) receptor complex promotes NMDA-dependent long-term potentiation and working memory. *Biological psychiatry* 67:246-254.

Naudon L, Dourmap N, Leroux-Nicollet I, Costentin J (1992) Kainic acid lesion of the striatum increases dopamine release but reduces 3-methoxytyramine level. *Brain research* 572:247-249.

Neve KA, Seamans JK, Trantham-Davidson H (2004) Dopamine receptor signaling. *Journal of receptor and signal transduction research* 24:165-205.

Nirenberg MJ, Chan J, Liu Y, Edwards RH, Pickel VM (1996) Ultrastructural localization of the vesicular monoamine transporter-2 in midbrain dopaminergic neurons: potential sites for somatodendritic storage and release of dopamine. *The Journal of neuroscience : the official journal of the Society for Neuroscience* 16:4135-4145.

O'Carroll CM, Martin SJ, Sandin J, Frenguelli B, Morris RG (2006) Dopaminergic modulation of the persistence of one-trial hippocampus-dependent memory. *Learning & memory* (Cold Spring Harbor, NY) 13:760-769.

O'Malley KL, Harmon S, Tang L, Todd RD (1992) The rat dopamine D4 receptor: sequence, gene structure, and demonstration of expression in the cardiovascular system. *The New biologist* 4:137-146.

O'Neill J, Pleydell-Bouverie B, Dupret D, Csicsvari J (2010) Play it again: reactivation of waking experience and memory. *Trends in neurosciences* 33:220-229.

Oak JN, Oldenhof J, Van Tol HH (2000) The dopamine D(4) receptor: one decade of research. *European journal of pharmacology* 405:303-327.

Opmeer EM, Kortekaas R, Aleman A (2010) Depression and the role of genes involved in dopamine metabolism and signalling. *Progress in neurobiology* 92:112-133.

Ouyang M, Young MB, Lestini MM, Schutsky K, Thomas SA (2012) Redundant catecholamine signaling consolidates fear memory via phospholipase C. *The Journal of neuroscience : the official journal of the Society for Neuroscience* 32:1932-1941.

Packard MG, White NM (1991) Dissociation of hippocampus and caudate nucleus memory systems by posttraining intracerebral injection of dopamine agonists. *Behavioral neuroscience* 105:295-306.

Peczely L, Ollmann T, Laszlo K, Kovacs A, Galosi R, Szabo A, Karadi Z, Lenard L (2014) Role of D1 dopamine receptors of the ventral pallidum in inhibitory avoidance learning. *Behavioural brain research* 270c:131-136.

Pei L, Lee FJ, Moszczynska A, Vukusic B, Liu F (2004) Regulation of dopamine D1 receptor function by physical interaction with the NMDA receptors. *The Journal of neuroscience : the official journal of the Society for Neuroscience* 24:1149-1158.

Pennartz CM, Ito R, Verschure PF, Battaglia FP, Robbins TW (2011) The hippocampal-striatal axis in learning, prediction and goal-directed behavior. *Trends in neurosciences* 34:548559.

Perreault ML, Hasbi A, Alijaniaram M, Fan T, Varghese G, Fletcher PJ, Seeman P, O'Dowd BF, George SR (2010) The dopamine D1-D2 receptor heteromer localizes in dynorphin/enkephalin neurons: increased high affinity state following amphetamine and in schizophrenia. *The Journal of biological chemistry* 285:36625-36634.

Pierce KL, Premont RT, Lefkowitz RJ (2002) Seven-transmembrane receptors. *Nature reviews Molecular cell biology* 3:639-650.

Probst WC, Snyder LA, Schuster DI, Brosius J, Sealfon SC (1992) Sequence alignment of the Gprotein coupled receptor superfamily. *DNA and cell biology* 11:1-20.

Puglisi-Allegra S, Cestari V, Cabib S, Castellano C (1994) Strain-dependent effects of posttraining cocaine or nomifensine on memory storage involve both D1 and D2 dopamine receptors. *Psychopharmacology* 115:157-162.

Restivo L, Vetere G, Bontempi B, Ammassari-Teule M (2009) The formation of recent and remote memory is associated with time-dependent formation of dendritic spines in the hippocampus and anterior cingulate cortex. *The Journal of neuroscience : the official journal of the Society for Neuroscience* 29:8206-8214.

Roesler R, McGaugh JL (2010) Memory Consolidation. Encyclopedia of Behavioral Neuroscience Koob, GF, LeMoal, M and Thompson, RF (Eds), 2010, volume 2, pp 206214 Oxford: Academic Press

Rondou P, Haegeman G, Van Craenenbroeck K (2010) The dopamine D4 receptor: biochemical and signalling properties. *Cellular and molecular life sciences : CMLS* 67:1971-1986.

Roozendaal B (2002) Stress and memory: opposing effects of glucocorticoids on memory consolidation and memory retrieval. Neurobiology of learning and memory 78:578-595.

Rossato JI, Radiske A, Kohler CA, Gonzalez C, Bevilaqua LR, Medina JH, Cammarota M (2013). Consolidation of object recognition memory requires simultaneous activation of dopamine D1/D5 receptors in the amygdala and medial prefrontal cortex but not in the hippocampus. *Neurobiology of learning and memory* 106:66-70.

Sahu A, Tyeryar KR, Vongtau HO, Sibley DR, Undieh AS (2009) D5 dopamine receptors are required for dopaminergic activation of phospholipase C. *Molecular pharmacology* 75:447-453.

Salamone JD, Cousins MS, Snyder BJ (1997) Behavioral functions of nucleus accumbens dopamine: empirical and conceptual problems with the anhedonia hypothesis. *Neuroscience and biobehavioral reviews* 21:341-359.

Santambrogio L, Lipartiti M, Bruni A, Dal Toso R (1993) Dopamine receptors on human T- and B-lymphocytes. *Journal of neuroimmunology* 45:113-119.

Sargolini F, Roullet P, Oliverio A, Mele A (1999) Effects of lesions to the glutamatergic afferents to the nucleus accumbens in the modulation of reactivity to spatial and nonspatial novelty in mice. *Neuroscience* 93:855-867.

Sawaguchi T, Goldman-Rakic PS (1991) D1 dopamine receptors in prefrontal cortex: involvement in working memory. *Science* 251:947-950.

Sawaguchi T, Goldman-Rakic PS (1994) The role of D1-dopamine receptor in working memory: local injections of dopamine antagonists into the prefrontal cortex of rhesus monkeys performing an oculomotor delayed-response task. *Journal of neurophysiology* 71:515528.

Schott BH, Frischknecht R, Debska-Vielhaber G, John N, Behnisch G, Duzel E, Gundelfinger ED, Seidenbecher CI (2010) Membrane-Bound Catechol-O-Methyl Transferase in Cortical Neurons and Glial Cells is Intracellularly Oriented. *Frontiers in psychiatry* 1:142.

Seeman P (2006) Targeting the dopamine D2 receptor in schizophrenia. *Expert opinion on therapeutic targets* 10:515-531.

Seeman P, Van Tol HH (1995) Deriving the therapeutic concentrations for clozapine and haloperidol: the apparent dissociation constant of a neuroleptic at the dopamine D2 or D4 receptor varies with the affinity of the competing radioligand. *European journal of pharmacology* 291:59-66.

Seeman P, Tallerico T, Ko F (2003) Dopamine displaces [3H]domperidone from high-affinity sites of the dopamine D2 receptor, but not [3H]raclopride or [3H]spiperone in isotonic medium: Implications for human positron emission tomography. *Synapse* 49:209-215.

Setlow B, McGaugh JL (1998) Sulpiride infused into the nucleus accumbens posttraining impairs memory of spatial water maze training. *Behavioral neuroscience* 112:603-610.

Setlow B, McGaugh JL (2000) D2 dopamine receptor blockade immediately post-training enhances retention in hidden and visible platform versions of the water maze. *Learning & memory* (Cold Spring Harbor, NY) 7:187-191.

Sibley DR (1999) New insights into dopaminergic receptor function using antisense and genetically altered animals. *Annual review of pharmacology and toxicology* 39:313-341.

Sidhu A (1998) Coupling of D1 and D5 dopamine receptors to multiple G proteins: Implications for understanding the diversity in receptor-G protein coupling. *Molecular neurobiology* 16:125-134.

Silva AJ, Rosahl TW, Chapman PF, Marowitz Z, Friedman E, Frankland PW, Cestari V, Cioffi D, Sudhof TC, Bourtchuladze R (1996) Impaired learning in mice with abnormal shortlived plasticity. *Current biology* : CB 6:1509-1518.

Simon H (1981) [Dopaminergic A10 neurons and frontal system (author's transl)]. *Journal de physiologie* 77:81-95.

Snyder SH, Taylor KM, Coyle JT, Meyerhoff JL (1970) The role of brain dopamine in behavioral regulation and the actions of psychotropic drugs. *The American journal of psychiatry* 127:199-207.

Soetens E, D'Hooge R, Hueting JE (1993) Amphetamine enhances human-memory consolidation. *Neuroscience letters* 161:9-12.

Soetens E, Casaer S, D'Hooge R, Hueting JE (1995) Effect of amphetamine on long-term retention of verbal material. *Psychopharmacology* 119:155-162.

Soeter M, Kindt M (2012) Stimulation of the noradrenergic system during memory formation impairs extinction learning but not the disruption of reconsolidation. *Neuropsychopharmacology : official publication of the American College of Neuropsychopharmacology* 37:1204-1215.

Sokoloff P, Giros B, Martres MP, Bouthenet ML, Schwartz JC (1990) Molecular cloning and characterization of a novel dopamine receptor (D3) as a target for neuroleptics. *Nature* 347:146-151.

Sokoloff P, Diaz J, Le Foll B, Guillin O, Leriche L, Bezard E, Gross C (2006) The dopamine D3 receptor: a therapeutic target for the treatment of neuropsychiatric disorders. *CNS & neurological disorders drug targets* 5:25-43.

Spano PF, Govoni S, Trabucchi M (1978) Studies on the pharmacological properties of dopamine receptors in various areas of the central nervous system. *Advances in biochemical psychopharmacology* 19:155-165.

Stam CJ, de Bruin JP, van Haelst AM, van der Gugten J, Kalsbeek A (1989) Influence of the mesocortical dopaminergic system on activity, food hoarding, social-agonistic behavior, and spatial delayed alternation in male rats. *Behavioral neuroscience* 103:24-35.

Stewart S, Cacucci F, Lever C (2011) Which memory task for my mouse? A systematic review of spatial memory performance in the Tg2576 Alzheimer's mouse model. *Journal of Alzheimer's disease : JAD* 26:105-126.

Sunahara RK, Guan HC, O'Dowd BF, Seeman P, Laurier LG, Ng G, George SR, Torchia J, Van Tol HH, Niznik HB (1991) Cloning of the gene for a human dopamine D5 receptor with higher affinity for dopamine than D1. *Nature* 350:614-619.

Sutherland GR, McNaughton B (2000) Memory trace reactivation in hippocampal and neocortical neuronal ensembles. *Current opinion in neurobiology* 10:180-186.

Svenningsson P, Nishi A, Fisone G, Girault JA, Nairn AC, Greengard P (2004) DARPP-32: an integrator of neurotransmission. *Annual review of pharmacology and toxicology* 44:269296.

Swanson LW (1982) The projections of the ventral tegmental area and adjacent regions: a combined fluorescent retrograde tracer and immunofluorescence study in the rat. *Brain research bulletin* 9:321-353.

Tang L, Todd RD, Heller A, O'Malley KL (1994) Pharmacological and functional characterization of D2, D3 and D4 dopamine receptors in fibroblast and dopaminergic cell lines. The *Journal of pharmacology and experimental therapeutics* 268:495-502.

Tarazi FI, Baldessarini RJ (1999) Brain dopamine D(4) receptors: basic and clinical status. *The international journal of neuropsychopharmacology / official scientific journal of the Collegium Internationale Neuropsychopharmacologicum* (CINP) 2:41-58.

Tiberi M, Jarvie KR, Silvia C, Falardeau P, Gingrich JA, Godinot N, Bertrand L, Yang-Feng TL, Fremeau RT, Jr., Caron MG (1991) Cloning, molecular characterization, and chromosomal assignment of a gene encoding a second D1 dopamine receptor subtype: differential expression pattern in rat brain compared with the D1A receptor. *Proceedings of the National Academy of Sciences of the United States of America* 88:7491-7495.

Tinsley CJ, Narduzzo KE, Ho JW, Barker GR, Brown MW, Warburton EC (2009) A role for calcium-calmodulin-dependent protein kinase II in the consolidation of visual object recognition memory. *The European journal of neuroscience* 30:1128-1139.

Trainor BC (2011) Stress responses and the mesolimbic dopamine system: social contexts and sex differences. *Hormones and behavior* 60:457-469.

Usiello A, Baik JH, Rouge-Pont F, Picetti R, Dierich A, LeMeur M, Piazza PV, Borrelli E (2000) Distinct functions of the two isoforms of dopamine D2 receptors. *Nature* 408:199203.

Vallar L, Muca C, Magni M, Albert P, Bunzow J, Meldolesi J, Civelli O (1990) Differential coupling of dopaminergic D2 receptors expressed in different cell types. Stimulation of phosphatidylinositol 4,5-bisphosphate hydrolysis in LtK- fibroblasts, hyperpolarization, and cytosolic-free Ca2+ concentration decrease in GH4C1 cells. *The Journal of biological chemistry* 265:10320-10326.

Van Tol HH, Bunzow JR, Guan HC, Sunahara RK, Seeman P, Niznik HB, Civelli O (1991) Cloning of the gene for a human dopamine D4 receptor with high affinity for the antipsychotic clozapine. *Nature* 350:610-614.

Varon J (2008) Treatment of acute severe hypertension: current and newer agents. *Drugs* 68:283297.

Ventura R, Cabib S, Puglisi-Allegra S (2001) Opposite genotype-dependent mesocorticolimbic dopamine response to stress. *Neuroscience* 104:627-631.

Verma A, Moghaddam B (1996) NMDA receptor antagonists impair prefrontal cortex function as assessed via spatial delayed alternation performance in rats: modulation by dopamine. *The Journal of neuroscience : the official journal of the Society for Neuroscience* 16:373379.

Volkow ND, Wang GJ, Fowler JS, Gatley SJ, Logan J, Ding YS, Hitzemann R, Pappas N (1998) Dopamine transporter occupancies in the human brain induced by therapeutic doses of oral methylphenidate. *The American journal of psychiatry* 155:1325-1331.

Wang H, Feng R, Phillip Wang L, Li F, Cao X, Tsien JZ (2008) CaMKII activation state underlies synaptic labile phase of LTP and short-term memory formation. *Current biology* : CB 18:1546-1554.

Wang SH, Redondo RL, Morris RG (2010) Relevance of synaptic tagging and capture to the persistence of long-term potentiation and everyday spatial memory. *Proceedings of the National Academy of Sciences of the United States of America* 107:19537-19542.

Wiig KA, Whitlock JR, Epstein MH, Carpenter RL, Bear MF (2009) The levo enantiomer of amphetamine increases memory consolidation and gene expression in the hippocampus without producing locomotor stimulation. *Neurobiology of learning and memory* 92:106113.

Williams GV, Goldman-Rakic PS (1995) Modulation of memory fields by dopamine D1 receptors in prefrontal cortex. *Nature* 376:572-575.

Wise RA, Rompre PP (1989) Brain dopamine and reward. *Annual review of psychology* 40:191225.

Wittmann BC, Schott BH, Guderian S, Frey JU, Heinze HJ, Duzel E (2005) Reward-related FMRI activation of dopaminergic midbrain is associated with enhanced hippocampusdependent long-term memory formation. *Neuron* 45:459-467.

Xu TX, Yao WD (2010) D1 and D2 dopamine receptors in separate circuits cooperate to drive associative long-term potentiation in the prefrontal cortex. *Proceedings of the National Academy of Sciences of the United States of America* 107:16366-16371.

Yang CR, Mogenson GJ (1987) Hippocampal signal transmission to the pedunculopontine nucleus and its regulation by dopamine D2 receptors in the nucleus accumbens: an electrophysiological and behavioural study. *Neuroscience* 23:1041-1055.

Yim CY, Mogenson GJ (1983) Response of ventral pallidal neurons to amygdala stimulation and its modulation by dopamine projections to nucleus accumbens. *Journal of neurophysiology* 50:148-161.

Zeeuws I, Soetens E (2007) Verbal memory performance improved via an acute administration of D-amphetamine. *Human psychopharmacology* 22:279-287.

Zhang K, Grady CJ, Tsapakis EM, Andersen SL, Tarazi FI, Baldessarini RJ (2004a) Regulation of working memory by dopamine D4 receptor in rats. *Neuropsychopharmacology : official publication of the American College of Neuropsychopharmacology* 29:16481655.

Zhang L, Lou D, Jiao H, Zhang D, Wang X, Xia Y, Zhang J, Xu M (2004b) Cocaine-induced intracellular signaling and gene expression are oppositely regulated by the dopamine D1 and D3 receptors. *The Journal of neuroscience: the official journal of the Society for Neuroscience* 24:3344-3354.

Zhou QY, Grandy DK, Thambi L, Kushner JA, Van Tol HH, Cone R, Pribnow D, Salon J, Bunzow JR, Civelli O (1990) Cloning and expression of human and rat D1 dopamine receptors. *Nature* 347:76-80.

Zintzaras E, Kitsios GD, Papathanasiou AA, Konitsiotis S, Miligkos M, Rodopoulou P, Hadjigeorgiou GM (2010) Randomized trials of dopamine agonists in restless legs syndrome: a systematic review, quality assessment, and meta-analysis. *Clinical therapeutics* 32:221-237.

In: Memory Consolidation
Editors: Manabu Sakakibara and Etsuro Ito
ISBN: 978-1-63482-596-2

Chapter 11

SLEEP'S ROLE IN THE CONSOLIDATION AND INTEGRATION OF DECLARATIVE MEMORIES

Enmanuelle Pardilla-Delgado, Stephen M. Mattingly, Sara E. Alger, Alexis M. Chambers, Tony J. Cunningham and Jessica D. Payne*
University of Notre Dame, Department of Psychology,
Notre Dame, IN, US

ABSTRACT

Sleep is a multi-stage, active state that benefits the consolidation of different types of memory in both animals and humans. However, this beneficial effect is not all encompasing. Many factors can contribute to the role sleep plays in memory consolidation, such as the type of stimuli and task direction used, timing and composition of sleep, retrieval methods, and subject population. A growing body of research demonstrates that memory is facilitated by post-learning sleep, with slow wave sleep (SWS) often considered to be more advantageous for emotionally neutral information and semantic memory, and rapid eye movement (REM) sleep more beneficial for emotional memory. This relationship, however, is far more complex than this simple dichotomy suggests. More specifically, distinct electrophysiological markers of each sleep stage often relate to the processing of different types of memory. During SWS, reactivation of hippocampus-dependent memory traces results in a redistribution of these traces to the neocortex through systems consolidation. During subsequent REM sleep, increased immediate-early gene plasticity aids synaptic consolidation in cortical sites, integrating new information into existing networks. The idea of consolidation has recently evolved from one in which newly learned information is solidified in its original form to one in which only the most salient and useful aspects of memories are selectively maintained (e.g., emotional salience, future relevance). This chapter will present contemporary behavioral, correlational, and physiological evidence for a role of sleep in the consolidation and integration of declarative memories, following the discussion of

* Corresponding author. Address: University of Notre Dame, Department of Psychology, Haggar Hall, Room 122B, Notre Dame, IN 46556. E-mail: jpayne7@nd.edu.

theoretical frameworks that might help understand these findings. Throughout the chapter we will discuss under what conditions sleep does and does not benefit consolidation processes.

INTRODUCTION

Sleep has repeatedly been found to benefit memory formation. Since the studies by Ebbinghaus in the late 19th century, which resulted in diminished forgetting of nonsense syllables after sleep, a vast amount of research has been accomplished to investigate how sleep-dependent processes aid memory consolidation (Ackerman & Rasch, 2014; Stickgold & Walker, 2013).

This chapter will discuss experimental evidence that points to a clear role of sleep in the formation of declarative memory, defined as memories for facts and experiences. To begin, a brief introduction into what physiological and neurochemical processes compose sleep will be presented. This will be followed by a discussion of studies and theories regarding the role of sleep in the consolidation of neutral and emotionally salient information, finishing with evidence supporting sleep-based assistance of the re-organization of this information and its integration into existing knowledge.

1. INTRODUCTION TO SLEEP PHYSIOLOGY AND NEUROCHEMISTRY AND THEIR RELATIONSHIP TO MEMORY FORMATION

Sleep is not a period of static recuperation for the brain. Rather, the sleeping brain is dynamically and systematically cycling through several distinct stages, each with different neurochemical and electrophysiological features. Broadly, sleep is dichotomized into two types; rapid eye movement (REM) sleep, named due to characteristic eye movements during this stage, and non-REM (NREM) sleep. NREM can be broken down into Stages 1 through 4 as defined by Rechtschaffen and Kales (1968), or N1 through N3 as more recently redefined by the American Academy of Sleep Medicine (2007).

The deepest sleep stage is N3 (or Stages 3 and 4, combined), more generally known as slow wave sleep (SWS), and in this stage is difficult to rouse people and leaves the greatest sleep inertia. In a standard night of sleep for a healthy population, a regular ~90-minute cycle begins with Stage 1, descends through stages 2, 3, and 4, and rapidly ascends from 4 to 1 and concludes with REM sleep. However, the distribution of these stages is not uniform throughout a regular night of sleep; the first half of the night is dominated by SWS with little REM sleep. With each successive ~90 minute cycle, SWS decreases and REM sleep increases. By the last ~90 minute cycle, REM sleep nearly doubles from what is found in first cycles (Figure 1; Payne, 2011).

The following is a more thorough explanation of the most relevant stages to this chapter: Stage 2, SWS, and REM sleep. Their neurochemical and electrophysiological features are critical to understanding how sleep modulates the consolidation of declarative memories.

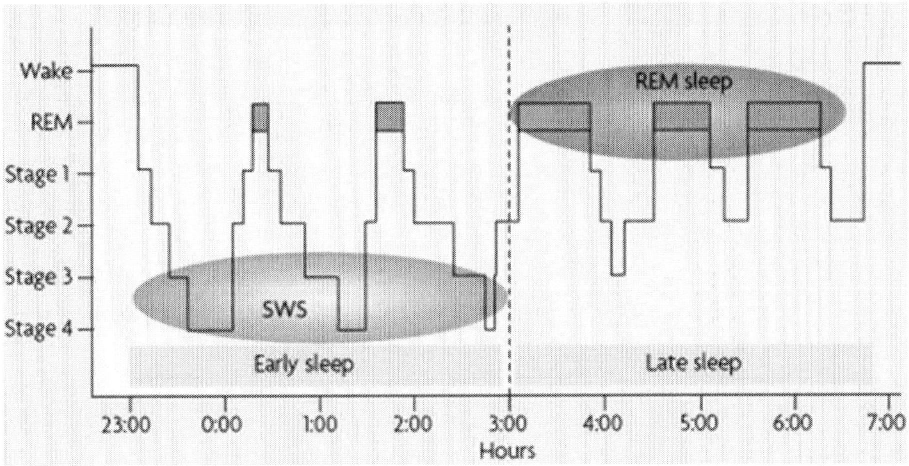

Figure 1. During the first part of the night, sleep is dominated by slow wave sleep (SWS), while the second part of the night is dominated by rapid-eye movement (REM) sleep (from Payne, 2010).

1.1. NREM Stage 2 Sleep

In general, Stage 1 sleep is a brief, transitional period to Stage 2. As the brain falls asleep, waking alpha frequency (8-13 Hz) transitions to a slower, 5-8 Hz activity. Within Stage 2, there are distinctive physiological markers including sleep spindles and K-complexes. Sleep spindles are defined by short, high frequency bursts of sigma activity (12-16 Hz), while K-complexes are brief negative peaks of high voltage/high amplitude followed by a slower positive complex, and another negative peak. While transitioning into Stage 2 sleep, acetylcholine (ACh), norepinephrine (NE), and serotonin levels decrease compared to wake and are similar to levels seen during SWS (for review see Stickgold et al., 2001; Smith et al., 2004).

Sleep spindles during Stage 2 come in two types, divided by originating region and frequency. Slow spindles (12-14 Hz) originate in frontal locations, whereas fast spindles (14-16 Hz) generate in centro-parietal regions (Sterman et al., 1978; Tanguay et al., 1975; Zeitlhofer et al., 2003). While the purpose of spindles is not fully understood, spindles are thought to contribute to the integration of new and old information, (Tamminen et al., 2010; 2012; see *section 4.2*) and in reactivating memories during sleep for synaptic modifications (Rosanova & Ulrich, 2005). Spindle quantity and density both increase following intensive declarative memory training, and have been shown to positively correlate with memory performance in several tasks (Gais et al., 2002; Clemens et al., 2005; Schabus et al., 2004; see *section 2.2*)

1.2. Slow-Wave Sleep

As the brain descends into SWS, brain activity shifts into delta waves, which are low-frequency, high-amplitude oscillations in the 0.5-4 Hz range. There is an overall decrease of regional cerebral blood flow during SWS compared to wakefulness in areas including limbic, midbrain, and cortical areas, such as the dorsolateral prefrontal cortex (Braun et al., 1997). During delta waves, fast neural oscillations (150-250 Hz) in the hippocampus occur during the up-states of slow oscillations, called sharp-wave ripples (Steriade et al., 1993; De Gennaro & Ferrara, 2003; Battaglia et al., 2004; Mölle et al., 2006). These sharp-wave ripples are temporally aligned with sleep spindles (Siapas & Wilson, 1998; Sirota et al., 2003), and constitute a primary component in a memory consolidation theory discussed in *section 2.3* (Buzsaki et al., 1983; Buzsaki, 1989).

	ACTIVE WAKE	QUIET WAKE	SWS	REM SLEEP
ACh	++	+	−	+++
NE 5-HT	++	+	+	−

Figure 2. Representation of the relative concentrations of neurotransmitters across sleep-related brain states. ACh = acetylcholine, NE = norepinehrine, 5-HT = serotonin, SWS = slow-wave sleep, REM = rapid eye movement.

In addition to delta activity, neurochemical changes during SWS include low firing levels of serotonergic, noradrenergic, and cholinergic connections, the latter occurring at their lowest (during active wakefulness ACh levels are tripled; see Figure 2). NE and ACh are critical neurotransmitters for the formation of declarative memories due to their direct influence on hippocampal activity. During NREM sleep (including SWS), NE is released from the locus coeruleus (LC), which moderates hippocampal and neocortical activity (Aston-Jones, 2004). In addition, LC activity is related to sleep spindles and slow oscillations during SWS (Gais et al., 2011). Given that NE from the LC influences the hippocampus, a critical memory structure, and is related to sleep spindles and slow oscillations thought to assimilate old and new information, it is likely that this key structure plays an important role in SWS's effects on declarative memory. Indeed, after a period of learning, NE and LC activity have been found to increase (Eschenko & Sara, 2008; Sara, 2009). Furthermore, by pharmacologically blocking NE activity during NREM sleep (but not during wakefulness), researchers were able to inhibit sleep-dependent memory improvement in an odor recognition task, while a NE reuptake inhibitor increased memory facilitation (Gais et al., 2011).

Acetylcholine's suppressive role in glutamate receptors in the hippocampus assists encoding during wakefulness by reducing communication between the hippocampus and long-term memory storage in the neocortex. The net effect is a reduction of interference from old, competing memories through suppression of reactivation in the hippocampus. During SWS, however, ACh is at its lowest levels, enabling hippocampal feedback to the neocortex by releasing glutamatergic connections from suppression, supporting long-term memory processing and systems consolidation of declarative information in humans (Hounsgaard, 1978; Valentino & Dingledine, 1981; Rovira et al., 1983; Hasselmo & Bower, 1992; Herreras et al., 1988; Hasselmo, 1999; Hasselmo & McGaughy, 2004).

1.3. Rapid Eye Movement Sleep

REM sleep brain activity is defined by low-amplitude, high-frequency oscillations. Additionally, there is a drastic relaxation of muscular tone (tonic event of REM sleep), as well as occasional muscle twitches and periodic rapid eye movements (phasic events of REM sleep; Carskadon & Dement, 1989). Global neural activity is equivalent to or greater than wakefulness, especially in the basal ganglia, brainstem, hippocampus, nearby limbic areas, and some cortical areas.

There is growing evidence that suggests a role of REM sleep in selectively assisting emotional declarative memory (see *section 3*). During REM sleep, there is increased activity and connectivity between critical emotional and memory centers of the brain, including the amygdala, hippocampus and cortical areas like the medial PFC (Braun et al., 1997; Maquet et al., 1996). This increased connectivity and activity is an ideal candidate for emotional memory consolidation. Considering the neurophysiology of the brain during REM sleep, NE synapses are highly inactive and cholinergic activity is about 60% lower than during wakefulness (though, see Vazquez & Baghdoyan, 2001). Conversely, in the hippocampus and parahippocampal areas, ACh levels are elevated to those observed during the waking state (Hasselmo, 1999). High cholinergic activity suppresses glutamatergic receptors, disrupting the communication between the hippocampus and the neocortex (Hasselmo, 1999), possibly aiding cortico-cortico interactions (Payne et al., 2004; Payne, 2011). Increases in ACh levels have also been related to the facilitation of emotional memory during REM sleep since cholinergic activation in the amygdala has been associated with memory enhancing effects (McGaugh, 2004).

2. The Role of Sleep on Declarative Memory

Here we provide evidence that sleep improves declarative memory consolidation across full nights of sleep, partial nights of sleep, and naps, as well as evidence that sleep deprivation impairs declarative memory formation. Given the same amount of time between learning and test, delays including sleep outperform delays spent awake, either by preserving memory or by actively enhancing memory.

2.1. The Role of Sleep on Declarative Memory: Behavioral Evidence

Payne and colleagues (2012a) had participants encode semantically unrelated word pairs in the morning or the evening and tested them 12 hours later after either a day filled with wakefulness or a night of sleep, respectively. They found that memory for the semantically unrelated word pairs, as measured by a cued recall test, was less eroded after a period of sleep compared to wake. Additionally, another group learned the same word pairs and were given a cued recall test 24 hours later. While both groups included equal amounts of wake and sleep, one group slept immediately following learning, while another slept after a wake-filled delay (i.e., waited ~16 hours to go to bed). For the immediate sleep group, memory was again better preserved compared to the delayed sleep group. As another example, Talamini and colleagues

(2008) trained participants on face-location pairings over a 24-hour delay including immediate or delayed sleep. Like Payne and colleagues (2012a), they reported that immediate sleep reduced forgetting occurring during a subsequent wake period compared to sleeping after a waking delay.

In addition to preserving memory, there is some evidence that sleep can restore previously degraded memories. For instance, Backhaus and colleagues (2007) entrained children on word pairs and tested at 12 and 24 hours. One group learned in the evening before bed (immediate sleep) or in the morning (delayed sleep). When comparing the 12-hour delay between groups, children who slept first had better memory than children who remained awake. When comparing the 12-hour delay to the 24-hour delay within groups, participants who slept first did not show memory improvement at 24 hours relative to 12 hours, while children who slept later after time spent awake showed improved memory at 24 hours compared to 12 hours, demonstrating sleep-based memory recovery. As another example, Fenn and colleagues (2003) had participants listen to synthetic speech then spell out the words (pre-test). Next, participants were trained on a novel set of synthetic speech words paired with correct word spellings. Finally, learning was assessed by presenting yet another novel synthetic speech word list and having participants spell the words (post-test) and comparing this performance to baseline performance. Participants who stayed awake for 12 hours between pre- and post- test had worse performance gains compared to an immediate test control condition, while those who slept between pre- and post- test had similar performance levels compared to the immediate test control condition. Interestingly, when participants who stayed awake between pre-and post- test were allowed a night of sleep and were issued a second post-test (24 hours after pre-test), their performance improved to be similar to the immediate test control condition, demonstrating a sleep-based memory recovery effect.

Naps have also been shown to improve declarative memory formation, reducing the potential for time-of-day confounds by allowing all subjects to be trained and tested at the same times. Tucker and colleagues (2006) had participants learn a verbal paired associates task and take a baseline memory test for all items. Following encoding, participants either napped for approximately an hour or remained awake. Six hours after encoding, participants were tested via cued recall on the learned word pairs a second time for all items. Participants who napped had better overall memory and greater improvement from baseline compared to those who did not nap. As another example, Lahl et al. (2008) had participants encode a word list and either remain awake, nap for 6 minutes, or nap for 35 minutes on average, before a recall memory test was administered an hour after encoding. They found that a nap as short as *6 minutes* improved memory compared to wake, and that a longer nap improved memory more than a shorter nap. Following on the ideas of Lahl et al. (2008), Alger and colleagues (2012) trained participants in a bimodal paired-associates task in which participants intentionally learned relationships between words that were associated with easily recognizable sounds. After training there was a 90-minute retention period, and participants were split into 3 experimental groups: wake, 10-minute nap, or 60-minute nap opportunity. Following the delay, participants were tested via audio cued-recall on a subset of the trained bimodal pairs to assess short-term memory. Next, all participants underwent an interference task in which a subset of the previously learned auditory stimuli was paired with novel words. After a 10-minute delay, this subset of the initially encoded paired associates was tested via audio cued-recall for both the original and interference paired word to assess the degree to which the newly learned words interfered with the originally paired words. Finally, participants returned to the lab a week later and were tested on the final subset of initially encoded pairs via cued recall to measure long-term retention. Alger et al. reported that napping for approximately 60 minutes improved memory for word-sound associations, and napping for 10 minutes marginally improved memory for associations, relative to a wake control, during the short-term memory test, essentially replicating the findings of Lahl et al. (2008). However, the longer nap, during which participants obtained SWS, maintained memory performance in the face of interference, while the shorter nap, containing only Stages 1 and 2 sleep, and the wake control conditions demonstrated degraded memory performance. For long-term memory, while all groups demonstrated degraded memory over a weeklong delay, a 60-minute nap resulted in less degradation compared to the 10-minute nap and wake conditions, who performed similarly to one another. Thus, while naps as short as 6-10 minutes can provide measurable improvements to short-term memory, longer naps containing deeper sleep stages may be needed to create long-term memory effects.

While sleep has been shown to improve memory, sleep deprivation has been shown to impair declarative memory formation. Gais and colleagues (2006) trained participants in English-German word pairs and administered an English word cued recall memory test immediately after encoding, and 48 hours later. Participants underwent 2 experimental conditions with counterbalanced order; sleep deprivation during the first night immediately followed by recovery sleep during the morning and recovery sleep the second night before retest, and two nights of uninterrupted sleep. Importantly, participants slept and remained awake across 48 hours in similar amounts. When participants were sleep deprived after learning they performed more poorly compared to when they were allowed to sleep naturally, demonstrating that sleep deprivation can weaken declarative memory relative to undisturbed sleep, even if they get a similar amount of sleep.

In a study by Plihal & Born (1997), participants learned word pairs before sleeping for the first half of the night (SWS-rich sleep) on one night and before sleeping the second half of the night (REM-rich sleep) a week later, with sleeping order counterbalanced between participants. On one night, participants encoded stimuli before bed and slept until woken up 3 hours later for a memory test and were allowed to sleep after, while the other night they slept 3 hours, woke up and encoded stimuli, and slept 3 more hours in the second part of the night before being woken up for a memory test. Another group underwent the same learning, but stayed awake between learning and test in the first half of the night in one night and the second half of the night of another night, order counterbalanced. This split-night procedure takes advantage of the natural sleep cycle (see *section 1.1*) resulting in a higher percentage of SWS in the early-night condition and a higher percentage of REM sleep in the late-night condition, while both conditions achieve equal amounts of Stages 1 and 2. Memory on the paired-associate task was improved over the wake condition only when the participants slept in the first, SWS-rich part of the night, but not when the same participants slept in the second, REM sleep heavy part of the night, suggesting SWS, not REM sleep, is responsible for sleep-based declarative memory benefits. In another, similar study, Plihal & Born (1999) trained participants on an object placement task (spatial-dependent memory) using the same split-night procedure. Again, they report that participants had better retention when they slept in the SWS-rich, first part of the night compared to when they slept in the later, REM-sleep rich part of the night. In agreement with these results, Schabus et al. (2005) found that only nappers who obtained SWS during an hour-long nap opportunity showed improvement in a paired associates task, compared to those who only had Stage 1 and 2.

2.2. The Role of Sleep in Declarative Memory: Correlational and Physiological Evidence

In addition to the observed behavioral effects of sleep on declarative memory formation, there is strong physiological evidence to support an active role of sleep in declarative memory consolidation. Several studies have found that specific sleep stages, generally slow wave sleep (SWS) and SWS physiological features, relate to memory modification. To demonstrate the role of SWS in declarative memory processing, Marshall and colleagues (2004; 2006) used transcranial direct current stimulation to increase slow-wave activity (SWA) after learning word pairs during sleep or during wake. They report that memory was improved for those who received transcranial direct current stimulation during NREM sleep (mostly SWS) that increased SWA, but no change in memory was found for the wake control condition, indicating that SWA during the state of sleep, not just SWA alone, is likely driving this memory-improving effect. Van Der Werf and colleagues (2009) instead suppressed SWS using auditory tones during sleep after encoding pictures. They reported impaired memory for those who obtained less SWS due to auditory suppression. Considering the experimental evidence that increasing SWS improves memory and decreasing SWS impairs memory, it is reasonable to assert that SWS plays a role in sleep's benefit to declarative memory formation.

However, while SWS may be a key contributor to declarative memory consolidation, Stage 2 sleep may also contribute. Clemens and colleagues (2005) measured sleep spindle density (i.e., amount of 12-16Hz bursts within Stage 2 per 30s epoch) during a baseline (no-task) night and after an intensive learning task, and found that spindle density increased after intensive learning compared to the baseline night. Further, this increase in spindle density predicted improved memory performance. Similar studies recorded sleep spindle density after a non-learning control task and an intensive learning control and reported increased spindle density after intensive learning that correlated with memory performance (Gais et al., 2002; Schabus et al., 2004). In another study, Mednick and colleagues (2013) used zolpidem to pharmacologically manipulate sleep. Zolpidem is a short-acting agonist for the neurotransmitter gamma-aminobutyric acid (GABA), responsible for inhibiting neural activity and able to increase sleep spindles and decrease REM sleep during a morning nap. Compared to a regular sleep with placebo condition, zolpidem naps, with increased spindle activity, resulted in improved memory for a word-pair associates task. In another study, Mander and colleagues (2011) trained and tested participants on face-name associations at two sessions 8 hours apart. Between session 1 and 2, one group was provided a 100-minute nap opportunity, while the other was kept awake. They report that participants who remained awake between sessions had lower memory performance comparing session 2 to session 1, while participants who napped had a non-significant increase in memory performance comparing session 2 to session 1. Further, the improvement between session 1 and session 2 for the nap group correlated with Stage 2 NREM sleep and with Stage 2 sleep spindles, supporting a role for sleep spindles in the formation of declarative memory.

Examinations of physiological evidence with neuroimaging methods also support a link between sleep, particularly SWS, and declarative memory. Peigneux and colleagues (2004) had participants learn a virtual navigation task while in an fMRI scanner. They report that regions important for navigating the task, including the right hippocampus and parahippocampal region, showed increased activation during the task and during subsequent NREM sleep, especially during SWS. Not only was there an increase in regional cerebral blood flow during sleep, this increased activity over a period of sleep correlated with better memory performance after sleep. When subjects were not presented with this task before sleep, SWS activity did not show additional activity in the right hippocampus and parahippocampal region, suggesting that the additional SWS activity in those who navigated the task was resultant from learning.

Orban et al. (2006) trained participants on a virtual town in an fMRI scanner, and either sleep deprived participants or allowed them to sleep naturally after encoding. Three days later (two nights of recovery sleep for the sleep deprived) memory was tested in the scanner. They reported that, while participants did not demonstrate behavioral memory differences, those who slept during the first night had differential activation compared to sleep deprived participants. Participants who slept had lower activation of the hippocampus and increased activity in the caudate nucleus, middle cingulate cortex, precuneus, supplementary parietal lobule, and temporal and frontal cortices compared to sleep deprived participants. This suggests that sleep may shift the neural organization of the brain from the hippocampus to a more response-based strategy. In contrast, sleep deprivation may impede this shift, creating greater demands for cognitive resources and making sleep-deprived participants rely more heavily on the hippocampus.

2.3. Theoretical Explanations for Evidence of a Role of Sleep in Declarative Memory Consolidation

Current theories that attempt to explain how sleep benefits declarative memory utilize neurochemical and electrophysiological properties of SWS and fall under two broad categories. The two-stage theory of consolidation (Buzsaki et al., 1983; Buzsaki, 1989; McClelland et al., 1995) suggests that waking experience is initially encoded and replayed during sleep to strengthen neural connections, while the synaptic downscaling theory suggests that global neural activity is suppressed during sleep, allowing only strongly encoded information to be preserved (Tononi & Cirelli, 2003; 2006).

The two-stage theory of consolidation (Buzsaki et al., 1983; Buzsaki, 1989; McClelland et al., 1995) proposes that during the first stage, new information is rapidly encoded through waking interaction with the environment. Sensory receptors (except olfactory) pass information through the thalamus to the entorhinal cortex and input layers of the CA3 region of the hippocampus directly and alternately through the dentate gyrus (Chrobak & Buzsaki, 1994). High levels of ACh during wake promote neuronal excitability, which can increase network connectivity (Hasselmo, 1999). During sleep (especially SWS), sharp-wave ripples assist in the reactivation of neuronal networks that were most recently fired during waking, namely those representing declarative information (Pavlides & Winson, 1989; Wilson & McNaughton, 1994; Skaggs & McNaughton, 1996; Kudrimoti et al., 1999; Nadasdy et al., 1999; Maquet et al., 2000; Louie & Wilson, 2001; Hoffman & McNaughton, 2002; Ji & Wilson, 2007). Evidence supporting the role of sharp-wave ripples in memory formation can be taken from animal models. Eschenko and colleagues (2008) found that sharp-wave ripple activity (frequency and density) in the rat hippocampus and PFC increased after encoding in an olfactory discrimination task, when compared to baseline levels. Furthermore, rats that learned to associate scent with food had drastically more ripple activity compared to rats that did not show associative learning. Additionally, Ramadan et al. (2009) showed that sharp-wave ripples during SWS also increased after rats were trained in a spatial discrimination task. Further, the increase in ripples correlated with successful learning, supporting the idea that ripples might have a causal role in memory formation (O'Neill et al., 2010).

In an elegant experimental design of the two-process hypothesis, Wilson & McNaughton (1994) examined neuronal firing patterns of place cells in rat hippocampi during waking exploration of an environment and during subsequent sleep. They observed a 'replay' of neural activity seen during wake in a time-compressed manner during SWS, providing evidence that waking experience was re-activated during SWS (however, see Hennevin et al., 2007, for a critique). This replay is thought to lead to long-term potentiation (LTP), the predominant possible mechanistic explanation for synaptic consolidation within a network (Sirota et al., 2003; Brehens et al., 2005; Whitlock et al., 2006; Diba & Buzsaki, 2007). LTP is a complex process, essentially leading to gene transcription and protein synthesis, which increases surface receptors, promotes new terminal branches and new synapses of the pre-synaptic neuron, and increases dendritic growth and channel conductance in the post-synaptic neuron (Sweatt, 1999; LeDoux, 2002). All of these changes work together to link pre- and post- synaptic cell firing, increasing connectivity in subsequent activations (Bliss & Lomo, 1973), similar to the Hebbian learning adage 'cells that fire together, wire together' (Hebb, 1949).

This two-stage theory also provides some support for the previously mentioned Stage 2 declarative memory benefits (e.g., Clements et al., 2005; Mander et al., 2011; Mednick et al., 2013). Spindles are temporally correlated with sharp-wave ripples (Siapas & Wilson, 1998; Sirota et al., 2003) and slow oscillations (Steriade & Timofeev, 2003, Mölle et al., 2009), and may be considered plastic events (Rosanova & Ulrich, 2005). Given the physiological relationship between spindles and slow wave activity, it is not unexpected that both may support declarative memory.

Another theory, "synaptic homeostasis" is proposed by Tononi & Cirelli (Tononi & Cirelli, 2003; 2006), in which SWS primarily serves a metabolic energy conservation role, and memory consolidation is incidentally benefitted. As neurons fire throughout waking hours (with focused learning especially increasing synaptic activity), there is a depletion of neuronal resources. This depletion increases neural thresholds for firing. Concurrently, time spent awake increases the homeostatic drive for SWS, and initial sleep is SWS heavy, with decreasing amounts of SWS as the night continues. The synaptic homeostasis theory suggests that SWS, in addition to replenishing neural resources and removing waste products (Xie et al., 2013), lowers neural firing thresholds to baseline, restoring plasticity and the associated ability to learn new information. This downscaling results from membrane potential changes due to alternating synchronous hyper- and de- polarization (i.e., slow oscillations) across cortical structures during SWS. In addition to widespread slow wave activity, there is also regionally specific, 'local' slow wave activity. This local activity correlates with intensive learning regional activation, allowing SWS to return neural activity to baseline both locally and globally (Huber et al., 2004).

While synaptic strength is downscaled across the cortex, strong synaptic connections (such as those involved in a learned memory trace) remain. Thus, synapses involved with mundane experience might be reduced below threshold after downscaling, while stronger synaptic connections are preserved, increasing the signal to noise ratio, allowing them to 'stand out from the crowd'.

Thus, according to synaptic homeostasis theory, the main purpose of SWS is to reset synaptic weights to baseline, and the improvement of the memory signal relative to noise indirectly results in improved memory. However, this signal to noise benefit may only be associated with cortical areas associated with the memory trace, as the hippocampus seems to maintain firing rate over time (Buzsaki et al. 2002).

While both theories can explain current observed memory improvements after sleep, it is important to realize that these theories are not mutually exclusive. It is reasonable to suggest that both processes can co-occur, with hippocampal replay strengthening neo-cortical memory traces of important information as global downscaling resets weaker memory traces.

2.4. The Role of Sleep in Declarative Memory Consolidation Is Not Absolute

While data supporting the enhancing effect of sleep on the consolidation and retrieval of all kinds of memory is prolific, there are instances in which sleep does not appear to actively augment memory performance. Initial studies demonstrating a "sleep effect" on memory suggested that sleep played a more passive role, protecting the veridicality of memories only by reducing the amount of interference experienced by the observer (Jenkins & Dallenbach, 1924). Subsequent investigation led to the development of designs that suggest a more complex picture in which sleep may play a more active role in memory processing (Benson & Feinberg, 1977; Grosvenor & Lack, 1984; Payne et al., 2008; Scullin and McDaniel, 2010). However, our understanding of exactly what kinds of memory are benefited by sleep and the best methodological practices to unearth these effects remain unclear.

For instance, we present three studies that investigated the retention of serial lists of nonsense syllables across a variety of different delays in which participants were allowed to sleep immediately after learning or had to remain awake after learning. Benson and Feinberg (1975) found that the sleep group performed better than the wake group after an 8-hour delay, but when tested again at 24 hours the groups performed similarly. Harlow et al. (1971) reported that the sleep group outperformed the wake group at both 8 and 24 hours. Richardson and Gough (1963) found that the sleep-first group and the wake-first group performed similarly at testing 24 and 48 hours after initial encoding, but those that were allowed to sleep immediately after testing performed significantly better than the wake-first group six days after initial encoding (they did not have an 8 hour testing session). As is evident here, despite designing similar studies with the intention of investigating the effect of sleep on the same type of neutral, declarative memory, very different results were produced making interpretation of the exact role of sleep in memory consolidation difficult.

Another study investigating the effect of sleep on declarative memory found that sleep only benefited memory for neutral word pairs compared to equivalent delays of daytime or nocturnal wakefulness if the participants expected or suspected subsequent testing (Wilhelm et al., 2011). Sleep did not benefit memory for subjects that did not anticipate a need for the information at a later time. The authors suggest that the expectation instruction manipulated the future relevance of the information, and that the sleep-dependent memory processing is able to explicitly consolidate information needed to guide future behavior. Using the split-night procedure, Plihal and Born (1997) found that participants showed significant improvement on recall of a paired-associates word list after early (SWS-rich) sleep (see *section 2.2 for details*). Critically, when tested on a procedural memory task the effects were reversed and participants performed better after the REM-rich sleep period compared to the earlier SWS-filled delay. The results from this study suggest that both the type of memory and the type of sleep may be critical factors when investigating the role of sleep on memory consolidation.

3. The Role of Sleep on Emotional Memory

Emotion is a potent contributor to memory. Emotion is generally considered on two dimensions: valence, or how positive or negative the stimulus is, and arousal, to what degree does the stimulus excite or agitate (Kensinger, 2004). Both dimensions can independently and additively contribute to memory formation. In general, the amygdala is considered a critical emotional brain structure, and it is intimately connected with the hippocampus, a critical memory structure (Phelps, 2004). Emotion is a useful and adaptive memory tag that helps distinguish important events that might require additional mental resources and attention, such as great successes or dangerous situations (Öhman, 2005). Emotional arousal has been demonstrated to benefit memory for a variety of stimuli types over neutral stimuli, including negative pictures (Hamann et al., 1999; Kensinger, 2009) and words (Kensinger & Corkin, 2003) without sleep. However, sleep can improve emotional memory above and beyond wake-filled delays.

3.1. The Role of Sleep on Emotional Memory: Behavioral Evidence

As an example of sleep improving emotional memory far more than sleep's improvement of neutral memory, Hu and colleagues (2006) had participants encode a mix of negative and neutral pictures and administered a subsequent memory test after a 12-hour delay on two separate occasions. In counterbalanced order separated by 1 week, the 12-hour delay included a night of sleep or daytime wakefulness. After sleep, participants considered emotional pictures to be more familiar than neutral pictures during a recognition memory test. For the wake group, there was no difference in familiarity between negative and neutral stimuli. In addition to improving memory for emotionally salient scenes, sleep has been shown to selectively improve memory for just the emotional components of scenes, at the expense of memory for the neutral contextual details. Payne and colleagues (2008) trained participants on negative and neutral objects placed within plausible neutral backgrounds (e.g., a snake in the woods, a taxi in the street) and had them undergo a delay including night time sleep or daytime wakefulness and a subsequent recognition memory test of objects and scenes presented separately. While both groups showed better memory for negative objects relative to neutral objects, and impaired memory for backgrounds associated with negative objects relative to backgrounds associated with neutral objects (i.e., the emotional memory tradeoff effect), the magnitude of this effect was greater for those who slept. Further, the enhancement of memory for negative objects over 12 hours was numerically superior to memory tested just 30 minutes after encoding (see figure 3).

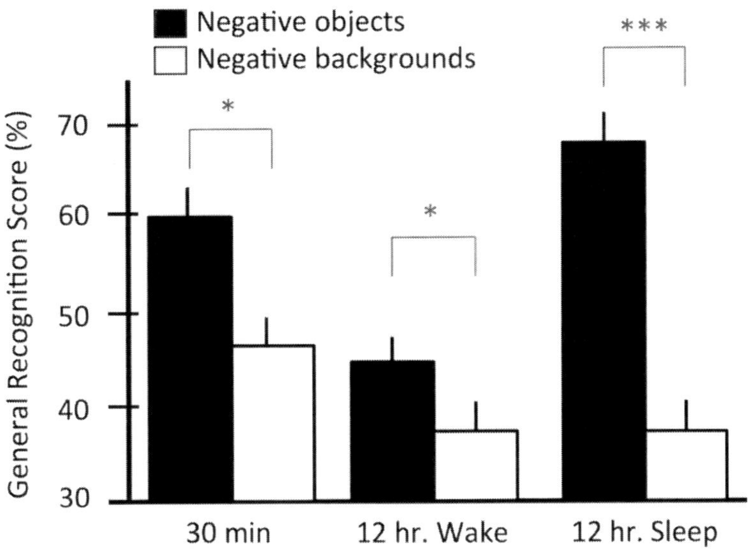

*p,<.05; ***p<.001.

Figure 3. Sleep seems to enhance the magnitude of the 'emotional memory tradeoff effect'. Participants encode scenes composed of emotionally neutral or negative objects within neutral backgrounds (e.g., a snake in the woods, a taxi in the street). The emotional memory tradeoff effect is defined by participants having better memory for negative objects at the expense of worse memory for their accompanying backgrounds (e.g., Payne et al., 2008).

In addition to studies that demonstrate whole night sleep benefits to emotional memory, studies using daytime naps and overnight sleep restrictions also show emotional memory benefits. Nishida and colleagues (2009) had participants encode negative and neutral images before and after a 90-minute nap opportunity or wake condition. A recognition memory test shortly after examined both sets of encoded stimuli. Within the nap group, pre-nap emotional stimuli were better remembered than post-nap emotional stimuli, even though the post-nap stimuli were learned only 15 minutes before test. For the wake condition, emotional memory performance did not differ between sessions in memory performance.

While periods of sleep seem to improve and consolidate neural networks underlying memory, it appears that sleep deprivation can impair this process for neutral and positive memory more than negative memory. Sterpenich and colleagues (2007) had participants learn emotionally salient (positive, negative) and neutral pictures and allowed them to sleep or totally deprived them of sleep the first night. Seventy-two hours later, after 2 nights of recovery sleep, participants underwent memory retrieval inside an fMRI scanner (imaging findings discussed in the next section). They report that while sleep deprivation, compared to regular sleep, deteriorated recollection for positive and neutral stimuli, recollection for negative stimuli was not different between the groups, suggesting that negative emotional memory is more resilient to sleep deprivation effects.

Even with sleep restrictions, sleep improves emotional memory. Wagner et al. (2001), using the split-night procedure explained in *section 2.1*, had participants encode emotional and neutral narratives and allowed them to sleep two nights in the lab, one week apart in counterbalanced order. Wagner and colleagues report that both early (SWS-rich) and late (REM-rich) sleep, compared to the same period of time spent awake, enhanced memory retention, and that sleep-dependent emotional memory benefits (but not neutral memory benefits) drove this effect. In addition to sleep improving memory, when participants slept in the REM-heavy second part of the night, they showed improved emotional memory compared to when they slept in the first half of the night. In a follow-up memory test performed four years later (Wagner et al., 2006), they found that sleep following learning improved emotional memory performance compared to a wake control, and that early and late sleep did equally well. No sleep-dependent benefit was found for neutral memory. Similar to Wagner et al. (2001), Groch and colleagues (2013) also used the split night procedure and found that REM-rich late sleep, compared to SWS-rich early sleep, resulted in higher memory for emotionally negative pictures. Moreover, emotional pictures were better remembered than neutral pictures in the late, but not early, sleep condition.

3.2. The Role of Sleep on Emotional Memory: Correlational and Physiological Evidence

While sleep has been shown to improve emotional memory, it is possible that different sleep stages may have different effects on emotional memory consolidation. Researchers have focused heavily on REM sleep (e.g., Greenberg et al., 1983; Nishida et al., 2009; Payne et al., 2012b; for review see Payne & Kensinger, 2010), though SWS may also contribute to emotional memory formation as well (e.g., Groch et al., 2011; Hauner et al., 2013; Kaestner et al., 2013). REM sleep has been examined with the emotional memory tradeoff discussed in the previous section (Payne et. al., 2008), in which sleep magnified the difference between negative objects and their associated backgrounds, and REM sleep achieved during the night correlated with the size of the improvement (Payne et al., 2012b). As discussed earlier, Nishida and colleagues (2009) allowed participants a 90+ minute nap and found better memory for emotional stimuli. Within the sleep group, several REM sleep measures were associated with the improved memory performance of pre-nap stimuli: amount of REM sleep (duration and percentage), reduced REM sleep latency (going into REM sleep earlier), and intensified theta rhythms during REM sleep.

Similar to previous studies (Payne et al., 2012b; Nishida et al., 2009), Groch et al. (2013) found a positive relationship between emotional memory and REM sleep in the late (REM-rich) sleep period. Groch and colleagues (2013), using event-related potentials (ERP) were also interested in the role of a specific ERP marker important for recognition accuracy (Rugg et al., 1988): the frontal late positive potential (300-500ms after stimulus onset). Successful recognition of emotional pictures was related with a significant increase in this positivity ERP over frontal areas, but only after the REM-rich late sleep condition. An increase in this late positive potential could reveal theta activity generating coherently in memory-relevant regions and the amygdala during previous REM sleep (Popa et al., 2010).

Payne & Kensinger (2011) used fMRI to examine negative and neutral memory after a period of wake or sleep. They found that participants who slept utilized a more concentrated and efficient neural network of the amygdala, ventromedial PFC, and cingulate gyrus to successfully recognize negative objects. Further, the connectivity between the amygdala, hippocampus, and ventromedial PFC was increased in the sleep group. In a sleep deprivation study discussed in the previous section, Sterpenich and colleagues (2007) had participants learn emotional and neutral pictures and allowed them to sleep or deprived them of sleep the first night and tested them 3 days later. While emotional memory was retained after sleep deprivation comparable to sleeping controls, how the brain represented the memory was different between sleep deprivation and control groups. Those who slept the first night recruited the hippocampus and medial PFC more extensively compared to the first-night sleep-deprived participants, while participants who were sleep deprived after encoding recruited more amygdalar activation, suggesting that even though both groups performed similarly on memory tests, the sleep group had undergone stronger memory consolidation and relied on a more permanent memory network. After a 6-month delay, participants were retested (Sterpenich et al., 2009). While emotional stimuli were better remembered than neutral stimuli, there was no additional benefit of sleep. However, the authors attribute this to the large amount of variability and low number of subjects due to fMRI usage. While no sleep benefits were present for memory tests, for those who slept, emotional memory network activation shifted to the amygdala, frontal cortical areas including the ventromedial PFC, and the middle occipital cortex, while sleep deprived subjects only recruited ventral medial prefrontal cortex, again suggesting less rich encoding for participants who were sleep deprived.

In addition to studies that examine the effects of REM sleep on emotional memory consolidation, other studies suggest that NREM sleep plays a role in this process. For example, Groch et al. (2011) had participants learn emotional and neutral stories and pictures (which they rated on valence and arousal) before a 3-hour sleep opportunity in the first part of the night (SWS heavy, similar to Wagner et., al. 2001). In 2 separate nights in counterbalanced order, participants underwent a drug (clonidine – norepinephrine suppressor) or placebo administration during sleep. With minimal changes in sleep architecture (REM was suppressed from 5% to 0.5 %) they report that clonidine reduced participant's ability to remember temporal aspects of the emotional story to similar levels as the neutral story, eliminating the benefit of emotion seen in the placebo condition. In addition, while only emotional picture ratings of valence and arousal decreased after sleep for the placebo, both neutral and emotional pictures received decreased ratings of valence after clonidine administration, and no consistent reduction in arousal. Thus, it may be that norepinephrine activity during SWS may help differentiate emotional from neutral processing. In another study by Kaestner et al. (2013), participants underwent 3 drug manipulations across 3 study days (each separated by 5-10 days), order counterbalanced. The drugs included zolpidem, which enhances sleep spindle density and decreases and/or delays REM sleep, sodium oxybate (SO), which increases SWS but decreases sleep spindle density, and a placebo. Each study day, participants slept in the lab, woke up to encode emotional (positive high arousal, positive low arousal, negative high arousal, and negative low arousal) and neutral picture stimuli, and were administered a drug (or placebo), and allowed to nap between 9AM and 11AM, followed by a retrieval session at 4PM. In the zolpidem condition, accurate recognition was better for negative stimuli and high arousal stimuli compared to placebo. The authors conclude that increasing SWS (via SO) did not improve emotional memory while zolpidem did because zolpidem increased spindles while SO suppressed them. This suggests that sleep spindles during stage 2 sleep may play a role in the consolidation of emotional memory.

3.3. Theoretical Explanations for Evidence of a Role of Sleep on Emotional Memory

Theories that examine emotional memory consolidation and sleep are less established than in other areas of research, and like declarative memory, both REM sleep and SWS likely play a role. However, most research has focused on the relationship between REM sleep and emotional memory. REM sleep increases activity in certain areas of the brain to levels higher than seen during wake; specifically the visual cortex, limbic lobe (including the amygdala) and hippocampus. During REM sleep, a cascade of neurotransmitters results in stimulation of cortical regions and reduced connectivity between the hippocampus and neocortex, though it does not suppress the hippocampus itself (Hasselmo, 1999, Maquet, 1997; McCarley, 2007). With increases in neocortical and amygdalar activity combined with a suppression of hippocampal connectivity to the neocortex, REM sleep is an ideal state to integrate novel, emotional information into existing neocortical networks.

The amygdala, a critical emotional center, modulates the suppression and dis-inhibition of other areas of the brain seen during REM sleep, including a selective up-regulation of the parietal operculum and other cortical regions (Maquet, 1997). This modulating activity of the amygdala during REM sleep may allow emotional cues to influence memory consolidation in long-term memory storage areas in the cortex. In addition to influencing up-regulation and suppression, the amygdala modulates activity in the hippocampus during REM sleep, which may in turn (through theta and gamma oscillations) affect the hippocampal – neocortical dialogue in subsequent sleep previously discussed in the two-stage model (Buzsaki et al., 1994; see *section 2.3*). Thus, the critical emotional brain structure of the amygdala could influence memory consolidation during REM sleep through up and down regulation of the neocortex both directly and indirectly.

3.4. The Role of Sleep in Emotional Memory Consolidation Is Not Absolute

Research on the impact of sleep on emotional memory consolidation has also had its share of mixed results. One major consideration in this field is that the measured impact of sleep on emotional memory may depend on the methods used for testing memory. This is exemplified by the differing effects of sleep on emotional vs. neutral picture memory when the Remember/Know recognition paradigm is used. During recognition testing, participants are asked to respond "Remember" for previously learned pictures they can remember with contextual details, "Know" for those they remember but lack contextual details, and "New" for images they did not see during the learning phase. This method allows the measurement of two memory processes suggested to occur during recognition: recollection, provided by the "Remember" responses, and familiarity, provided by the "Know" responses (Yonelinas, 2002). While Hu et al. (2006) found that a period of nocturnal sleep as opposed to daytime wakefulness enhanced memory for emotional pictures given "Know" responses during testing, no effects of sleep were obtained for pictures given "Remember" responses. This study suggests that sleep does not benefit emotional recollection processes. However, other studies have found the opposite, with sleep benefiting memory for measures of recollection, but not familiarity. Specifically, Atienza and Cantero (2008) found that recollection memory (as assessed by "Remember" responses) for emotional pictures was attenuated less than memory for neutral pictures by 40 hours of sleep deprivation, while familiarity memory (as assessed by "Know" responses) was not impacted by sleep or sleep deprivation manipulations. Similarly, Chambers and Payne (2014) found nocturnal sleep to enhance memory for positive components of scenes over a period of daytime wakefulness for recollection memory, but not familiarity memory. The discrepancies between the Hu et al. (2006) study and these latter two are likely due to the differences in methodology, such as utilizing a sleep deprivation design (Atienza & Cantero, 2008) or focusing on specifically positive emotional information (Chambers & Payne, 2014), which differ from the focus on nocturnal sleep vs. daytime wakefulness and negative emotional picture memory in the Hu et al. (2006) study.

Additionally, the emotional memory trade-off effect, which finds sleep to benefit memory for negative foreground objects at the expense of their associated neutral backgrounds to a greater extent than wakefulness (Payne et al., 2008), is not impacted by sleep when the emotional characteristics of backgrounds (as opposed to the objects) are manipulated.

Lewis and colleagues (2011) found that sleep did not impact memory performance for negative vs. neutral backgrounds that were paired with neutral objects differently than a period of wakefulness, regardless of whether sleep occurred over the course of a daytime nap or night of sleep. It is unclear why sleep provided no emotional memory benefits over wakefulness here, though it may be that attention processes, which are focused on negative vs. neutral *objects* in previous studies (Payne et al., 2008) do not focus in a similar way on negative vs. neutral *backgrounds*. Using a similar emotional memory trade-off paradigm, Cunningham et al. (2014) recently found that when expectation of a future memory test was established prior to the consolidation phase, the magnitude of the emotional memory tradeoff for negative objects over their paired neutral backgrounds was equivalent across periods of sleep and wake.

This suggests that certain cues during encoding and prior to consolidation may drive the extent to which successful consolidation and later retrieval occurs, regardless of whether sleep happens during the delay.

Finally, investigations into disorders or across age groups have also produced the conditions for null sleep effects. Prehn-Kristensen et al. (2013) tested memory for neutral vs. negative pictures after nocturnal sleep and daytime wakefulness in healthy children, healthy adults, and children with ADHD.

Only the healthy children showed the typical emotional memory bias for negative over neutral pictures after sleep relative to wake. ADHD children had comparable performance for emotional and neutral stimuli, which may be associated with the purported dysfunction of oscillatory activity over frontal EEG sites during SWS and REM, as suggested by negative correlations between memory performance and frontal EEG slow oscillations (during SWS) and theta (during REM) in this group. It should be noted that this study additionally failed to find an emotional memory bias in the group of healthy adults, but the authors propose that this could be due to the negative pictures used, which were toned down so that they were appropriate for use with children. It is possible that sufficient oscillatory activity during sleep as well as sufficiently arousing material is necessary improve emotional memory following sleep.

4. The Role of Sleep in Gist Extraction, Insight, and Integration of Information

While beneficial effects of sleep have been shown to apply to the consolidation of veridical memory and selective enhancement of emotionally salient information, recent evidence suggests that sleep also facilitates the flexible use of learned information to mold future behavior. Such memory transformations include processes like gist extraction, insight, and the integration of newly learned material into existing stores, all of which will be discussed below.

4.1. The Role of Sleep in Gist Extraction, Insight, and Integration of Information: Behavioral Evidence

Sleep may not only support the consolidation of single events, but also the extraction of the overall meaning, or "gist", of a certain set of information. In the Deese-Roediger-McDermott (DRM) false memory task, participants study lists of semantically related words (e.g., nurse, hospital, medicine, etc.) that have an unpresented, critical 'gist' word in common (in this case, doctor; Deese, 1959; Roediger & McDermott, 1995). When asked to retrieve these words, participants will remember the studied words (e.g., nurse) and falsely remember the gist words (e.g., doctor) with equally high probability and confidence (Roediger & McDermott, 1995; Seamon et al., 2002). Payne and colleagues (2009) found, in two separate experiments using a 12hr delay, that participants who slept were able to freely recall more gist words than those who spent the time awake (Figure 4). In a third experiment, they replicated this finding, but instead of a full night of sleep, participants either napped or stayed awake in a 90-min period.

Recent studies support the idea that sleep enhances gist extraction, at least in the DRM task. A similar study to Payne et al. (2009) was conducted by Diekelmann et al. (2010) that resulted in increased gist word recall after sleep, when compared to a similar amount of wakeful activity. Nonetheless, this effect was driven by correct memory performance (recall of studied words), as it was only observed in participants that recalled fewer studied words. The authors divided their sample into low and high performers, applying a median split to their corrected recall measure (i.e., unstudied intrusions that are not the gist words subtracted from studied words). A closer look at their data implies that general performance in the Payne et al. (2009) study was in fact more similar to this "low", rather than "high", performance of the Diekelemann et al. (2010) study. McKeon and colleagues (2012) investigated the role of emotional (negative) content in the DRM after a period of sleep or wake activity. Similarly, recall of gist words, both neutral and negative, was increased after a 12 hr period of sleep, but not wake.

The research involved in assessing the role of sleep in gist memory, at least with the DRM task, is far from clear, as several studies have found contrasting results to the ones presented above. Sleep has been found to *decrease* gist-based false memory, but only when tested by recognition, instead of recall (Fenn et al., 2009; Lo et al., 2014). In one study, this effect was found when participants encoded the list words aurally (experiment 1) and visually (experiment 2; Fenn et al., 2009). Similarly, another study using the Remember/Know paradigm discussed in *section 3.4* found a decrease in gist word recollection ('Remember' responses) after sleep in an elderly population (mean age 67; Lo et al., 2014).

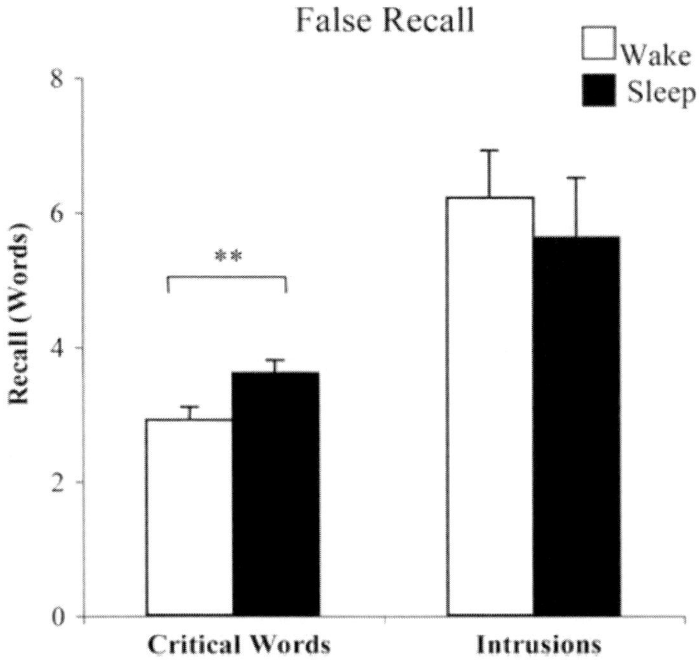

Figure 4. Participants encoded lists of semantically related words (DRM task), which had an unpresented critical 'theme' word in common, and 12 hours later had a free recall test. Those whom slept during those 12 hours were able to recall more gist (critical) words, compared to those who did not (from Payne et al., 2009).

Full-night sleep deprivation seems to also increase recognition of gist words, compared to a similar time awake and a full night of sleep (experiment 1, Diekelmann et al., 2008). Follow-up experiments were conducted to assess whether sleep deprivation during consolidation or retrieval contributed to this effect. In one experiment (experiment 2, Diekelmann et al. 2008), participants encoded the DRM lists in the night of day 1, went to sleep normally that night, and then were either sleep deprived or not on the second night. They found that when tested the following morning (day 3; after a full night of sleep or sleep deprivation), participants showed increased recognition of gist words. In contrast, if the sleep deprivation manipulation occurred (or not) in the first night and all participants had a recovery night of sleep on day 2, targeting only consolidation processes, no effect was found when memory was tested the morning of day 3 (experiment 3). The authors concluded that sleep deprivation increases gist-based false memories by affecting retrieval processes, not consolidation processes. Different results were found by another group of researchers when participants were either sleep deprived or not, given two nights of recovery sleep, and tested on day 4 (Darsaud et al., 2010). This is comparable to experiment 3 of Diekelmann et al. (2008), except that in that study participants only got one night of recovery sleep. Darsaud and colleagues (2010) found, unlike the null effect of Diekelmann et al., that sleep deprivation on day 1, when compared to regular sleep, marginally decreased recognition of gist words on day 4, but only with recollection and not with familiarity judgments.

Cai and colleagues (2009) used the remote associates task (RAT) to investigate sleep-dependent item integration. In this task, participants are presented with word triplets (e.g., heart, sixteen, cookies) and must find the word that associates all three (in this case, 'sweet'). In this study, participants completed a morning RAT, napped or stayed awake, and finished with an afternoon RAT. Between the morning RAT and the nap/wake interval, participants completed analogies that either contained or did not contain the answers to the afternoon RAT, i.e., half the RAT answers were primed. The authors found that only when the items were primed, napping resulted in an improvement from morning to afternoon RAT. Importantly, only participants whose naps contained REM sleep, but not those whose naps only contained NREM sleep, showed this improvement when compared to wake controls. Interestingly, these results cannot be explained by better memory for primed RAT answers, as all three groups (REM, NREM, wake) had comparable memory as measured by recognition and cued recall tests.

There is also evidence that indicates that sleep plays a role in the integration of novel information into pre-existing networks (Dumay & Gaskell, 2007; Tamminen et al., 2010; 2013). When a new word (e.g., 'cathedruke') is integrated into the mental lexicon, it will compete with similar-sounding items (e.g., 'cathedral') in a way that will make recognition of the old word slower. Dumay and Gaskell (2007) tested this lexical competition effect immediately after exposure to the new words, at 12 hours after a period of sleep or wake, and at 24 hours when all participants had slept. Although participants had good memory for the new words in the first test, no lexical competition had occurred yet. After 12 hours, only participants whose delay contained sleep showed integration of the new words, and after 24 hours, when "wake" participants at 12hr had slept, all participants showed similar magnitudes of the lexical competition effect. Unlike Cai et al.'s (2009) results which integration's benefit may be independent of memory for the items, Dumay and Gaskell (2007) found that participants showing improvement in memory for new words also showed stronger lexical competition (i.e., better integration). Additionally, Tamminen and colleagues (2010; 2013) have recently established a role for NREM sleep spindle and slow-wave activity (SWA; < 4 Hz) in lexical and semantic integration, discussed in the next section.

Sleep may also aid the achievement of insight. Wagner and colleagues (2004) used the number reduction task (NRT) and found that participants who had a full night of sleep, compared to a period of wakefulness, were more likely to discover a shortcut to finish the task faster. In the NRT participants are supposed to reduce a series of eight numbers into seven, by a simple rule that reduces two numbers into one. The objective of the task is to find the final number in the series as fast as possible. Unbeknownst to the participants, there was a hidden rule that allowed them to find this last number faster: the last three digits in a series were a mirror image of the previous three, so that after solving the second number in a series, the participant was able to know the last number (see Figure 5). Those who slept were twice as likely to find this hidden rule, compared to those who spent their time awake. This supports the idea that sleep may be an ideal state for the development of insight after a problem has been introduced. Ellenbogen and colleagues (2007) also found evidence that point to a beneficial role of sleep in building inferential relationships based on individual premises. In this study participants learned five "premise pairs" (A>B, B>C, C>D, D>E, and E>F) represented with abstract shapes. However, the participants did not know that these pairs were built in a larger hierarchy (A>B>C>D>E>F, so that A>F). Participants were tested after a 20-minute, 12-hour (including sleep or wake), or 24-hour delay on the basic premise pairs, but also in higher-order pairs that they were not explicitly trained on (i.e., A>C and the even higher order B>E). After 20 minutes, participants were at chance-level performance. Participants in the 12- and 24-hour delays showed significant enhanced inference skills. The greatest difference was between the 12-hour delay containing wakefulness vs. the 12-hour delay containing sleep, in which participants were more able to develop the most distant inferential relationship (i.e., B>E). This improvement in performance was not correlated with an improvement in subjective confidence, suggesting that while sleep may assist relational memory binding and inferential abilities, this process may transpire outside of consciousness (Ellenbogen et al., 2007).

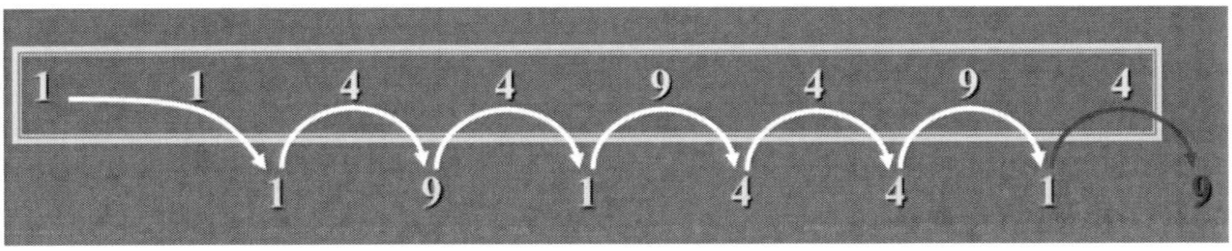

Figure 5. A visual representation of the number reduction task used in Wagner et al., 2004. In this task, participants are supposed to solve puzzles composed of 3 numbers (here 1's, 4's, and 9's) by applying two simple rules. If the numbers are the same, starting from the left, then participants enter that number. In this example, the first two numbers are 1's, so the correct answer would be 1. If the numbers are different, like the second pair here (1 and 4), then the correct answer is the third number (9). The goal of the task is to keep doing this until they reach the last number in the series of digits (in red). This task has a hidden rule that is unknown to the participants: the last three digits (419) are always a mirror image of the previous three digits (914), making the first digit of the mirror (here, 9) always the same as the final answer (the red 9).

Lau and colleagues (2010; 2011) have provided similar evidence that sleep aids the useful reorganization of information to form associative inferences. Lau et al. (2010) used a task in which one object (B) is paired with one face (A) in a first trial but paired to a different face (C) on a second trial (AB-BC). Participants who napped after encoding, when compared to those who stayed awake, were able to remember better the direct pairings (AB and BC) and formed the larger inference of the two faces paired with the same object (AC). Demonstrating another form of inferential memory, Lau et al. (2011) presented Chinese characters to participants that could be categorized into groups by characters that included the same radical; that is, the element of the characters with a shared semantic meaning within the group (i.e., characters for 'sister' or 'mother' have the same radical 'female'). Participants whose retention period contained a nap, compared to a wake period, were more capable to identify the meaning of a new character that shared a radical of one of the previously learned character groups, as well as to declare the radical's meaning, demonstrating an inferential extraction of the meaning that was never explicitly stated.

4.2. The Role of Sleep in Gist Extraction, Insight, and Integration of Information: Correlational and Physiological Evidence

In *section 2,* evidence is discussed regarding the beneficial role of SWS in episodic memory (e.g., Plihal & Born, 1997; Marshall et al., 2006: see Diekelmann & Born, 2010), but

SWS might have a negative role for gist processing (Payne et al., 2009; Lo et al., 2014). Payne and colleagues (2009) found an inverse relationship between SWS and memory for the DRM list words (e.g., nurse, hospital) that might be explained by the semantic nature of this task. Unlike the other mainly episodic tasks used in this literature (e.g., word-pair associates), the DRM paradigm heavily relies on semantic/gist processing due to the strong relatedness of the list words. It was suggested that SWS, while assisting episodic memory consolidation, might impede semantic processing.

Darsaud and colleagues (2011) examined functional brain activity during encoding and recognition of the DRM task. For recognition of both list and gist words, participants who were sleep deprived for the first night after encoding, compared to regular sleep, relied on a broader, more distributed cortical network, suggesting that sleep deprivation shifts the brain to more effortful strategies to retrieve information. Additionally, they found that regardless of experimental condition, hippocampal activity during encoding was the primary factor for the subsequent recollection of list rather than gist words. Across conditions, hippocampus activity at encoding for those lists that did not produce gist word recollection was greater than for lists that later produced gist word recollection.

SWS has been proposed to aid the reorganization of memory traces to establish associations between items not directly linked during learning. Lau and colleagues (2010) found a positive relationship between the number of correctly matched face-face associations (AC pair; relational memory) and SWS duration. A follow-up regression analysis showed that, although direct associative (AB and BC pairs; face-object) memory was strongly associated with relational (face-face) memory, the SWS-relational memory positive correlation was still significant when accounting for the correlation between direct and relational memory. These findings suggest that while sleep plays a role in strengthening what is explicitly learned (e.g., face-object pairs, word pairs; see *section 2*), it may have a distinct role in the integration of information.

Tamminen et al., 2010 used the same lexical competition task ('cathedral' vs. 'cathedruke'; see previous section) as Dumay and Gaskell (2007) to investigate whether sleep spindle activity was associated with the integration of newly learned information to existing knowledge. Similar to Dumay and Gaskell, there was no lexical integration immediately after exposure to the novel words ('cathedruke'). In contrast to Dumay and Gaskell, participants showed integration of the new words (i.e., a significant lexical competition effect) regardless of whether their retention interval (9 hours) was filled with sleep or wakefulness. Although both groups achieved integration, showing that this process is not exclusively sleep-dependent, the authors found a strong positive correlation between sleep spindle count and the lexical competition effect, suggesting a definitive role for this NREM sleep physiological feature in integration processing.

Sleep spindles and SWA (<4 HZ) may also play a role in the integration of novel information into semantic memory. In the Tamminen et al. (2010) study, new words ('cathedruke') were presented without an actual meaning. In a more recent study, Tamminen et al. (2013) assessed the role of sleep physiology in the integration of novel words with novel meanings. In this study, semantic neighborhood density (the number of semantically associated concepts) was manipulated by constructing novel semantic concepts using a pre-existing concept and adding a new semantic feature to it, producing a novel variation of an existing concept (e.g., 'bee whose sting feels pleasant'). These novel variations had either many semantic associates (dense semantic neighborhood) or few associates (sparse semantic neighborhood). Semantic neighborhood density affects memory, with stronger, fast access memories for newly learned information with few semantic neighbors. That is, memory traces for novel words in a sparse semantic neighborhood should be accessed faster because of less interference from competing neighbors. Participants were trained at night in both sparse and dense neighborhood novel concepts with at least one week apart (within-subject design). After encoding, participants went to sleep and various tests of recognition tests were completed the next morning. The sparse condition resulted in higher spindle density (spindle count/30s) and SWA than the dense condition. Although no wake controls were used in this experiment, these results suggest that sleep spindle and SWA have a role in the preferential (faster) integration of new information that lacks semantic competition (i.e., sparse semantic neighborhood).

4.3. Theoretical Explanations for a Role of Sleep in Gist Extraction, Insight, and Integration of Information

While the role of sleep in declarative memory is explained with models explaining that reactivation of memory traces during sleep aids the transfer of memories from a short-term store (i.e., hippocampus) to long-term neocortical areas (e.g., two-stage model; Buzsaki, 1989; McClelland et al., 1995), the role of sleep in gist extraction and integration of newly learned material may be explained with a model that builds upon schema theory (Howard, 1987; Mandler, 1984; Rumelhart, 1980) and an additional role for SWS reactivation: the information overlap to abstract, or iOtA, model developed by Lewis and Durant (2011).

The iOtA model aims to describe how the overlapping replay of multiple memories and succeeding synaptic downscaling during SWS could underpin both schema formation and integration of new knowledge into an existing schema. A schema is a cognitive concept that represents organized information acquired on the basis of experience (Howard, 1987; Mandler, 1984; Rumelhart, 1980) and is formed by the abstraction of regularities within an environment while simultaneously discarding the differences.

iOtA explains that overlap in memory reactivation may occur simultaneously or sequentially to create a schema. In simultaneous reactivation, neurons that represent shared elements of a memory are activated at the same time, resulting in facilitation of Hebbian learning. Memories that are replayed simultaneously will result in a stronger connection between areas that are associated with both memories compared to areas associated with a single memory. Then, during synaptic downscaling, the weaker connections of the non-overlapping areas will perish, whereas the overlapping traces survive. Similarly, in sequential reactivation, shared elements of memories will be more strongly activated than elements belonging to a single memory. Activation from subsequent reactivations will build upon activation from previous reactivations. With downscaling occurring throughout SWS bouts throughout the night, connections may be downscaled in a more subtle way in between spaced sequential reactivations. In general, the selective strengthening by overlap may depend on the extent of this downscaling and the number of reactivations during a night, providing a more controlled mechanism to extract similarities and create a schema.

iOtA builds upon the finding that, in rats, memories that fit into an existing schema become hippocampus-independent and are assimilated into the neocortex faster (Tse et al., 2007; van Kesteren et al., 2012) to explain how newly learned information is added to an existing schema. Findings from the Tamminen et al. studies (2010; 2013, discussed in the previous section) also support this idea. Reactivation of a new memory elicits responses in schemata that are associated to the new memory. The simultaneous activation of 'schema neurons' and 'new-memory neurons' leads to Hebbian learning, linking the latter to the neocortical representation of the schemata. However, initial reactivation and Hebbian learning are unlikely to be enough to add the new information into a schema. In iOtA, the new memory needs to overlap with the schema over multiple occasions, so that the 'new memory-schema' connections are strong enough to survive downscaling. Multiple reactivations of the same 'new' memory are key so that only memories associated with an existing schema are integrated, resulting in shared elements being incorporated while less relevant information is not. This process, similar to schema formation, depends on the number of reactivations of the 'new' memory, the degree of overlap between the memory and the schema, and the initial memory strength. That is, assimilation of a new memory that has little overlap with the schema will take a higher number of reactivations, especially if it has been weakly encoded.

Some of the findings discussed in *section 3.1* could be explained by iOtA. Overlapping replay of premise pairs should lead to the construction of the abstracted representation of the overall hierarchy (i.e., B>C + C>D + D>E = B>E; Ellenbogen et al., 2007). In the number reduction task, overlapping reactivation of the completed problems should lead to the insightful realization of the hidden rule (Wagner et al., 2004). Similarly, in the DRM task, because the gist word of each list is highly related to the learned words, sequential activation of these words leads to a strengthening of the gist words. During sleep, the semantic link between list words and gist words would be reinforced by overlapping reactivation, thus leading to increased gist-based false memories (Payne et al., 2009, Diekelmann et al., 2010). However, this explanation counters the negative correlation between SWS, in which most iOtA processes occur, and recall of list words in the Payne et al. study.

Findings that support a detrimental role of sleep in gist memory formation (Fenn et al., 2009; Lo et al., 2014) cannot be explained by iOtA. However, these studies used recognition testing to probe gist memory. While sleep may aid in gist extraction, increasing false *recall*, it may also enhance contextual details of the list words, enhancing source-monitoring processes, reducing gist memory when the list words and the gist words are encountered in a *recognition* test (Fenn et al., 2009; Lo et al., 2014).

This account can also explain how sleep deprivation increases gist extraction. Sleep deprivation, by impairing PFC functioning (Horne, 1993), might worsen monitoring skills, reducing the ability to discriminate between encountered (list words) and non-encountered (gist) words (Diekelmann et al., 2008).

REM sleep might also play a role in the development of insight. Cai and colleagues (2009) found that a nap that contained REM sleep improved performance for primed items in the remote associate task. The authors argue that activation, generated by the memory trace for the target item (primed answer), is spread through a network of other relevant stored information. REM sleep might be an optimal stage for this because, as discussed in *section 1.1*, high ACh levels in the hippocampus interrupt feedback to the neocortex, while low ACh levels in the neocortex might facilitate spreading activation within cortical areas, possibly leading to a reorganization of associative networks and a subsequent reinterpretation of information in the hippocampus.

4.4. The Role of Sleep in Gist Extraction, Insight, and Integration of Information Is Not Absolute

Although most evidence suggests that sleep supports the reorganization and integration of information, along with memory consolidation (*sections 2 and 3*), a handful of studies have found otherwise. In the Diekelmann et al. (2008) study discussed in *section 4.1*, sleep deprivation was found to increase gist-based false memory recognition, compared to a regular night of sleep and regular daytime wakefulness. However, there was no significant difference between memory performance in the overnight sleep and wake groups. It is possible that no "sleep effect" was found due to the retrieval method used (recognition, instead of recall), although this is unlikely since other studies found a significant reduction in gist word recognition after sleep (e.g., Fenn et al., 2009). Another possibility is that a significant effect was not observable in their data because, at least with that population (Diekelmann et al., 2010), recall of studied words was driving sleep's role in gist memory formation. That is, in the Diekelmann et al., 2010 study, the authors divided their sample using a median split with list word memory performance and only found a significant effect in low performers. In their 2008 study, the group did not report this median split in their data.

Tamminen and colleagues (2010) found a positive relationship between sleep spindle count and lexical integration, but behaviorally, no differences were found between the sleep and the wake group (i.e., the degree of lexical integration was similar between the groups after the retention interval). These findings may require an alternate explanation for the integration processes that occur during wakefulness, that according to the authors, are probably based in different neural bases than the (sleep) processes investigated in their study.

CONCLUSION

This chapter considered the wealth of evidence demonstrating an active role of sleep in the preservation, enhancement, and integration of information. We discussed how a full or partial night of sleep or a daytime nap could have beneficial effects on memory performance, whereas sleep deprivation is generally detrimental. Furthermore, recent evidence has been able to discern that it may not be sleep as a unitary process that improves these memory processes, but that specific stages of sleep, and even specific physiological features within the stages, are more important for the processing of certain types of information.

While there is an abundance of evidence suggesting that sleep plays a critical role in the successful consolidation of memory, some studies allude to a model with a level of complexity that we do not yet fully understand and no contemporary experiment, hypothesis or theory has been able to completely address. When designing research experiments it is important to take into consideration a host of factors that could influence the results such as the type of memory task, composition of the stages of sleep, multiple stages of post-encoding memory processing (i.e., consolidation vs. retrieval effects), instructions during and after the acquisition phase, rehearsal effects, and expectation effects, just to name a few. While an examination of the influences of each of these factors is beyond the scope of this chapter, several exhaustive reviews detail these considerations and propose numerous potential models based on contemporary research (see Ackermann & Rasch, 2014; Conte & Ficca, 2013;

Diekelmann & Born, 2010 Diekelmann et al., 2009; Ellenbogen et al., 2006; Payne, 2011). It is almost undeniable that sleep is crucially involved in memory consolidation and integration, but many of the specifics have not yet been answered, promising an abundant road for future researchers. There is still much to be learned by probing deeper into the electrophysiological phenomena that occur during human sleep, beyond behavioral and correlational findings, in an attempt to gain greater understanding of the active role for sleep in memory processing. While research has focused on the parsing of sleep into different stages to discover the contributions of these states to memory consolidation, sleep itself may be more than the sum of its parts. Future research should not only continue to refine our knowledge of sleep stages individually but should also integrate across these findings to better understand the big picture of the role of sleep in memory consolidation.

REFERENCES

Ackermann, S., & Rasch, B. (2014). Differential effects of non-REM and REM sleep on memory consolidation?. Current neurology and neuroscience reports,14(2), 1-10.

Alger, S. E., Lau, H., & Fishbein, W. (2012). Slow wave sleep during a daytime nap is necessary for protection from subsequent interference and long-term retention. Neurobiology of Learning and Memory, 98(2), 188–196. doi:10.1016/j.nlm.2012.06.003

Aston-Jones, G. (2004). The locus coeruleus, A5 and A7 noradrenergic cell groups. In G. Paxinos (Ed.), The rat nervous system (pp. 259–294). San Diego, CA: Elsevier Academic Press.

Atienza, M., & Cantero, J. L. (2008). Modulatory effects of emotion and sleep on recollection and familiarity. *Journal of sleep research*, *17*(3), 285-294.

Backhaus, J., Hoeckesfeld, R., Born, J., Hohagen, F., & Junghanns, K. (2007). Immediate as well as delayed post learning sleep but not wake enhances declarative memory consolidation in children. *Neurobiology of Learning and Memory*; 89, 76-80.

Battaglia, F. P., Sutherland, G. R., & McNaughton, B. L. (2004). Hippocampal sharp wave bursts coincide with neocortical "up-state" transitions. *Learning & Memory*, *11*(6), 697-704.

Benson, K., & Feinberg, I. (1975). Sleep and memory: retention 8 and 24 hours after initial learning. *Psychophysiology*, *12*(2), 192-195.

Benson, K., & Feinberg, I. (1977). The beneficial effect of sleep in an extended Jenkins and Dallenbach paradigm. *Psychophysiology*, *14*(4), 375-384.

Bliss, T. V. & Lomo, T. (1973). Long-lasting potentiation of synaptic transmission in the dentate area of the anaesthetized rabbit following stimulation of the perforant path. *Journal of Physiology,* 232, 331-356.

Braun, A. R., Balkin, T. J., Wesenten, N. J., Carson, R. E., Varga, M., Baldwin, P., et al. (1997). Regional cerebral blood flow throughout the sleep-wake cycle. An H2 (15) O PET study. *Brain*, 120(7), 1173-1197.

Brehens, C. J., van den Boom, L. P., de Hoz, L., Friedman, A., & Heinemann, U. (2005). Induction of sharp-wave ripple complexes in vitro and reorganization of hippocampal networks. *Nature Neuroscience*; 8(11), 1560-1567.

Buzsaki, G. (1989). Two-stage model of memory trace formation: A role for "noisy" brain states. *Neuroscience*; 31, 551-70.

Buzsáki, G., Bragin, A., Chrobak, J. J., Nadasdy, Z., Sik, A., Hsu, M., & Ylinen, A. (1994). Oscillatory and intermittent synchrony in the hippocampus: relevance to memory trace formation. In *Temporal coding in the brain* (pp. 145-172). Springer Berlin Heidelberg.

Buzsaki, G., Csicsvari, J., Dragoi, G., Harris, K., Henze, D., & Hirase, H. (2002). Homeostatic maintenance of neuronal excitability by burst discharges in vivo. *Cereb Cortex*; 12, 893-899.

Buzsaki, G., Leung, L. W., & Vanderwolf, C. H. (1983). Cellular bases of hippocampal EEG in the behaving rat. *Brain Research*; 287(2), 139-171.

Cai, D. J., Mednick, S. A., Harrison, E. M., Kanady, J. C., & Mednick, S. C. (2009). REM, not incubation, improves creativity by priming associative networks. *Proceedings of the National Academy of Sciences*, *106*(25), 10130-10134.

Carskadon, M. A. & Dement, W. C. (1989). Normal Human Sleep: An Overview. In M. H. Kryger, T. Roth, & W. C. Dement (Eds.). *Principles and Practice of Sleep Medicine* (3-13). Philadelphia: W.B. Saunders.

Chambers, A. M., & Payne, J. D. (2014). Laugh yourself to sleep: memory consolidation for humorous information. *Experimental brain research*, *232*(5), 1415-1427.

Chrobak, J. J., & Buzsaki, G. (1994). Selective activation of deep layer (V-VI) retrohippocampal cortical neurons during hippocampal sharp waves in the behaving rat. *The Journal of Neuroscience*; 14, 6160-6170.

Clemens, Z., Fabo, D., & Halasz, P. (2005). Overnight verbal memory retention correlates with the number of sleep spindles. *Neuroscience*; 132, 529-535.

Conte, F., & Ficca, G. (2013). Caveats on psychological models of sleep and memory: a compass in an overgrown scenario. *Sleep medicine reviews*, *17*(2), 105-121.

Cunningham, T. J., Chambers, A. M., & Payne, J. (2014). Prospection and emotional memory: how expectation affects emotional memory formation following sleep and wake. *Cognitive Science*,*5*, 862.

Darsaud, A., Dehon, H., Lahl, O., Sterpenich, V., Boly, M., Dang-Vu, T., ... & Collette, F. (2011). Does sleep promote false memories?. *Journal of cognitive neuroscience*, *23*(1), 26-40.

De Gennaro, L., & Ferrara, M. (2003). Sleep spindles: an overview. *Sleep medicine reviews*, 7(5), 423-440.

Deese, J. (1959). On the prediction of occurrence of particular verbal intrusions in immediate recall. *Journal of experimental psychology*, 58(1), 17.

Diba, K., & Buzsaki, G. (2007). Forward and reverse hippocampal place-cell sequences during ripples. *Nature Neuroscience*; 10(10), 1241-1242.

Diekelmann, S., & Born, J. (2010). The memory function of sleep. *Nature Reviews Neuroscience*, *11*(2), 114-126.

Diekelmann, S., Born, J., & Wagner, U. (2010). Sleep enhances false memories depending on general memory performance. *Behavioural brain research*, 208(2), 425-429.

Diekelmann, S., Landolt, H. P., Lahl, O., Born, J., & Wagner, U. (2008). Sleep loss produces false memories. *PLoS One*, 3(10), e3512.

Diekelmann, S., Wilhelm, I., & Born, J. (2009). The whats and whens of sleep-dependent memory consolidation. *Sleep medicine reviews*, *13*(5), 309-321.

Dumay, N., & Gaskell, M. G. (2007). Sleep-associated changes in the mental representation of spoken words. *Psychological Science*, *18*(1), 35-39.

Ellenbogen, J. M., Hu, P. T., Payne, J. D., Titone, D., & Walker, M. P. (2007). Human relational memory requires time and sleep. *Proc Natl Acad Sci U S A.*; 104(18), 7723-8.

Ellenbogen, J. M., Payne, J. D., & Stickgold, R. (2006). The role of sleep in declarative memory consolidation: passive, permissive, active or none? *Current Opinion in Neurobiology*; 16, 716–722.

Eschenko, O., & Sara, S. J. (2008). Learning-Dependent, Transient Increase of Activity in Noradrenergic Neurons of Locus Coeruleus during Slow Wave Sleep in the Rat: Brain Stem–Cortex Interplay for Memory Consolidation?. *Cerebral Cortex*, 18(11), 2596-2603.

Eschenko, O., Ramadan, W., Mölle, M., Born, J., & Sara, S. J. (2008). Sustained increase in hippocampal sharp-wave ripple activity during slow-wave sleep after learning. *Learning & Memory*, 15(4), 222-228.

Fenn, K. M., Gallo, D. A., Margoliash, D., Roediger, H. L., & Nusbaum, H. C. (2009). Reduced false memory after sleep. *Learning & Memory*, 16(9), 509-513.

Fenn, K. M., Nusbaum, H. C., & Margoliash, D. (2003). Consolidation during sleep of perceptual learning of spoken language. *Nature*, 425(6958), 614–616.

Gais, S., Lucas, B. & Born, J. (2006). Sleep after learning aids memory recall. *Learning & Memory*, 13, 259-262.

Gais, S., Mölle, M., Helms, K., & Born, J. (2002). Learning-dependent increases in sleep spindle density. *Journal of Neuroscience*; 22(15), 6830-4.

Gais, S., Rasch, B., Dahmen, J. C., Sara, S., & Born, J. (2011). The memory function of noradrenergic activity in Non-REM sleep. *Journal of cognitive neuroscience*, 23(9), 2582-2592.

Greenberg, R., Pearlman, C., Schwartz, W., & Grossman, H. (1983). Memory, emotion, and REM sleep. *Journal of abnormal psychology*, 92 (3), 378-381.

Groch, S., Wilhelm, I., Diekelmann, S., & Born, J. (2013). The role of REM sleep in the processing of emotional memories: evidence from behavior and event-related potentials. *Neurobiology of learning and memory*, *99*, 1-9.

Groch, S., Wilhelm, I., Diekelmann, S., Sayk, F., Gais, S., & Born, J. (2011). Contribution of norepinephrine to emotional memory consolidation during sleep. *Psychoneuroendocrinology*, 36(9), 1342–1350. doi:10.1016/j.psyneuen.2011.03.006

Grosvenor, A., & Lack, L. C. (1984). The effect of sleep before or after learning on memory. *Sleep*, *7*(2), 155-167. Scullin, M. K., & McDaniel, M. A. (2010). Remembering to execute a goal sleep on it!. *Psychological Science*, *21*(7), 1028-1035.

Hamann, S. B., Ely, T. D., Grafton, S. T., & Kilts, C. D. (1999). Amygdala activity related to enhanced memory for pleasant and aversive stimuli. *Nature Neuroscience*, 2(3), 289-293.

Harlow, H., McGaugh, J. L., & Thompson, R. F. (1971). Psychology. San Francisco. *CA: Albion Publication Company*.

Hasselmo, M. E. (1999). Neuromodulation: Acetylcholine and memory consolidation. *Trends in Cognitive Sciences;* 3(9), 351-359.

Hasselmo, M. E., & Bower, J. M. (1992). Cholinergic suppression specific to intrinsic not afferent fiber synapses in rat piriform (olfactory) cortex. *Journal of Neurophysiology*; 67, 1222–1229.

Hasselmo, M. E., McGaughy, J. (2004). High acetylcholine levels set circuit dynamics for attention and encoding and low acetylcholine levels set dynamics for consolidation. *Progress in Brain Research*; 145, 207-31.

Hauner, K. K., Howard, J. D., Zelano, C., & Gottfried, J. A. (2013). Stimulus-specific enhancement of fear extinction during slow-wave sleep. *Nature neuroscience*.

Hebb, D. O. (1949). The Organization of Behavior. New York: John Wiley and Sons.

Hennevin, E., Huetz, C., & Edeline, J. M. (2007). Neural representations during sleep: from sensory processing to memory traces. *Neurobiology of Learning and Memory*; 87(3), 416-440.

Herreras, O., Solis, J. M., Munoz, M. D., Martin del Rio, R., & Lerma, J. (1988). Sensory modulation of hippocampal transmission: I. Opposite effects on CA1 and dentate gyrus synapsis. *Brain Research*; 461, 290–302.

Hoffman, K. L., & McNaughton, B. L. (2002). Coordinated reactivation of distributed memory traces in primate neocortex. *Science*, 297(5589), 2070-2073.

Horne, J. A. (1993). Human sleep, sleep loss and behaviour: Implications for the prefrontal cortex and psychiatric disorder. *The British Journal of Psychiatry*.

Hounsgaard, J. (1978). Presynaptic inhibitory action of acetylcholine in area CA1 of the hippocampus. *Experimental Neurology*; 62, 787–797.

Howard, R. W. (1987). *Concepts and schemata: An introduction*. London: Cassell.

Hu, P., Stylos-Allan, M., & Walker, M. P. (2006). Sleep facilitates consolidation of emotional declarative memory. *Psychological Science*; 17(10), 891-898.

Huber, R., Ghilardi, M. F., Massimini, M., & Tononi, G. (2004). Local sleep and learning. *Nature*; 430, 78-81.

Iber, C., Ancoli-Israel, S., Chesson, A. L., & Quan, S. F. (Eds.). (2007). The AASM Manual for the Scoring of Sleep and Associated Events: Rules, Terminology, and Technical Specifications (1st ed.). Westchester, IL: American Academy of Sleep Medicine.

Jenkins, J. G., & Dallenbach, K. M. (1924). Obliviscence during sleep and waking. *American Journal of Psychology*; 35, 605-612.

Ji, D. & Wilson, M. A. (2007). Coordinated memory replay in the visual cortex and hippocampus during sleep. *Nature Neuroscience*; 10(1), 100-7.

Kaestner, E. J., Wixted, J. T., & Mednick, S. C. (2013). Pharmacologically Increasing Sleep Spindles Enhances Recognition for Negative and High-arousal Memories. *Journal of Cognitive Neuroscience*, 25(10), 1597–1610. doi:10.1162/jocn_a_00433

Kaestner, E. J., Wixted, J. T., & Mednick, S. C. (2013). Pharmacologically increasing sleep spindles enhances recognition for negative and high-arousal memories. *Journal of cognitive neuroscience*, *25*(10), 1597-1610.

Kensinger, E. A. (2004). Remembering emotional experiences: The contribution of valence and arousal. *Reviews in the Neurosciences*, 15, 241–252.

Kensinger, E. A. (2004). Remembering emotional experiences: The contribution of valence and arousal. *Reviews in the Neurosciences*, *15*(4), 241-252.

Kensinger, E. A. (2009). What factors need to be considered to understand emotional memories?. *Emotion Review*, 1(2), 120-121.

Kensinger, E. A., & Corkin, S. (2003). Memory enhancement for emotional words: Are emotional words more vividly remembered than neutral words?. *Memory & Cognition*, 31(8), 1169-1180.

Kudrimoti, H. S., Barnes, C. A., McNaughton, B. L. (1999). Reactivation of hippocampal cell assemblies: effects of behavioral state, experience, and EEG dynamics. *Journal of Neuroscience;* 19, 4090-4101.

Lahl, O., Wispel, C., Willigens, B., & Pietrowsky, R. (2008). An ultra short episode of sleep is sufficient to promote declarative memory performance. *Journal of Sleep Research*; 17, 3-10.

Lau, H., Alger, S. E., & Fishbein, W. (2011). Relational memory: A daytime nap facilitates the abstraction of general concepts. *PLoS ONE 6*(11): e27139. doi:10.1371/journal.pone.0027139

Lau, H., Tucker, M. A., Fishbein, W. (2010). Daytime napping: Effects on human direct associative and relational memory. *Neurobiology of Learning and Memory*; 93(4), 554-60.

LeDoux, J. (2002). *Synaptic Self: How Our Brains Become Who We Are*. New York: Penguin Putnam Inc.

Lewis, P. A., & Durrant, S. J. (2011). Overlapping memory replay during sleep builds cognitive schemata. *Trends in cognitive sciences*, *15*(8), 343–51. doi:10.1016/j.tics.2011.06.004

Lewis, P. A., Cairney, S., Manning, L., & Critchley, H. D. (2011). The impact of overnight consolidation upon memory for emotional and neutral encoding contexts. *Neuropsychologia*, *49*(9), 2619-2629.

Lo, J. C., Sim, S. K., & Chee, M. W. (2014). Sleep reduces false memory in healthy older adults. *Sleep*, 37(4), 665-U247.

Louie, K., & Wilson, M. A. (2001). Temporally structured replay of awake hippocampal ensemble activity during rapid eye movement sleep. *Neuron*; 29(1), 145-56.

Mander, B. A., Santhanam, S., Saletin, J. M., & Walker, M. P. (2011). Wake deterioration and sleep restoration of human learning. *Current Biology*, 21(5), R183–R184.

Mander, B. A., Santhanam, S., Saletin, J. M., & Walker, M. P. (2011). Wake deterioration and sleep restoration of human learning. *Current biology*, *21*(5), R183-R184.

Mandler, J. M. (2014). *Stories, scripts, and scenes: Aspects of schema theory*. Psychology Press.

Maquet, P. (1997). Positron emission tomography studies of sleep and sleep disorders. *Journal of neurology*, *244*(1), S23-S28.

Maquet, P., Laureys, S., Peigneux, P., Fuchs, S., Petiau, C., Phillips, C., et al. (2000). Experience-dependent changes in cerebral activation during human REM sleep. *Nature Neuroscience*, 3, 831–836.

Maquet, P., Péters, J. M., Aerts, J., Delfiore, G., Degueldre, C., Luxen, A., & Franck, G. (1996). Functional neuroanatomy of human rapid-eye-movement sleep and dreaming. *Nature*, 383(6596), 163-166.

Marshall, L., Helgadottir, H., Molle, M., & Born, J. (2006). Boosting slow oscillations during sleep potentiates memory. *Nature*; 444, 610-613.

Marshall, L., Molle, M., Hallschmid, M., Born, J. (2004). Transcranial direct current stimulation during sleep improves declarative memory. *The Journal of Neuroscience*; 24(44), 9985–9992.

McCarley, R. W. (2007). Neurobiology of REM and NREM sleep. *Sleep medicine*, *8*(4), 302-330.

McClelland, J. L., McNaughton, B. L., & O'Reilly, R. C. (1995). Why there are complementary learning systems in the hippocampus and neocortex: Insights from the successes and failures of connectionist models of learning and memory. *Psychological Review*; 102(3), 419-457.

McGaugh, J. L. (2004). The amygdala modulates the consolidation of memories of emotionally arousing experiences. *Annual Review Neurosci.*, 27, 1-28.

McKeon, S., Pace-Schott, E. F., & Spencer, R. M. (2012). Interaction of sleep and emotional content on the production of false memories. *PloS one*, 7(11), e49353.

Mednick, S. C., McDevitt, E. A., Walsh, J. K., Wamsley, E., Paulus, M., Kanady, J. C., & Drummond, S. P. A. (2013). The Critical Role of Sleep Spindles in Hippocampal-Dependent Memory: A Pharmacology Study. *Journal of Neuroscience*, 33(10), 4494–4504. doi:10.1523/JNEUROSCI.3127-12.2013

Mednick, S. C., McDevitt, E. A., Walsh, J. K., Wamsley, E., Paulus, M., Kanady, J. C., & Drummond, S. P. (2013). The critical role of sleep spindles in hippocampal-dependent memory: a pharmacology study. *The Journal of Neuroscience*, *33*(10), 4494-4504.

Mölle, M., Eschenko, O., Gais, S., Sara, S. J., & Born, J. (2009). The influence of learning on sleep slow oscillations and associated spindles and ripples in humans and rats. *European Journal of Neuroscience*, 29(5), 1071-1081.

Mölle, M., Yeshenko, O., Marshall, L., Sara, S. J., & Born, J. (2006). Hippocampal sharp wave-ripples linked to slow oscillations in rat slow-wave sleep. *Journal of neurophysiology*, 96(1), 62-70.

Nadasdy, Z., Hirase, H., Czurko, A., Csicsvari, J., & Buzsaki, G. (1999). Replay and time compression of recurring spike sequences in the hippocampus. *Journal of Neuroscience;* 19, 9497-9507.

Nishida, M., Pearsall, J., Buckner, R. L., & Walker, M. P. (2009). REM sleep, prefrontal theta, and the consolidation of human emotional memory. *Cerebral Cortex*; 19, 1158-1166.

O'Neill, J, Pleydell-Bouverie, B., Dupret, D., & Csicsvari, J. (2010). Play it again: reactivation of waking experience and memory. *Trends in Neurosciences*. 33 (5), 220-229.

Öhman, A. (2005). The role of the amygdala in human fear: Automatic detection of threat. *Psychoneuroendocrinology*, 30(10), 953–958. doi:10.1016/j.psyneuen.2005.03.019

Öhman, A. (2005). The role of the amygdala in human fear: automatic detection of threat. *Psychoneuroendocrinology*, *30*(10), 953-958.

Orban, P., Rauchs, G., Balteau, E., Degueldre, C., Luxen, A., Maquet, P., & Peigneux, P. (2006). Sleep after spatial learning promotes covert reorganization of brain activity. *Proceedings of the National Academy of Sciences*, 103(18), 7124–7129.

Pavlides, C., & Winson, J. (1989). Influences of hippocampal place cell firing in the awake state on the activity of these cells during subsequent sleep episodes. *Journal of Neuroscience;* 9, 2907-2918.

Payne, J. D. (2011). Learning, memory, and sleep in humans. *Sleep Medicine Clinics*, 6(1), 15-30.

Payne, J. D., & Kensinger, E. A. (2010). Sleep's role in the consolidation of emotional episodic memories. *Current Directions in Psychological Science*, 19(5), 290-295.

Payne, J. D., & Kensinger, E. A. (2011). Sleep leads to changes in the emotional memory trace: evidence from fMRI. *Journal of Cognitive Neuroscience*, 23(6), 1285-1297.

Payne, J. D., Chambers, A. M., & Kensinger, E. A. (2012). Sleep promotes lasting changes in selective memory for emotional scenes. *Frontiers in integrative neuroscience, 6.*

Payne, J. D., Nadel, L., Britton, W. B., Jacobs, W. J. (2004). *The biopsychology of trauma and memory.* In Reisberg, D. & Hertel, P. (Eds.), Memory and emotion (76-128). New York: Oxford University Press.

Payne, J. D., Schacter, D. L., Propper, R. E., Huang, L. W., Wamsley, E. J., Tucker, M. A., ... & Stickgold, R. (2009). The role of sleep in false memory formation. *Neurobiology of learning and memory*, 92(3), 327-334.

Payne, J. D., Stickgold, R., Swanberg, K., & Kensinger, E. A. (2008). Sleep preferentially enhances memory for emotional components of scenes. *Psychological Science*; 19, 781-788.

Payne, J. D., Tucker, M. A., Ellenbogen, J. M., Wamsley, E. J., Walker, M. P., Schacter, D. L., & Stickgold, R. (2012a). Memory for Semantically Related and Unrelated Declarative Information: The Benefit of Sleep, the Cost of Wake. *PLoS ONE*, 7(3), e33079. doi:10.1371/journal.pone.0033079

Peigneux, P., Laureys, S., Fuchs, S., Collette, F., Perrin, F., Reggers, J., Phillips, C., Degueldre, C., Del Foire, G., Aerts, J., Luxen, A., & Maquet, P. (2004). Are spatial memories strengthened in the human hippocampus during slow-wave sleep? *Neuron*; 44, 535-545.

Phelps, E. (2004). Human emotion and memory: interactions of the amygdala and hippocampal complex. *Current Opinion in Neurobiology*, 14(2), 198–202. doi:10.1016/j.conb.2004.03.015

Phelps, E. A. (2004). Human emotion and memory: interactions of the amygdala and hippocampal complex. *Current opinion in neurobiology*, *14*(2), 198-202.

Plihal, W., & Born, J. (1997). Effects of early and late nocturnal sleep on declarative and procedural memory. *Journal of Cognitive Neuroscience*; 9, 534–547.

Plihal, W., & Born, J. (1999). Effects of early and late nocturnal sleep on priming and spatial memory. *Psychophysiology*, 36(5), 571–582.

Popa, D., Duvarci, S., Popescu, A. T., Léna, C. M., & Paré, D. (2010). Coherent amygdalocortical theta promotes fear memory consolidation during paradoxical sleep. Proceedings of the National Academy of Sciences, 107, 6516–6519.

Prehn-Kristensen, A., Munz, M., Molzow, I., Wilhelm, I., Wiesner, C. D., & Baving, L. (2013). Sleep promotes consolidation of emotional memory in healthy children but not in children with attention-deficit hyperactivity disorder. *PloS one*, *8*(5), e65098.

Ramadan, W., Eschenko, O., & Sara, S. J. (2009). Hippocampal sharp wave/ripples during sleep for consolidation of associative memory. *PLoS One*, 4(8), e6697.

Rechtschaffen, A., & Kales, A., editors (1968). *A manual of standardized terminology, techniques and scoring system for sleep stages in human subjects.* Los Angeles: Brain Information Service/Brain Research Institute, University of California.

Richardson, A., & Gough, J. E. (1963). The long range effect of sleep on retention. *Australian Journal of Psychology*, *15*(1), 37-41.

Roediger, H. L., & McDermott, K. B. (1995). Creating false memories: Remembering words not presented in lists. Journal of experimental psychology: *Learning, Memory, and Cognition*, 21(4), 803.

Rosanova, M., & Ulrich, D. (2005). Pattern-specific associative long-term potentiation induced by a sleep spindle-related spike train. *The Journal of Neuroscience*, 25(41), 9398-9405.

Rovira, C., Ben-Ari, Y., Cherubini, E., Krnjevic, K., & Ropert, N. (1983). Pharmacology of the dendritic action of acetylcholine and further observations on the somatic disinhibition in the rat hippocampus in situ. *Neuroscience*; 8, 97–106.

Rugg, M. D., Mark, R. E., Walla, P., Schloerscheidt, A. M., Birch, C. S., & Allan, K. (1998). Dissociation of the neural correlates of implicit and explicit memory. *Nature*, 392, 595–598.

Rumelhart, D. E. (1980). Schemata: The building blocks of cognition. In; RJ Spiro/BC Bruce/WF Brewer (Eds.): Theoretical Issues in Reading Comprehension.

Sara, S. J. (2009). The locus coeruleus and noradrenergic modulation of cognition. *Nature reviews neuroscience*, 10(3), 211-223.

Schabus, M., Gruber, G., Parapatics, S., Sauter, C., Klösch, G., Anderer, P., Klimesch, W., Saletu, B., & Zeitlhofer, J. (2004). Sleep spindles and their significance for declarative memory consolidation. *Sleep*; 27(8), 1479-85.

Schabus, M., Hodlmoser, K., Pecherstorfer, T., & Klosch, G. (2005). Influence of midday naps on declarative memory performance and motivation. *Somnologie*; 9(3), 148-153.

Seamon, J. G., Luo, C. R., Kopecky, J. J., Price, C. A., Rothschild, L., Fung, N. S., & Schwartz, M. A. (2002). Are false memories more difficult to forget than accurate memories? The effect of retention interval on recall and recognition. *Memory & Cognition*, 30(7), 1054-1064.

Siapas, A.G., & Wilson, M.A. (1998). Coordinated interactions between hippocampal ripples and cortical spindles during slow-wave sleep. *Neuron*; 21, 1123-1128.

Sirota, A., Csicsvari, J., Buhl, D., & Buzsáki, G. (2003). Communication between neocortex and hippocampus during sleep in rodents. *Proc Natl Acad Sci U S A*; 100(4), 2065-9.

Skaggs, W.E., & McNaughton, B.L. (1996). Replay of neuronal firing sequences in rat hippocampus during sleep following spatial experience. *Science*; 271, 1870–1873.

Smith, C. T., Aubrey, J. B., & Peters, K. R. (2004). Different roles for REM and stage 2 sleep in motor learning: A proposed model. *Psychologica Belgica*, 44(1-2), 81-104.

Steriade, M., & Timofeev, I. (2003). Neuronal Plasticity in Thalamocortical Networks during Sleep and Waking Oscillations. *Neuron*; 37(4), 563-576.

Steriade, M., Nuñez, A., & Amzica, F. (1993). Intracellular analysis of relations between the slow (< 1 Hz) neocortical oscillation and other sleep rhythms of the electroencephalogram. *The Journal of Neuroscience*, 13(8), 3266-3283.

Sterman, M. B., Goodman, S. J., & Kovalesky, R. A. (1978). Effects of sensorimotor EEG feedback training on seizure susceptibility in the rhesus monkey. *Experimental Neurology*, 62(3), 735-747.

Sterpenich, V., Albouy, G., Boly, M., Vandewalle, G., Darsaud, A., et al. (2007). Sleep-related hippocampal-cortical interplay during emotional memory recollection. *PloS ONE* 5(11), e282.

Sterpenich, V., Albouy, G., Darsaud, A., Schmidt, C., Vandewalle, G., et al. (2009). Sleep promotes the neural reorganization of remote emotional memories. *The Journal of Neuroscience*; 29(16), 5143-5152.

Stickgold, R., & Walker, M. P. (2013). Sleep-dependent memory triage: evolving generalization through selective processing. *Nature neuroscience*,16(2), 139-145.

Stickgold, R., Hobson, J. A., Fosse, R., & Fosse, M. (2001). Sleep, learning, and dreams: Off-line memory reprocessing. *Science*, 294, 1052-1057.

Sweatt, J. D. (1999). Toward a molecular explanation for long-term potentiation. *Learning & Memory*, 6, 399-416.

Talamini, L. M., Nieuwenhuis, I. L., Takashima, A., & Jensen, O. (2008). Sleep directly following learning benefits consolidation of spatial associative memory. *Learning & Memory,* 15(4), 233-237.

Tamminen, J., Payne, J. D., Stickgold, R., Wamsley, E. J., & Gaskell, M. G. (2010). Sleep spindle activity is associated with the integration of new memories and existing knowledge. *The Journal of Neuroscience*, 30(43), 14356-14360.

Tanguay, P. E., Ornitz, E. M., Kaplan, A., & Bozzo, E. S. (1975). Evolution of sleep spindles in childhood. *Electroencephalography and Clinical Neurophysiology*, 38(2), 175-181.

Tononi, G., & Cirelli, C. (2003). Sleep and synaptic homeostasis: a hypothesis. *Brain Research Bulletin;* 62, 143-150.

Tononi, G., & Cirelli, C. (2006). Sleep function and synaptic homeostasis. *Sleep Medicine Review*; 10, 49-62.

Tse, D., Langston, R. F., Kakeyama, M., Bethus, I., Spooner, P. A., Wood, E. R., ... & Morris, R. G. (2007). Schemas and memory consolidation. *Science*,*316*(5821), 76-82.

Tucker, M. A., Hirota, Y., Wamsley, E. J., Lau, H., Chaklader, A., & Fishbein, W. (2006). A daytime nap containing solely non-REM sleep enhances declarative but not procedural memory. *Neurobiology of Learning & Memory*; 86(2), 241-7.

Valentino, R.J., & Dingledine, R. (1981). Presynaptic inhibitory effect of acetylcholine in the hippocampus. *Journal of Neuroscience*; 1, 784-792.

Van Der Werf, Y. D., Altena, E., Schoonheim, M. M., Sanz-Arigita, E. J., Vis, J. C., De Rijke, W., & Van Someren, E. J. (2009). Sleep benefits subsequent hippocampal functioning. *Nat Neurosci.*; 12, 122–123.

van Kesteren, M. T., Fernández, G., Norris, D. G., & Hermans, E. J. (2010). Persistent schema-dependent hippocampal-neocortical connectivity during memory encoding and postencoding rest in humans. *Proceedings of the National Academy of Sciences*, *107*(16), 7550-7555.

Vazquez, J. & Baghdoyan, H. A. (2001). Basal forebrain acetylcholine release during REM sleep is significantly greater than during waking. *American Journal of Physiology-Regulatory, Integrative and Comparative Physiology*, 280(2), R598-R601.

Wagner, U., Gais, S., & Born, J. (2001). Emotional memory formation is enhanced across sleep intervals with high amounts of rapid eye movement sleep. *Learning & Memory;* 8, 112–119.

Wagner, U., Gais, S., Haider, H., Verleger, R., & Born, J. (2004). Sleep inspires insight. *Nature*; 427(6972), 352-5.

Wagner, U., Hallschmid, M., Rasch, B., & Born, J. (2006). Brief sleep after learning keeps emotional memories alive for years. *Biological psychiatry*, *60*(7), 788-790.

Whitlock, J. R., Heynen, A. J., Shuler, M. G., Bear, M. F. (2006). Learning induces long-term potentiation in the hippocampus. *Science*; 313, 1093-1097.

Wilhelm, I., Diekelmann, S., Molzow, I., Ayoub, A., Mölle, M., & Born, J. (2011). Sleep selectively enhances memory expected to be of future relevance. *The Journal of Neuroscience*, *31*(5), 1563-1569.

Wilson, M.A., & McNaughton, B.L. (1994). Reactivation of hippocampal ensemble memories during sleep. *Science*; 265(5172), 676-679.

Xie, L., Kang, H., Xu, Q., Chen, M. J., Liao, Y., Thiyagarajan, M., … Iliff, J. J. (2013). Sleep Drives Metabolite Clearance from the Adult Brain. *Science*, 342(6156), 373–377.

Xie, L., Kang, H., Xu, Q., Chen, M. J., Liao, Y., Thiyagarajan, M., ... & Nedergaard, M. (2013). Sleep drives metabolite clearance from the adult brain. *Science*, *342*(6156), 373-377.

Yonelinas, A. P. (2002). The nature of recollection and familiarity: A review of 30 years of research. *Journal of memory and language*, *46*(3), 441-517.

Zeitlhofer, J., Gruber, G., Anderer, P., Asenbaum, S., Schimicek, P., & Saletu, B. (2003). Topographic distribution of sleep spindles in young healthy subjects. *Journal of sleep research,* 6(3), 149-155.

In: Memory Consolidation
Editors: Manabu Sakakibara and Etsuro Ito
ISBN: 978-1-63482-596-2

Chapter 12

EMOTIONS AND EYE MOVEMENTS: EYE TRACKER AND MNESTIC PARAMETERS

***Rosa Angela Fabio*[1], *Iconia Gullà*[1] *and Antonino Errante*[2]**
[1]Department of Cognitive Science, University of Messina, Italy
[2]Department of Humanities and Social Sciences, Division of Psychology, University of Messina, Italy

ABSTRACT

The purpose of this chapter is to illustrate the mains mechanisms implicate in memory consolidation for stimuli and objects with emotive contents. Emotions are a fundamental part in human life. For this reason emotions are in close connection with other cognitive functions, such as attention and memory. In particular, attention influences memory in different ways, mainly by capture of attentive sources. Emotions "retain" attention and direct it toward emotionally salient stimuli. In the first part of the chapter the relationship between attention and eye movement is explored. The study of eye movements is an important instrument to understand mechanisms of selective attention and switching it in the space. In fact, examination of the eye movement pattern of a viewer on a scene demonstrates that viewers do not fixate every part of the scene. For this reasons the use of eye tracker is aimed at understanding the mechanisms underlying the shift of attention in the space. After, the relation between eye movement and emotive responses is described. Pattern of different movements are linked to selective emotive responses modulated by attention. Furthermore, the capture of selective attention by stimuli with emotive contents is displayed, as an example of asymmetry in evaluation of emotive and neutral stimuli. Since attention precedes eye movements, indirectly attention is the vehicle by which information is stored in the working memory. The brain networks responsible for the planning and execution of eye movements are describe as partially overlapping with those involved in spatial attention and spatial working memory. Also, we describe the main evidence on gender differences in emotion recognition and visual scanning, because a great number of studies showing that male and female using different strategy to perceptual identify emotion expressions. The impact of strategy of scanning, together attentive and working memory mechanisms were explored based on sex differences. In the last part of the chapter, we present an experimental study, carry out to analyze the differences between male and female in relation to visual task containing neutral and emotional stimuli. More over the study explore gender difference in relation to the mnestic index, specifically to check the correlation between memory, FL indexes (fixations length) and the emotional or neutral stimuli. Finally, a third aim of the study was measuring FL parameters only on facial expression of the stimuli, to investigate the differences between males and females on the fixation time of the faces.

INTRODUCTION

Eye Movement and Attention

The visual mode is, among the different sensory modalities, one of the most used to build consistent representations of the surrounding world, to acquire knowledge and information, to interact with the others. The direct relationship between eye movements and cognitive tasks has been widely studied. Memory and emotions are related too. A first type of relationship is based on the mediation of the attention. It is well known that viewers can allocate attention independent of eye position in simple tasks in which they must hold their eyes on a fixation target while other stimuli are presented elsewhere in the visual field (Posner, 1980). In many tasks, both (a) eye location (overt attention) and covert attention are overlapping and at the same location or (b) attention disengagement is a product of a saccade program (wherein attention precedes the eyes to the next saccade target). More researchers have started to use eye tracking techniques in the last few years. In large part, this is because systems to track the eyes have become more readily available and more user friendly. The two basic components of eye movements in the various tasks under consideration here are the movements themselves (called saccades) and the fixations (the period of time when the eyes remain fairly still and new information is acquired from the visual array). Since vision is suppressed during a saccade (Matin, 1974), new information is only acquired during fixations. There are special situations wherein information can be acquired during a saccade (see Uttal & Smith, 1968), but under most normal circumstances we do not obtain new information during a saccade because the eyes are moving so quickly across the stable visual stimulus that only a blur would be perceived. While new information is not encoded during saccades, cognitive processing does continue in most situations during the saccade (Irwin, 1998). Eye movements are motor responses that take time to plan and execute. Saccade latency, the time needed to encode the location of a target in the visual field and initiate an eye movement, is of the order of 175–200 ms (Rayner et al., 1983). The average fixation duration during scene perception tends to be closer to 300 ms (but varies as a function of the task and the characteristics of the scene). Examination of the eye movement pattern (or the scan path) of a viewer on a scene demonstrates that viewers do not fixate every part of the scene. Most fixations tend to fall on the informative parts of the scene. Viewers obtain preview benefit from objects that they have not yet fixated (Henderson, 1992; Pollatsek, Rayner and Henderson, 1990) with the amount of the preview benefit of the order of 100ms. Interestingly, as is now rather well known, viewers are rather insensitive to changes in scenes. Grimes and McConkie (1995) and McConkie and Currie (1996) asked observers to view scenes with the task of memorizing what they saw. They were also told that changes could be made to the image while they were examining it, and they were instructed to press a button if they detected the changes. While observers viewed the scenes, changes were made during a saccade (when vision is suppressed). Remarkably, observers were unaware of most changes, which included the appearance and disappearance of large objects and the changing of colors. Other studies found that viewers were unable to detect changes when there was a corresponding disruption to processing, such as with the simultaneous onset of patches covering portions of the scene (O'Regan, Rensink and Clark, 1999). However, this lack of awareness of changes does not mean that there is no recollection of any visual details, but rather that the likelihood of remembering visual information is highly dependent on the processing of that information (Henderson and Hollingworth, 2003). A strong link exists between spatial attention and eye movements. Attention precedes an eye movement, and attention is the vehicle by which information is stored in the working memory. Early theories of trans-saccadic memory (Jonides, Irwin, & Yantis, 1982; McConkie & Rayner, 1976b) proposed that information is integrated across saccades in an integrative visual buffer (with properties like iconic memory). There is evidence that visual short-term memory (VSTM), which is thought to be at a higher level than a visual buffer, serves a primary role in integrating information across saccades (Hollingworth, Richard and Luck, 2008). Thus, memory across saccades during scene perception appears to be due to higher level visual codes (Carlson-Radvansky, 1999; Irwin and Gordon, 1998), which are abstracted away from precise sensory representations, with VSTM as the basis for integration. It is clear that viewers' eyes are drawn to important aspects of the visual scene and that their goals in looking at the scene very much influence their eye movements. It is also clear that the saliency of different parts of the scene influence what part of the scene is fixated (Parkhurst and Niebur, 2003). Neider and Zelinksy (2006a) examined the influence that context has on the placement of eye movements in search of certain objects within pseudo-realistic scenes. Viewers were asked to look for target objects that are typically constrained to certain parts of the scene (i.e., jeep on the ground, hot air balloon in the sky). When a target was present, fixations were largely limited to the area where one would expect to find the target object (i.e., ground or sky), while, when the target was absent, search was less restricted to these areas. More results suggest that the initial scene representation is based on abstract, visual information that is useful across changes in spatial scales. Thus, the information used to guide eye movements has two sources: the saliency of the scene and the information in memory about that scene and scene type. To bring aspects of the visual world in the focus of attention, eye fixations are directed to those visual aspects of interest. This makes a shift of gaze between the visual stimuli and the overt sign of the attention orienting.

Eye Movement and Emotive Responses

Harris et al. (2008) introduced emotional, yet somewhat unusual, aspects into scenes; for example, in a control scene people are tossing a beach ball at a beach, while in the emotional scene the beach ball is replaced by a baby. They found that viewers looked earlier at the emotional aspect of the scene. A number of other recent studies (Calvo and Lang, 2005; Calvo and Nummenmaa, 2007; Kirchner and Thorpe, 2006) have likewise reported that the eyes quickly move to emotional objects or scenes (though in these studies, the object/scene is usually presented in parafoveal vision, and the latency of a saccade from a central fixation point is measured for an emotional scene/object versus a neutral scene/object). Many studies have shown that the meaning of a hedonic input perceptual is, in fact, automatically evaluated, implying fast and pre-attentive estimates, irrespective of awareness or intention by the subject (Robinson, 1998), but these processes do not appear to be independent from the attention (Pessoa et al., 2005). The evidences of the emotional events processing outside the attentional focus are contradictory. Some studies have reported that the emotional responses are not modulated by attention (Vuillemier et al., 2001; Anderson et al., 2003). Others have found opposite results (Pessoa et al., 2002; Eimer et al., 2003; Bishop et al., 2004; Ishai et al., 2004; Krolak-Salmon and coll., 2004), showing a strong interaction between emotion and attention characterized by a reduction of emotional response during the performance of particularly challenging tasks. The reasons for this divergence could be traceable in different experimental paradigms used, due to the different tasks and perceptual characteristics of the stimuli used. A first type of studies has shown that the manipulation of attentive load by varying cognitive processing demands may alter responses to emotional stimuli (Hariri et al., 2000; Banich et al., 2009; Luo et al., 2010). These studies demonstrated that increasing attentional demands (for example by varying the difficulty of the task) may attenuate response to emotional stimuli, even if the task does not require diverting attention from the emotional aspect of the stimulus set. A second strand of research has used neuroimaging to investigate the neural correlates of asking individuals to attend to their own internal emotional state and modify it in a specific direction (Ochsner and Gross, 2005; Beauregard, 2007; Wager et al., 2008). This form of explicit emotional control is often referred to as "voluntary" or "deliberate" emotion regulation (Gross and Thompson, 2007). The insight emerging from functional neuroimaging has been the recognition of the existence of distinct brain circuits that are activated on the one hand by the perceptual encoding of emotional stimuli, and on the other hand by regulatory processes. The amygdala (a brain structure part of the limbic system located in the medial anterior temporal lobe) has been consistently involved in the early perception of emotionally arousing stimuli (Davis and Whalen, 2001; Adolphs and Spezio, 2006). The control of emotional stimuli interfering with cognitive tasks is associated with activation of the prefrontal areas that are recruited during cognitive control and mnemonic encoding of general stimuli (Compton et al., 2003; Banich et al., 2009). These areas have been shown by studies of cognition to be involved together with medial prefrontal and parietal areas in a dorsal cortical network critical to cognitive control (Owen et al., 2005). Interestingly, theorists of emotion regulation also mention the existence of "automatic emotion regulation" forms (Mauss et al., 2007; Phillips et al., 2008) or of unconscious varieties of emotion regulation (Bargh and Williams, 2007), which contrast with the voluntary form just mentioned. The term automatic refers here to processes evoked by the stimulus and running without monitoring, or initiated without awareness and not subject to strong capacity limitations (Williams et al., 2009). However, almost all existing studies have focused on forms of emotion regulation mediated by top-down regulatory processes. The resulting account of emotion regulation is based on a dual-process model (Barrett et al., 2004) that opposes automatic sensory encoding of emotionally arousing stimuli on the one hand, and on the other an integration of attentive mechanisms and prefrontal function envisaged to account more generally for cognitive control processes (Posner and Rothbart, 1998; Posner et al., 2002; Hofmann et al., 2012). Emotionally arousing stimuli in bottom-up perceptual channels may be viewed as particularly effective in competing for access to short-term memory (Stanley et al., 2009), requiring strong bias by top-down processes to maintain cognitive control. An important debate has involved the influence of top-down processes at early stages of sensory encoding of emotional stimuli, or their relative automaticity (Pessoa and Adolphs, 2010; Dolcos et al., 2011). These influences suggest that the form of automaticity of early sensory encoding in the amygdala and visual pathways consists of the low level of control minimally required for its processing. Another important issue is the capacity of the amygdala and subcortical processing to prime visual cortex for processing emotionally salient stimuli, allowing the early perception of emotional stimuli to refocus attention effectively (Dolan, 2002; Bach et al., 2011).

Selective Attention Captured by Stimuli with Emotional Contents

An adaptive cognitive system is likely to be complemented by a perceptual mechanism that is biased to readily detect and process emotional stimuli among other, competing stimuli. If so, it may be assumed that affective stimuli are especially likely to capture attention. Three paradigms are used to investigate whether emotional stimuli capture attention. In the *dot-probe paradigm* (e.g., Mogg and Bradley, 1999) two pictures—one neutral and one emotional—are briefly displayed simultaneously side by side. When they disappear, a dot probe replaces one of them, and the participants' task is to press a button as soon as they detect the dot. A shorter reaction time for an emotional than a neutral picture indicates that the observer was attending to the emotional picture at the time when the dot probe was presented. Anxious individuals have been found to react particularly quickly when the dot replaces an angry face (Bradley et al., 1998; Mogg and Bradley, 1999). A second paradigm that has been frequently used with emotional pictorial stimuli is the visual search task (Ohman, Flykt and Esteves, 2001). Participants are required to search a pre-specified target stimulus (e.g., a spider) embedded in an array of no target stimuli (e.g., flowers). When more complex stimuli such as pictures of animals have been used, the results have been less clear cut. Oman, Flit and Estevez (2001) observed faster identification of phobic animals among no phobic stimuli than vice versa. In contrast, a search detection advantage has also been observed for pictures of both threatening and pleasant animals (Tipples et al., 2002) and for any animal regardless of their fear relevance (Lip et al., 2004). A third experimental paradigm is the exogenous cueing task (Posner, 1980), in which a picture is presented very briefly as a cue to the right or left of a fixation point and a target (e.g., a square) is presented at the same (valid trials) or opposite (invalid trials) location where the cue had appeared. Faster reaction times for the target in valid trials are assumed to indicate attention orienting to the cued picture, whereas slower responses in invalid trials reflect attentional engagement on the cued picture. Fox et al. (2001) found no reaction time benefit for trials validly cued with angry or happy expressions when compared with trials validly cued with neutral expressions. In other words, this study found no bias in attentional orienting to emotional faces. Instead, participants were slower to respond on trials invalidly cued with angry expressions, which suggests that emotional content engaged attention, especially for anxious individuals. This finding has been replicated by other studies using both pictures of angry faces and of threat-related scenes (Fox, Russo and Dutton, 2002; Yiend and Mathews, 2001). Recently, Nummenmaa, Hyona and Calvo (2006) using the eye-tracking method to assess attentional orienting and engagement on emotional visual scenes. In Experiment 1, unpleasant, neutral, or pleasant target pictures were presented simultaneously with neutral control pictures in peripheral vision under instruction to compare pleasantness of the pictures. The probability of first fixating an emotional picture, and the frequency of subsequent fixations were greater than those for neutral pictures. In Experiment 2, participants were instructed to avoid looking at the emotional pictures, but these were still more likely to be fixated first and gazed longer during the first-pass viewing than neutral pictures. Low-level visual features cannot explain the results. The author concluded that overt visual attention is captured by both unpleasant and pleasant emotional content. Relatively few neuroimaging studies in the healthy have combined spatial attention paradigms with emotional stimuli. Two studies using emotional stimuli conditioned to aversive events found activations that extended from the IPS into the supramarginal gyrus and anteriorly toward the secondary sensory cortex (Fredrikson et al., 1995; Armony and Dolan, 2002). In contrast, a recent study in which the cue was conveying information about the magnitude of reward reported only weak or no effect on ventral parietal or temporal-parietal regions (Tosoni et al., 2013). A larger number of studies have investigated the effect of attention to emotion in tasks originally devised to study cognitive interference. In these studies, emotional stimuli appear as distracters. The most obvious use of emotional material is as a spatial distracter in a selection or identification task (Eriksen and Eriksen, 1974). Temporal distracters have an interfering effect on reports on the identity of a stimulus if they immediately precede or follow the target stimulus (Broadbent and Broadbent, 1987).

Interaction between Attention, Working Memory and Eye Movement

One of the most influential conceptualizations of working memory is the multiple component model of Baddeley and colleagues (e.g., Baddeley, 1986; Baddeley and Hitch, 1974). In the original model, Baddeley and colleagues suggested two independent buffers for the storage of verbal and visuospatial information. In a later version of the model, an additional buffer was added called ''episodic buffer" which was assumed to link information across domains to form integrated units of visual, spatial, and verbal information (Baddeley, 2000). The Central Executive makes it possible to manage information in and between the different buffers. Baddeley (1986) suggested that, analogous to articulator rehearsal of verbal material, the oculomotor system may play a role in keeping visuospatial information "in mind" during a delay period between study and test. This suggestion that eye movements support memory echoed similar sentiments expressed by Noton and Stark (1971, a-b), who theorized that eye movements are a critical aspect of perception and the subsequent activation of the corresponding internal image during recognition.

A great number of evidence has indicated that the allocation of spatial attention may support spatial working memory (see Awh and Jonides, 2001, Pearson and Sahraie, 2003). Given that the brain networks responsible for the planning and execution of eye movements are partially overlapping with those involved in spatial attention and spatial working memory (Awh, Armstrong and Moore, 2006; Belopolsky and Theeuwes, 2009). When a location is kept in working memory, processing of stimuli at the memorized location is facilitated relative to other locations (Awh and Jonides, 2001), just like attending to a location improves the processing of information at that location (Posner, 1980). Conversely, when attention to memorized locations is interrupted, the ability to remember these locations is impaired. Brain imaging studies of working memory confirm the notion that rehearsal of spatial information modulates early sensory areas (Awh et al., 1999), and that the same fronto-parietal network is involved as in attention-related tasks (Kastner and Ungerleider, 2000). Studies investigating visuospatial working memory use a procedure that is similar to the exogenous cueing task used in attention research. Typically, a peripheral onset cue is presented with abrupt onset somewhere in the visual field (see, for example, Awh et al.'s 1998). However, instead of responding to a target, the main task of the participant is to remember the location of the abrupt onset cue. During the retention interval (which could last up to 5 sec), participants are required to make a speeded response to a probe stimulus presented somewhere in the visual field. In 25% of the trials, the probe stimulus happens to appear at the location that was kept in memory. The crucial finding is that when the probe happened to match the location kept in memory, probe reaction time (RT) was about 15 ms faster then when the probe was presented elsewhere in the visual field.

The interpretation of these findings is that during the retention interval attention is endogenously shifted to the location of the abrupt onset cue. This active shift of focal spatial attention to a specific location in space is assumed to allow an active maintenance of this location in working memory. The work of Awh and colleagues suggests that spatial attention is used to keep information in working memory. Research conducted by Schmidt and colleagues (2002) showed that focusing attention exogenously or endogenously on a spatial location increases the probability that information at that location will be transferred into visual working memory.

There appears to be a strong link between visual working memory and attention. In turn, an equally strong link exists between spatial attention and eye movements. Previous research has shown that the eyes typically move to the location where attention is allocated, and, vice versa, attention is allocated at possible saccade targets (Deubel and Schneider, 1996; Hoffman and Subramaniam, 1995). The relationship between eye movements and memory has been explored with tasks that require the maintenance and/or retrieval of spatial information. For example, during mental imagery and memory retrieval of spatial information, eye fixations were directed toward locations that were initially studied (Johansson et al., 2012; Johansson and Johansson, 2013).

Tremblay, Saint-Aubin and Jalbert (2006) assessed the contribution of delay period eye movements to visuospatial memory performance during the recall of studied locations in serial order, which requires both the formation of spatial memory representations as well as memory for the correct temporal sequence of the study item presentations. Their findings indicated that eye movements during the delay period to previously studied locations benefitted memory performance for serially presented visuospatial information, but questions remain about whether memory for the spatial and/or the temporal information was affected by these overt shifts of attention. A number of studies have shown that eye movements disrupt spatial working memory (e.g., Baddeley and Lieberman, 1980; Smyth and Scholey, 1994). For example, Smyth and Scholey (1994) demonstrated that additional tasks that require shifts of spatial attention disrupt performance of the *Corsi Blocks Task*, a typical clinical measure used to determine spatial memory span.

The underlying notion is that attention is needed both to maintain locations active in spatial working memory and for the execution of eye movement. Because both processes make use of the same attentional mechanism, working memory performance deteriorates when eye movements have to be executed.

Visual Scanning and Emotion Recognition: Gender Differences

Neuroimaging studies have found that males and females use different frontal and subcortical regions when perceiving facial expressions of emotion. For example, gender differences in amygdale activation appear to depend on the valence of the emotion, with males showing greater right amygdala activation when perceiving happy expressions, and both genders showing increased activation in the left amygdala for sad ones (Killgore and Yurgelun-Todd, 2001). Behavioral studies provide further evidence for gender differences in emotion perception, with females performing better than males when recognizing facial emotion (Rahman, Wilson and Abrahams, 2004), vocal emotion (Voyer & Rodgers, 2002) and non-emotional faces (McBain, Norton and Chen, 2009). Females appear to excel in facial emotion recognition when face identification and discrimination are more difficult (McBain et al., 2009) and emotions are faint and subtle (Hoffmann et al., 2010). Evidence also suggests that males and females use different perceptual strategies when identifying emotional expressions. For example, Vassallo, Cooper and Douglas (2009) found out that males and females showed different scanning patterns when discriminating facial emotions, with both genders more often fixating the eye region, but with males making more fixations on nose and mouth. Vassallo et al. (2009) suggest that the extra time spent by males fixating non-salient regions, such as the nose, may cause them to be less efficient at recognizing emotional expressions if compared to females. However, since it has been found that the nose and the mouth are important for recognizing ''disgust'' and ''happy'' expressions, respectively, and the eyes are particularly important in the fear recognition (Schyns, Petro and Smith, 2007), Vassallo et al.'s results (2009) also indicate that males spend longer scanning additional facial regions that are important for the identification of particular expressions. While some researchers have reported that females demonstrate greater accuracy overall in decoding facial affect (Nowicki and Hartigan, 1988), others have found that sex differences only exist for certain emotional expressions. For example, Mandal and Palchoudhury (1985) found that females were highly accurate in their perception of sad faces, while males were better at recognizing anger. This trend has also been noted in other studies (Biele and Grabowska, 2006), while Goos and Silverman (2002) reported that females were more sensitive to angry and sad expressions specifically when the poser was female. Rahman et al. (2004) reported no overall sex differences in accuracy in their large sample of participants (N = 240; 120 males) but noted a sex difference in the speed of identifying the emotion. Females were significantly faster than males at identifying happy, sad, and neutral faces, though these were the only expressions used in the study. In the same year, Palermo and Coltheart (2004) created a large normative database of information about facial affect processing, when they repeatedly showed seven different universal facial expressions of emotion (N = 336 stimuli) to 12 males and 12 females. There were no significant differences in relation to observer gender (or stimulus gender) when reaction time was recorded using a voice key. Miyahira et. al. (2000) found sex differences in exploratory eye movements made to different schematic pictures but did not assess pictures of facial affect. The manner in which a face is viewed can be assessed in relation to the frequency and duration of fixation to an area of interest. For example, Vassallo, Cooper and Douglas (2009) assessed whether differences exist in the way males and females overtly orient their visual attention to salient facial features while viewing static emotional facial expressions. Eye movements were recorded while fifty healthy participants viewed a series of six universal facial expressions. The number and duration of foveal fixations to four predefined facial areas of interest (AOIs) each eye, nose, mouth were also recorded.

There were no significant group differences with respect to accuracy, though females were significantly faster than males in correctly identifying expressions. While both groups spent more time and looked more frequently at the eye region, males spent significantly more time viewing the nose and mouth. Interestingly, gender differences become much more apparent when facial stimuli are presented at the edge of consciousness (Hoffmann et al., 2010).

Thus, women appear to recognize facial emotions better than men in particular under conditions of minimal stimulus information, i.e., when facial expression is shown for less than a second. Moreover, women appear to display more emotional contagion to both positive and negative expressions than men in a dyadic interaction setting concerning subjective experience and facial response (Doherty et al., 1995). Murphy and Zajonc (1993) exposed participants to subliminal facial expressions of emotions that varied in valence (angry, happy, and neutral) and were masked by ideographs. It was observed that valence of expression influenced judgments of subsequent ideographs. In the happy prime condition ideographs were evaluated more positively than those shown in the neutral prime condition, whereas in the angry prime condition ideographs were evaluated more negatively than those presented in the neutral control condition. This phenomenon of affect congruent influence of emotional facial expression on subsequent judgments is called *affective priming effect*. Recently, Donges , Kersting and Suslow (2012) investigated the effect of gender on affective priming based on negative and positive facial expression. In a priming experiment sad, happy, neutral, or no facial expression was briefly presented (for 33 ms) and masked by neutral faces which had to be evaluated. Subjects had no subjective awareness of emotional primes. Women did not differ from men with regard to age, education, intelligence, trait anxiety, or depression. In the whole sample, happy but not sad facial expression elicited valence congruent affective priming. Furthermore, women manifested greater affective priming due to happy faces than men. Women seem to have a greater ability to perceive and respond to positive facial emotion at an automatic processing level compared to men. High perceptual sensitivity to minimal social-affective signals may contribute to women's advantage in understanding other persons' emotional states.

An Experimental Study

The purpose of the present study was to analyze differences between male and female in relation to visual task containing neutral and emotional stimuli through the parameters used in Eye tracker : FC (Fixation count) and FL (fixation length) with neutral and emotional stimuli. Another aim was to study gender difference in relation to the mnestic index, specifically to check the correlation between the mnestic and FL indexes (fixations length) and the emotional or neutral stimuli. The third aim, measuring FL parameters only on facial expression of the stimuli, was to investigate the differences between males and females on the fixation time of the faces.

METHOD

Participants

The project involved 56 students, 38 females and 18 males, aged from 22 to 24 (mean age 22.9 ± 3,5) years old of the Cognitive Science and Education Department of the University of Messina. Participants completed a general test in which it was requested to write general information about sex, age, and educational qualifications.

Materials

An unobtrusive eye tracking device, a model "iAble© - MyTobii®" ("D10" version) was used in this study. Before starting the eye tracking recording, all the subjects were taken through a calibration procedure. During such procedure, the eye tracker measured the features of the user's eyes and used them together with an internal, physiological 3D eye model to calculate the gaze data. This model includes information about shapes, light refraction and reflection properties of the different parts of the eyes (e.g., cornea, placement of the fovea, etc.).

During the calibration the user was asked to look at specific points on the screen, also known as calibration dots. During this period, several images of the eyes were collected and analyzed. The resulting information was then integrated in the eye model and the gaze point for each image sample was calculated. When the procedure was at the end, the quality of the calibration has been illustrated by green lines of varying length. The length of each line represents the offset between each sampled gaze point and the center of the calibration dot. My Tobii instruments have been able to define the Areas Of Interest (AoI) within the stimuli chosen for the statistical analysis of the eye tracking. AOI cluster refers to selected specific areas that will be used for the recalling details of the images. Six neutral and six emotional images were presented.

Design

It was used a mixed factorial design with repeated measures: 2 (gender: male vs. female) x 2 (category: Emotional content and Neutral content) x 6 (Specific emotional content: sadness, anger, fear, happiness, disgust, surprise; Specific neutral content: students, girls, girl at a computer, painter, man watching TV, children in the kitchen).

Procedure

In a preparation phase, two independent observers analyzed each image. For each image ten peripheral details were selected. The inter-rater agreement was at minimum 70%. For this reason, three were deleted and seven details were included for each image. A Total of 84 AoI were prepared. With reference to each selected AoI, the related item to be remembered was also prepared. So, the total list of the items was made up of 84 questions: 42 about neutral content and 42 about emotional content.

Each subject was tested individually in a quiet room. The images were presented on a computer monitor equipped with the eye tracker technology. All the images were presented in randomized order for 15 second each. After each presentation the subject had to reply to seven open questions on the images just watched. Then the following image was presented, and so on. The experimenter registered all the correct answers.

Figure 1. An example of an image presented in the study. The rectangles with different colors are the areas of interest (AoI) which correspond to specific statistics and specific questions.

RESULTS

With reference to the first aim related to the eye tracker parameters, gender show no main significant effect, both with fixation count (FC) and fixation length (FL). Only interaction effect "Gender x Specific content" was significant, with reference to fixation length, $F(1, 54) = 22.49$ $p<.001$. As it is shown in table 1, male fixation length is higher in some emotional contents: sadness and fear.

As regards to the second aim related to mnemonic index, gender show no significant effect. Interaction "Category x Specific content" was significant, $F(5, 270) = 17.56$ $p<.001$ the details of some images remain longer in memory. Correlation between FL and Mnemonic index were evaluated. There were correlations between mnemonic index and fixation lengths with reference to anger ($r = .21$, $p=.06$) and surprise ($r = .25$, $p=.05$), while other specific contents showed no statistically significant correlations.

To give value to the third hypothesis, the AoI was focused only on the face of the people in the images with an emotional and neutral content (table 2).

The variable "category" shows significant effects, F (1, 54) = 17.02, p <0.01. The face is more gazed if the images evoke emotions, while images with neutral content are gazed for shorter time. With regard to the interaction "category of content x specific content" there are significant effects, F (5, 270)= 10.64 p<0.01. Also the interaction "gender x category" shows a significant effect, F(1, 270)= 18,31, p<0.01. Women tend to look longer emotionally involving situations. In Figure 2 it is possible to analyze an example of a woman gaze faces.

Table 1. Means and standard deviation of the seven peripheral details for each condition

Category	Man		Woman	
Fixation count				
Emotional Content:	Mean	SD	Mean	SD
Sadness	30,88	7,24	28,78	8,31
Anger	19,55	7,22	19,71	6,67
Fear	24,5	9,13	23,21	9,74
Happiness	15	7,28	13,15	6,06
Disgust	27,28	14,6	28,78	10,82
Surprise	15,33	10,44	17,73	6,47
Neutral content				
Students	26,61	9,17	25,5	6,79
Girls	18,66	9,34	19,89	5,94
Girl at pc	23,27	10,4	25,15	8,07
Painter	21	11,76	20,74	8,08
Man Tv	17,22	11,11	17,84	7,15
Children	20,50	10,00	21,31	8,18
Fixation Length				
Emotional Content	Mean	SD	Mean	SD
Sadness	11,62	3,54	9,84	4,93
Anger	7,56	5,21	5,62	3,37
Fear	11,11	4,69	9,11	5,57
Happiness	6,46	3,41	5,49	3,18
Disgust	8,73	5,03	9,13	5,43
Surprise	6,23	4,36	6,20	3,42
Neutral content				
Students	**8,89**	3,87	7,23	3,15
Girls	**6,88**	4,40	6,42	3,18
Girls at pc	**8,55**	4,27	8,09	3,98
Painter	**8,09**	4,45	7,05	3,92
Man Tv	**5,92**	3,82	6,54	4,04
Children	**8,21**	3,78	8,29	3,56

Table 2. Means and Standard Deviation of FL on the AOI on the faces

Category	Man		Woman	
Fixation Length				
Emotional Content	Mean	SD	Mean	SD
Sadness	4,01	1,78	5,12	1,76
Anger	6,12	1,32	8,25	1,78
Fear	3,74	1,44	4,63	1,32
Happiness	5,22	1,62	6,27	1,16
Disgust	3,21	2,13	3,53	1,74
Surprise	3,52	1,54	4,28	1,44
Neutral content				
Students	**2,77**	1,32	3,01	1,67
Girls	**2,64**	1,61	2,78	1,05
Girls at pc	**2,15**	1,06	2,65	1,61
Painter	**3,01**	1,43	3,08	1,53
Man Tv	**2,44**	2,02	2,81	2,21
Children	**2,87**	1,15	3,05	1,23

Figure 2. Gaze plot mode: a view of the motion sequence, order and duration of the gaze fixation for each interviewed, separately (in this it is presented a typical women gaze).

CONCLUSION

The aim of this chapter was to explore the relationship between cognitive functions and eye movements. Behavioral research has shown that contextual information plays an important role in object detection. Changes in real-world scenes are noticed more quickly for objects and regions of interest (Rensink et al., 1997), and scene context can influence the detection of a change (Hollingworth and Henderson, 2000), suggesting a preferential deployment of attention to these parts of a scene. Experimental results suggest that the selection of these regions is governed not merely by low-level saliency but also by scene semantics (Henderson and Hollingworth, 1999). The global scene representation delivers contextual information in parallel with the processing of local features, providing a formal realization of an efficient feed-forward mechanism for the guidance of attention. An early impact of scene context is also compatible with the reverse hierarchy theory (Hochstein and Ahissar, 2002) in which properties that are abstracted late in visual processing (such as object shapes, categorical scene description) rapidly feed back into early stages and constrain local processing. Recently, there has been considerable interest in using eye movements in tasks that presumably share many components with scene perception and visual search. Here the focus is primarily on research using eye movements to examine how viewers search through arrays to find specific targets. Research shows that many phenomena related to visual attention taking place when selecting relevant information from the environment are similar to processes needed to keep information active in working memory. New data show that when retrieving information from working memory, people may allocate visual spatial attention to the empty location in space that used to contain the information that has to be retrieved. Moreover, we show that maintaining a location in working memory not only may involve attention rehearsal, but might also recruit the oculomotor system. The idea that spatial attention plays a functional role in the active maintenance of location information is also known as the attention-based rehearsal hypothesis (Awh, Armstrong and Moore, 2006). Note, however, that a recent behavioral study challenged the classic attention-based rehearsal hypothesis of Awh et al. (2006). Facilitation of processing comes from the maintenance of endogenous attention (attention-based rehearsal), while inhibition is most likely a result of suppression of an oculomotor program. Spatial attention controls the transfer of perceptual representations into visual working memory (see, Belopolsky, Kramer and Godijn, 2008). As described above, there appears to be a strong link between visual working memory and attention. In turn, an equally strong link exists between spatial attention and eye movements. In particular, attention precedes an eye movement, and attention is the vehicle by which information is stored in working memory (as concluded by Schmidt et al., 2002). According, face recognition of emotional expressions can be linked to attentive and visual working memory capacity. Indeed, the pattern of eye movements when viewing a complex scene is driven by both bottom-up and top-down factors. When looking at complex scenes, such as photographs or pictures, participants typically look around the image through an initial exploratory stage with many short fixations that are largely driven by bottom-up salience, followed by a later stage with a slower pace of fewer saccades per second and an increasing influence of the task. In addition, emotional content in the stimulus can modulate memory for scene details, perhaps by influencing where participants look in the image (Calvo and Lang, 2004). Eye-fixation monitoring has recently been used to investigate preferential attention to emotional pictures. In the Calvo and Lang's study (2004) eye movement measures were collected to differentiate between attentional orienting and engagement, and complex visual scenes were presented to healthy individuals. The results indicated that an emotional picture, either pleasant or unpleasant, was more likely to be fixated than a neutral picture, and that emotional pictures were also fixated longer during the first 500 ms of stimulus exposure. This suggests that emotional pictures capture attention and also engage attention during the early stages of picture processing. Furthermore evidence showed that eye movement patterns have been linked to performance on immediate (IM) and long-term memory (LTM) tests. In the real world the capacity of orienting attention to relevant stimuli and recognition those with emotive contents is fundamental. The ability to interpret emotion through facial expression plays an important role in interpersonal interactions. In the normal adult population, the majority of research findings have supported the notion that females outperform males in their ability to accurately decode emotional facial expressions regardless of whether the viewed faces are presented in static or dynamic form. Basic expressions of emotion include anger, contempt, sadness, happiness, surprise, disgust, and fear/anxiety (Matsumoto and Ekman, 2004). While some researchers have reported that females demonstrate greater accuracy overall in decoding facial affect, others have found that sex differences only exist for certain emotional expressions. For example, Mandal and Palchoudhury (1985) found that females were highly accurate in their perception of sad faces, while males were better at recognizing anger. In our experiment, to observe the differences between males and females in relation to visual behavior, eye tracker parameters (FC, FL) have been used to analyze neutral images and images that are an expression of emotions such as anger, fear, sadness, happiness, surprise, disgust. The interaction "gender x content" was significant. In fact, males look at emotional content images longer: sadness, fear and happiness. The findings seem to be reflected in a study conducted by Mumenthaler et al. (2012) based on the notion of *social appraisal*. Appraisal theories of emotions agree that emotions are elicited by the subjective evaluation (appraisals) of events and situations (Roseman and Smith, 2001). Even if the importance of contextual information is considered to arise from the basic principles of these theories, appraisal processes have commonly been studied independently of the social context. In the said work linked to the social theory of Aaronson (2005), the author argues that the social factors facilitate the recognition of happiness and fear. Various factors that depend either on the perceiver or on the producer can influence the perception of emotional facial expression. For example, personality traits, such as the attachment style of the perceiver, affect the ability to recognize the displayed expression (see Fraley et al., 2006). The gaze of the producer is also a very important component of the face that can provide information about mental states and can interact with emotional expression. Recent works has also demonstrated that an identical facial expression was perceived differently depending on the accompanying body expression. Studies also indicate that contextual effects go beyond the influence of body expression. For example, Righart and De Gelder (2008) showed that facial expressions were categorized more quickly and more accurately if presented together with congruent as compared with incongruent background pictures. Furthermore, in our study no significant differences between males and females in relation to the mnestic indexes were found. The lack of difference in mnestic indexes reports is due to the fact that they were taken into considerations peripheral indexes. In fact, many data in the literature state that not all the details of the emotional situations are remembered well. Specifically, the central details of an emotional event are relatively good, while memory for irrelevant information, the so-called *peripheral details*, is relatively poor (Christianson, 1992; Kensinger et al., 2005; Pickel, Franch and Betts, 2003). Therefore, the results of the present study seem to confirm further that neutral pictures attract less the attention of the subjects as already reported in literature (Calvo and Lang, 2004, 2005; Nummenmaa, Hyönä and Calvo, 2006) and suggest the effect of thematic content as an additional significant aspect of visual stimuli influencing the exploratory visual behavior. The cognitive representations of the emotional stimuli in short-term memory may persist longer than those of the neutral ones. In several studies it has been also shown that the detection of angry facial expressions is prioritized by our cognitive system, because the detection of facial threat is an evolutional advantage (Calvo, Avero and Lundqvist, 2006; Fox et al., 2000). Therefore, according to their standpoint, angry faces do not capture attention only because they represent a danger, but also because they show negative effects. Looking at gender difference, our results are partially in line with the study of Hall and Matsumoto (2004), where women recognized more facial emotions than man when the stimuli was presented at the edge of consciousness. Also our eye tracker data suggest that female fixation length were longer in emotion such sadness, anger, happiness and fear. Explanations for the sex difference range from sexual inequalities in power and social status (e.g., see Hall, 1984) to evolutionary perspectives based on women's near-universal responsibility for child-rearing (e.g., Babchuk, Hames and Thompson, 1985). The primary caretaker hypothesis proposed by Babchuk et al. (1985) contends that females, as a result of their evolutionary role as primary caretakers, will display evolved adaptations that enhance the probability of survival of their offspring. In humans, these adaptations are hypothesized to include the fast and accurate decoding of facial affect, an important means of communication especially in preverbal infants. A second interpretation, the fitness threat hypothesis, assigns a special status to negative emotions. It predicts a female superiority that is limited to expressions of negative emotion including fear, disgust, sadness, and anger. Although both sexes have a stake in infant survival, the ability to swiftly and accurately identify potential threat is a basic adaptation to the role of primary caretaker and would be maximized in the sex having the largest investment in each offspring. Finding a female advantage that is selective to negative emotions would constitute support for the fitness threat hypothesis. The possibility that the sex difference does not apply equally to all emotions is plausible from a neurobiological perspective. Perception of facial identity is separate from perception of facial emotion and can be differentially impaired by brain lesions. The neural regions and networks employed in facial affect recognition are vast, complex, and intertwined. They have been further described in recent years via the use of fMRI, or PET. For example, bilateral frontal and left parietal cortices are activated during the viewing of happy faces, while the amygdala predominantly subserves negative emotional recognitions. Sex differences in neural activation during the viewing of facial affect have been noted, especially for negative facial expressions (e.g., sad, fearful). Using chimeric faces, Bourne (2005) examined sex differences in hemispheric dominance to a happy facial expression. With accuracy as the dependent measure, males were found to be more lateralized (i.e., right hemisphere dominant) than females in their processing of positive facial affect. A study of Vassallo, Cooper and Douglas (2009) show that males tend to make more fixations and fixate for a longer period at the lower part of the face, especially the nose, when compared with females. Overall, however, the method of extraction of facial information by both groups involved predominantly orienting visual attention to the eyes rather than the nose or mouth. This pattern of visual search is a typical finding in normal healthy populations. In conclusion, the relationship between selective attention and working memory is a fundamental prerequisite to investigate the mnestic and recognition of emotive expression capacity. The role of eye movements is central to understanding the cognitive process to memory consolidation.

REFERENCES

Adolphs, R. & Spezio, M. (2006). Role of the amygdala in processing visual social stimuli. *Prog. Brain. Res.* 156, 363–378.

Anderson, A. K., Christoff, K., Panitz, D., De Rosa, E., Gabrieli, J. D. (2003). Neural correlates of the autonomic processing of treat facial signals. *Journal of Neuroscience*, 23, 5627-5633.

Armony, J. L., & Dolan, R. J. (2002). Modulation of spatial attention by fear-conditioned stimuli: an event-related fMRI study. *Neuropsychologia*, *40*(7), 817-826.

Aronson, E., Wilson, T.D., & Akert, R.M. (2005). Social psychology, 7th ed. Upper Saddle River, NJ: Pearson Education, Inc.

Awh, E., & Jonides, J. (2001). Overlapping mechanisms of attention and spatial working memory. *Trends in Cognitive Sciences*, 5(3), 119–126.

Awh, E., Armstrong, K. M., & Moore, T. (2006). Visual and oculomotor selection: Links, causes and implications for spatial attention. *Trends in Cognitive Sciences*, 10(3), 124–130.

Awh, E., Jonides, J., Smith, E. E., Buxton, R. B., Frank, L. R., Love, T., et al. (1999). Rehearsal in spatial working memory: Evidence from neuroimaging. *Psychological Science*, 10(5), 433–437.

Babchuk, W. A., Hames, R. B., & Thompson, R. A. (1985). Sex differences in the recognition of infant facial expressions of emotion: The primary caretaker hypothesis. Ethology and Sociobiology, 6, 89–101.

Bach, D. R., Talmi, D., Hurlemann, R., Patin, A., & Dolan, R. J. (2011). Automatic relevance detection in the absence of a functional amygdala. *Neuropsychologia*, *49*(5), 1302-1305.

Baddeley, A. D. (1986). Working memory. Oxford, UK: Oxford University Press.

Baddeley, A. D., & Hitch, G. (1974). Working memory. In G. Bower (Ed.), The psychology of learning and motivation, VIII (pp. 47–89). New York: Academic Press.

Baddeley, A. D., & Lieberman, K. (1980). Spatial working memory. In R. Nickerson (Ed.), Attention and performance VIII (pp. 521–539). Hillsdale, NJ: Erlbaum.

Baddeley, A. D. (2000). The episodic buffer: A new component of working memory? *Trends in Cognitive Science*, 4, 417–423.

Banich, M., Milham, M., Atchley, R., Cohen, N., Webb, A., Wszalek, T., ... & Magin, R. (2000). fMRI studies of Stroop tasks reveal unique roles of anterior and posterior brain systems in attentional selection. *Cognitive Neuroscience, Journal of*, *12*(6), 988-1000.

Banich, M. T., Mackiewicz, K. L., Depue, B. E., Whitmer, A. J., Miller, G. A., & Heller, W. (2009). Cognitive control mechanisms, emotion and memory: a neural perspective with implications for psychopathology. *Neuroscience & Biobehavioral Reviews*, *33*(5), 613-630.

Barrett, L. F., Tugade, M. M., & Engle, R. W. (2004). Individual differences in working memory capacity and dual-process theories of the mind. *Psychological bulletin, 130*(4), 553.

Beauregard, M. (2007). Mind does really matter: evidence from neuroimaging studies of emotional self-regulation, psychotherapy, and placebo effect. *Progress in neurobiology, 81*(4), 218-236.

Belopolsky, A. V., & Theeuwes, J. (2009). When are attention and saccade preparation dissociated? *Psychological Science*, 20(11), 1340–1347.

Belopolsky, A. V., Kramer, A. F., & Godijn, R. (2008). Transfer of information into working memory during attentional capture. *Visual Cognition,* 16(4), 409–418.

Biele, C., & Grabowska, A. (2006). Sex differences in perception of emotion intensity in dynamic and static facial expressions. *Experimental Brain Research,* 171, 1–6.

Bishop, S. J., Duncan, J., Lawrence, A. D. (2004). State anxiety modulation on the amygdala response to unattended threat-related stimuli. *Journal of Neuroscience*, 24, 10364-10368.

Blair, R. J., Morris, J. S., Frith, C. D., Perrett, D. I., and Dolan Bourne, V. J. (2005). Lateralised processing of positive facial emotion: Sex differences in strength of hemispheric dominance. *Neuropsychologia,* 43, 953–956.

Bradley, B. P., Mogg, K., Falla, S., & Hamilton, L. R. (1998). Attentional bias for threatening facial expressions in anxiety: Manipulations of stimulus duration. *Cognition and Emotion, 12,* 737–753.

Broadbent, D. E., & Broadbent, M. H. (1987). From detection to identification: Response to multiple targets in rapid serial visual presentation. *Perception & psychophysics, 42*(2), 105-113.

Brown, J. M. (2003). Eyewitness memory for arousing events: Putting things into context. *Applied Cognitive Psychology,*17, 93–106.

Calvo, M. G., & Lang, P. J. (2004). Gaze patterns when looking at emotional pictures: Motivationally biased attention. *Motivation and Emotion, 28*, 221–243.

Calvo, M. G., & Lang, P. J. (2005). Parafoveal semantic processing of emotional visual scenes. *Journal of Experimental Psychology: Human Perception and Performance,31*, 502-519.

Calvo, M. G., & Nummenmaa, L. (2007). Processing of unattended emotional visual scenes. Journal of Experimental Psychology: General, 136, 347–369.

Calvo, M.G., Avero, P., & Lundqvist, D. (2006). Facilitated detection of angry faces: Initial orienting and processing efficiency. *Cognition and Emotion, 20,* 785-811.

Carlson-Radvansky, L. A. (1999). Memory for relational information across eye movements. *Perception & Psychophysics*, 61, 919–934.

Christianson, S.-Å. (1992). Do flashbulb memories differ from other types of emotional memories? In E. Winograd, & U. Neisser (Eds.), *Affect and accuracy in recall* (pp. 191-211). Cambridge: Cambridge University Press.

Compton, R. J., Banich, M. T., Mohanty, A., Milham, M. P., Herrington, J., Miller, G. A., ... & Heller, W. (2003). Paying attention to emotion. *Cognitive, Affective, & Behavioral Neuroscience*, *3*(2), 81-96.

Davis M., Whalen P. J. The amygdala: vigilance and emotion. *Molecular Psychiatry.* 2001;6:13–34.

Deubel, H., & Schneider, W. X. (1996). Saccade target selection and object recognition: Evidence for a common attentional mechanism. *Vision Research,* 36(12), 1827–1837.

Doherty RW, Orimoto L, Singelis TM, Hatfield E, Hebb J (1995) Emotional contagion. Gender and occupational differences. *Psychol Women* Q 19: 355–371.

Dolan, R.J.(2002). Emotion, cognition, and behaviour. *Science* 298, 1191–1194.

Dolcos, F., Iordan, A. D., & Dolcos, S. (2011). Neural correlates of emotion–cognition interactions: a review of evidence from brain imaging investigations. *Journal of Cognitive Psychology*, *23*(6), 669-694.

Donges U-S, Kersting A, Suslow T (2012) Women's Greater Ability to Perceive Happy Facial Emotion Automatically: Gender Differences in Affective Priming. *PLoS ONE* 7(7): e41745.

Eimer, M., Holmes, A., McGlone, F. P. (2003). The role of spatial attention in the processing of facial expression : an ERP study of rapid brain responses to six basic emotions. *Cognitive and Affective Behavioral Neuroscience*, 3, 97-110.

Eriksen, B. A., & Eriksen, C. W. (1974). Effects of noise letters upon the identification of a target letter in a nonsearch task. *Perception & psychophysics*, *16*(1), 143-149.

Fox, E., Lester, V., Russo, R., Bowles, R.J., Pichler, A., & Dutton, K. (2000). Facial expressions of emotion: Are angry faces detected more efficiently? *Cognition & Emotion, 14,* 61-92.

Fox, E., Russo, R., Bowles, R., & Dutton, K. (2001). Do threatening stimuli draw or hold visual attention in subclinical anxiety? *Journal of Experimental Psychology: General, 130,* 681–700.

Fox, E., Russo, R., & Dutton, K. (2002). Attentional bias for threat. Evidence for delayed disengagement from emotional cues. *Cognition and Emotion, 16,* 335–379.

Fraley, R. C., Niedenthal, P. M., Marks, M., Brumbaugh, C., & Vicary, A. (2006). Adult attachment and the perception of emotional expressions: Probing the hyperactivating strategies underlying anxious attachment. *Journal of Personality, 74,* 1163–1190.

Fredrikson, M., Wik, G., & Andersson, J. (1995). Affective and attentive neural networks in humans: a PET study of Pavlovian conditioning. *Neuroreport*, *7*(1), 97-101.

Goos, L. M., & Silverman, I. (2002). Sex related factors in the perception of threatening facial expressions. *Journal of Nonverbal Behavior*, 26, 27–41.

Grimes, J., & McConkie, G. W. (1995). On the insensitivity of the human visual system to image changes made during saccades. In K. Atkins (Ed.), Problems in perception. Oxford, UK: Oxford University Press.

Gross, J. J., & Thompson, R. A. (2007). Emotion regulation: Conceptual foundations. *Handbook of emotion regulation*, *3*, 24.

Hariri, A.R., Bookheimer, S.Y., andMazziotta, J.C.(2000). Modulatingemotional responses: Effects of a neocortical network on the limbic system. *Neuroreport* 11, 42–48.

Hall, J. A. (1984). Nonverbal sex differences: Communication accuracy and expressive style. Baltimore Johns Hopkins University Press.

Hall, J. A., & Matsumoto, D. (2004). Gender differences in judgments of multiple emotions from facial expressions. *Emotion,* 4(2), 201−206.

Harris, C. R., Kaplan, R. L., & Pashler, H. (2008). Alarming events in the corner of your eye: Do they trigger early saccades? Manuscript submitted for publication.

Henderson, J. M. (1992). Identifying objects across saccades: Effects of extrafoveal preview and flanker object context. *Journal of Experimental Psychology: Learning, Memory, and Cognition,* 18, 521–530.

Henderson, J. M., & Hollingworth, A. (2003). Global transsaccadic change blindness during sceneperception. *Psychological Science*, 14, 493–497.

Heuer, F., & Reisberg, D. (1992). Emotion, arousal, and memory for detail. In S. Christianson (Ed.) Handbookof memory and emotion: Theory, research and practice (pp.151–180). Hillsdale, NJ. Lawrence Erlbaum Associates Inc.

Hochstein, S., & Ahissar, M. (2002). View from the top: Hierarchies and reverse hierarchies in the visual system. *Neuron, 36,* 791–804.

Hofmann, W., Schmeichel, B.J., and Baddeley, A.D.(2012). Executive functions and self-regulation. *TrendsCogn.Sci.* 16, 174–180.

Hoffman H., Kessler H., Eppel T., Rukavina S., Traue H. C. (2010). Expression intensity, gender and facial emotion recognition: women recognize only subtle facial emotions better than men. *Acta Psychol.* (Amst.) 135, 278–283.

Hoffman, J. E., & Subramaniam, B. (1995). Saccadic eye movements and visual selective attention. *Perception & Psychophysics*, 57, 787–795.

Hollingworth, A., Richard, A. M., & Luck, S. J. (2008). Understanding the function of visual short-term memory: Transsaccadic memory, object correspondence, and gaze correction. *Journal of Experimental Psychology*: General, 137, 163–181.

Hollingworth, A., & Henderson, J. M. (2000). Semantic informativeness mediates the detection of changes in natural scenes. *Visual Cognition, 7,*213–235.

Henderson, J. M., & Hollingworth, A. (1999). High level scene perception. *Annual Review of Psychology, 50,* 243–271.

Irwin, D. E. (1998). Lexical processing during saccadic eye movements. *Cognitive Psychology*, 36, 1–27.

Irwin, D. E., & Gordon, R. D. (1998). Eye movements, attention, and transsaccadic memory. *Visual Cognition*, 5, 127–156.

Ishai, A., Pessoa, L., Bikle, P. C., Ungerleider, L.G. (2004). Repetition suppression faces is modulated by emotion. *Proceeding of the National Academy of Sciences, U. S. A.* 101, 9827-9832.

Johansson, R., Holsanova, J., Dewhurst, R., & Holmqvist, K. (2012). Eye movements during scene recollection have a functional role, but they are not reinstatements of those produced during encoding. *Journal of Experimental Psychology: Human Perception & Performance*, 38(5), 1289–1314.

Johansson, R., & Johansson, M. (2013). Look here, eye movements play a functional role in memory retrieval. *Psychological Science*, (E-pub ahead of print).

Jonides, J., Irwin, D. E.,&Yantis, S. (1982). Integrating information from successive fixations. *Science*, 215,192–194.

Lipp, O. V., Derakshan, N., Waters, A. M., & Logies, S. (2004). Snakes and cats in the flower bed: Fast detection is not specific to pictures of fear-relevant animals. *Emotion, 4,* 233–250.

Luo, Q., Mitchell, D., Jones, M., Mondillo, K., Vythilingam, M., & Blair, R. J. R. (2007). Common regions of dorsal anterior cingulate and prefrontal–parietal cortices provide attentional control of distracters varying in emotionality and visibility. *Neuroimage, 38*(3), 631-639.

Kastner, S., & Ungerleider, L. G. (2000). Mechanisms of visual attention in the human cortex. *Annual Review of Neuroscience*, 23, 315–341.

Kensinger, E. A., Piguet, O., Krendl, A. C., & Corkin, S.(2005). Memory for contextual details: Effects of emotionand aging. *Psychology and Aging,* 20, 241–250.

Killgore, W. D. S., & Yurgelun-Todd, D. A. (2001). Sex differences in amygdale activation during the perception of facial affect. *Neuroreport,* 12, 2543–2547.

Kirchner, H., & Thorpe, S. J. (2006). Ultra-rapid object detection with saccadic eye movements: Visual processing speed revisited. *Vision Research*, 46, 1762–1776.

Kowler, E., Anderson, E., Dosher, B.,& Blaser,E.(1995). The role of attention in the programming of saccades. *Vision Research*, 35(13), 1897-1916.

Krolak-Salmon, P., Henaff, M. A., Vighetto, A., Bertrand, O., Mauguiere, F. (2004). Early amygdala reaction to fear spreading in occipital, temporal, and frontal cortex: A depth electrode ERP study in human. *Neuron* 42, 665-676.

Mandal, M. K., & Palchoudhury, S. (1985). Perceptual skill in decoding facial affect. *Perceptual and Motor Skills*, 60, 96–98.

Matsumoto, D., & Ekman, P. (2004). Japanese and Caucasian facial expressions of emotion (JACFEE) and neutral faces (JACNeuF). Berkeley, CA: Paul Ekman & Associates.

McBain, R., Norton, D., & Chen, Y. (2009). Females excel at basic face perception. *Acta Psychologica*, 130, 168–173.

Matin, E. (1974). Saccadic suppression: A review and an analysis. *Psychological Bulletin*, 81, 899–917.

McConkie, G. W., & Rayner, K. (1976b). Identifying the effective stimulus during a fixation in reading: Literature review and theories of reading. In H. Singer & R. N. Ruddell (Eds.), Theoretical models and processes of reading (2nd ed.). Newark, DE: International Reading Association.

Miyahira, A., Morita, K., Yamaguchi, H., Nonaka, K., & Maeda, H. (2000). Gender differences of exploratory eye movements: A life span study. *Life Sciences,* 68, 569–577.

Mogg, K., & Bradley, B. P. (1999). Orienting of attention to threatening facial expressions presented under conditions of restricted awareness. *Cognition and Emotion, 13,* 713–740.

Mumenthaler, Christian; Sander, David. (2012). Social Appraisal influences recognition of emotion. *Journal of Personality and Social Psychology*, Vol 102(6), Jun 2012, 118-1135.

Murphy ST, Zajonc RB (1993) Affect, cognition, and awareness: affective priming with optimal and suboptimal stimulus exposures. *J. Pers. Soc. Psychol.* 64: 723–739.

Neider, M. B., & Zelinsky, G. J. (2006a). Scene context guides eye movements during search. *Vision Research*, 46, 614–621.

Noton, D., & Stark, L. (1971a). Scanpaths in eye movements during pattern perception. *Science,* 171(3968), 308–311.

Noton, D., & Stark, L. (1971b). Scanpaths in saccadic eye movements while viewing and recognizing patterns. *Vision Research*, 11(9), 929–942.

Nowicki, S., Jr., & Hartigan, M. (1988). Accuracy of facial affect recognition as a function of locus of control orientation and anticipated interpersonal interaction. *Journal of Social Psychology*, 128, 363–372.

Nummenmaa, L., Hyönä, J., & Calvo, M. G. (2006). Eye movement assessment of selective attentional capture by emotional pictures. *Emotion, 6*, 257–268.

Ochsner, K. N., & Gross, J. J. (2005). The cognitive control of emotion. *Trends in cognitive sciences*, *9*(5), 242-249.

Ohman, A., Flykt, A., & Esteves, F. (2001). Emotion drives attention: Detecting the snake in the grass. *Journal of Experimental Psychology: General, 130,* 466–378.

O'Regan, J. K., Rensink, R. A., & Clark, J. J. (1999). Change blindness as a result of "mudsplashes". *Nature*, 398, 34.

Owen, A. M., McMillan, K. M., Laird, A. R., & Bullmore, E. (2005). N-back working memory paradigm: A meta-analysis of normative functional neuroimaging studies. *Human brain mapping*, *25*(1), 46-59.

Palermo, R., & Coltheart, M. (2004). Photographs of facial expression: Accuracy, response times, and ratings of intensity. *Behavior Research Methods, Instruments, & Computers*, 36, 634–638.

Pearson, D. G., & Sahraie, A. (2003). Oculomotor control and the maintenance of spatially and temporally distributed events in visuo-spatial working memory. *Quarterly Journal of Experimental Psychology Section A: Human Experimental Psychology*, 56A(7), 1089–1111.

Parkhurst, D., & Niebur, E. (2003). Scene context selected by active vision. Spatial Vision, 16, 125–154.

Pessoa, L., & Adolphs, R. (2010). Emotion processing and the amygdala: from a'low road'to'many roads' of evaluating biological significance. *Nature Reviews Neuroscience*, *11*(11), 773-783.

Pessoa, L., McKenna, M., Gutierrez, E., Ungerleider, L. G. (2002). Neural processing of emotional faces requires attention. *Proceedings of the National Academy of Sciences, U. S. A.* 99,11458-11463.

Pessoa, L., Padmala, S., and Morland, T. (2005). Fate of unattended fearful faces in the amygdala is determined by both attentional and cognitive modulation. *NeuroImage,* 15; 28(1): 249-255.

Pollatsek, A., Rayner, K., & Henderson, J. H. (1990). Role of spatial location in integration of pictorial information across saccades. *Journal of Experimental Psychology: Human Perception and Performance,* 16, 199–210.

Posner, M. I. (1980). Orienting of attention. *Quarterly Journal of Experimental Psychology*, 32, 3–25.

Posner, M. I., & Rothbart, M. K. (1998). Attention, self–regulation and consciousness. *Philosophical Transactions of the Royal Society of London. Series B: Biological Sciences*, *353*(1377), 1915-1927.

Posner, M. I., Rothbart, M. K., Vizueta, N., Levy, K. N., Evans, D. E., Thomas, K. M., & Clarkin, J. F. (2002). Attentional mechanisms of borderline personality disorder. *Proceedings of the National Academy of Sciences*, *99*(25), 16366-16370.

Pickel, K. L., French, T. A., & Betts, J. M. (2003). A crossmodal weapon focus effect: The influence of a weapons presence on memory for auditory information. *Memory,* 11 (3), 277-292.

Rahman, Q., Wilson, G. D., & Abrahams, S. (2004). Sex, sexual orientation, and identification of positive and negative facial affect. *Brain and Cognition,* 54, 179–185.

Rayner, K., Carlson, M., & Frazier, L. (1983). The interaction of syntax and semantics during sentence processing: Eye movements in the analysis of semantically biased sentences. *Journal of Verbal Learning and Verbal Behavior,* 22, 358–374.

Rensink, R. A., O'Regan, J. K., & Clark, J. J. (1997). To see or not to see: The need for attention to perceive changes in scenes. *Psychological Science, 8,* 368–373.

Righart, R., & De Gelder, B. (2008). Recognition of facial expressions is influenced by emotional scene gist. *Cognitive, Affective, & Behavioral Neuroscience, 8,* 264–272.

Robinson, M. D. (1998). Running from William James' bear: A review of preattentive mechanisms and their contribution to emotional experience. *Cognition and Emotion*, 12, 667-696.

Roseman, I. J., & Smith, C. A. (2001). Appraisal theory: Overview, assumptions, varieties, controversies. In K. R. Scherer, A. Schorr, & T. Johnstone (Eds.), *Appraisal processes in emotion: Theory, method, research* (pp. 3–19). New York, NY: Oxford University Press.

Schmidt, B. K., Vogel, E. K., Woodman, G. F., & Luck, S. J. (2002). Voluntary and involuntary attentional control of visual working memory. *Perception & Psychophysics*, 64, 754-763.

Smyth, M. M., & Scholey, K. A. (1994). Interference in immediate spatial memory. *Memory & Cognition*, 22(1), 1–13.

Schyns, P. G., Petro, L. S., & Smith, M. L. (2007). Dynamics of visual information integration in the brain for categorizing facial expressions. *Current Biology*, 17, 1580–1585.

Schmidt, B. K., Vogel, E. K., Woodman, G. F., & Luck, S. J. (2002). Voluntary and automatic attentional control of visual working memory. *Perception & Psychophysics*, 64(5), 754–763.

Vassallo, S., Cooper, S. L., & Douglas, J. M. (2009). Visual scanning in the recognition of facial affect: Is there an observer sex difference?. *Journal of Vision*, *9*(3), 11.

Stanley, D., Ferneyhough, E., & Phelps, E. A. (2009). Neural perspectives on emotion: impact on perception, attention, and memory. *Handbook of Neuroscience for the Behavioral Sciences*, 2, 829.

Tipples, J., Young, A. W., Quinlan, P., Broks, P., & Ellis, A. (2002). Searching for threat. *Quarterly Journal of Experimental Psychology, 55A,* 1007–1026.

Tosoni, A., Shulman, G. L., Pope, A. L., McAvoy, M. P., & Corbetta, M. (2013). Distinct representations for shifts of spatial attention and changes of reward contingencies in the human brain. *Cortex*, *49*(6), 1733-1749.

Tremblay, S., Saint-Aubin, J., & Jalbert, A. (2006). Rehearsal in serial memory for visual-spatial information: Evidence from eye movements. *Psychonomic Bulletin & Review*, 13(3), 452–457.

Uttal, W. R., & Smith, P. (1968). Recognition of alphabetic characters during voluntary eye movements. *Perception & Psychophysics*, 3, 257–264.

Voyer, D., & Rodgers, M. A. (2002). Reliability of laterality effects in a dichotic listening task with nonverbal material. *Brain and Cognition*, 48, 602–606

Vuilleumier P. (2005). How brains beware: Neural mechanisms of emotional attention. *Trends in Cognitive Sciences*, 9(12), 585-594.

Vuilleumier, P., and Schwartz, S. (2001). Beware and be aware: capture of spatial attention by fear-related stimuli in neglect. *Neuroreport,* 12, 1119-1122.

Wager, T. D., Davidson, M. L., Hughes, B. L., Lindquist, M. A., & Ochsner, K. N. (2008). Prefrontal-subcortical pathways mediating successful emotion regulation. *Neuron*, *59*(6), 1037-1050.

Williams, L. E., Bargh, J. A., Nocera, C. C., & Gray, J. R. (2009). The unconscious regulation of emotion: nonconscious reappraisal goals modulate emotional reactivity. *Emotion*, *9*(6), 847.

Williams, J. M. G., Watts, F. N., MacLeod, C., & Mathews, A. M. (1997). Cognitive psychology and emotional disorders (2nd ed.). Chichester, United Kingdom: Wiley.

Yiend, J., & Mathews, A. (2001). Anxiety and attention to threatening pictures. *Quarterly Journal of Experimental Psychology, 54A,* 665–681.

In: Memory Consolidation
Editors: Manabu Sakakibara and Etsuro Ito
ISBN: 978-1-63482-596-2

Chapter 13

MEMORY IN FOUR DIMENSIONS

Richard A. Sieb*
Edmonton, Alberta, Canada

ABSTRACT

Episodic memory appears to be four-dimensional. Episodic memory is memory of personal past experiences. The encoding, retention, and retrieval of episodic memories utilizes a variant of synaptic timing dependent plasticity (LTP) operating through the hippocampus and adjacent structures. This short term encoding becomes long term via new protein synthesis and structural change in the active cortical connections. Conscious recollection is required to demonstrate episodic memory. Conscious recollection indicates that episodic memory is organized in four dimensions (three dimensions of space, one dimension of time). During recollection, some conscious experiences making up an episodic memory occur simultaneously, but at different locations in space; while other conscious experiences occur at the same location in space, but at different times. This shows that episodic memories are organized in both space and time (four dimensions). Consciousness (awareness) at any one time is organized in a similar manner (some conscious experiences occur simultaneously at different locations, while other conscious experiences occur at the same location, but at different times) and appears therefore also to be organized in four dimensions. Both consciousness and episodic memory retrieval have the same contents (conscious experiences) and both originate from working memory. Consciousness and episodic memory are linked by working memory and the four dimensional organization of both appears to be determined by working memory. Einstein's special theory of relativity describes the organization and relationships of events in four-dimensional spacetime. A spacetime interval is the separation of two events in four-dimensional spacetime. A spacetime interval has both spatial and temporal coordinates. The special theory of relativity predicts the occurrence of three types of invariant spacetime intervals between events. Space-like spacetime interval separation between events (the spatial separation of two events is greater than the temporal separation of the events) gives rise to frames of reference (observational perspectives) in which events occur simultaneously, but in different locations; time-like spacetime interval separation between events (the temporal separation of two events is greater than

* E-mail: siebr@shaw.ca.

the spatial separation of the events) gives rise to frames of reference in which events occur at the same location, but at different times. The organization of events in four-dimensional spacetime is therefore similar to the organization of conscious experiences in consciousness at any one time and episodic memory retrieval. Hence consciousness and episodic memory may be frames of reference organized by space-like and time-like spacetime interval separations as predicted by Einstein's special theory of relativity. Consciousness and episodic memory are in other words four-dimensional. Consciousness and episodic memory therefore have a level of intricacy and flexibility not otherwise considered. They also have a personal perspective quality. In addition, this might be considered a different avenue of support for the validity of Einstein's theories of relativity.

Keywords: episodic, memory, consciousness, Einstein, relativity, four-dimensions, hippocampus, place cells, long term potentiation

INTRODUCTION

Memory is the ability to store, retain, and recall information. Memory of personal past experiences of a particular time and place ("what", "where", and "when") with reference to the person as a participant is known as episodic memory (Munoz-Lopez et al., 2010; Schacter et al., 2011; Tulving, 1985, 2001, 2002). Such memory is normally thought of as being three dimensional. However, utilizing relativistic principles, episodic memory can be shown to be four-dimensional (three dimensions of space, one dimension of time).

Declarative (explicit) memory refers to memory which can be consciously recalled (Tulving, 1983; Ullman, 2004). There are two categories of declarative memory (Graf & Schacter, 1985; Tulving, 1972; Tulving & Schacter, 1990). Semantic memory is a structured record of facts, concepts, and skills that is independent of personal experience. Episodic memory is a record of personal past experiences. Episodic memory may provide basic support for semantic memory and semantic information may be derived from accumulated episodic memory. Episodic memory may be thought of as a map that ties together items in semantic memory. Episodic memory and semantic memory each represent different parts of a complete picture. As such, something that affects episodic memory can also affect semantic memory. Anterograde amnesia is an impairment of declarative memory that affects both episodic and semantic memory (Howard & Kahana, 2002; Tulving & Markowitsch, 1998).

Episodic memory requires no training and is mostly automatic or incidental. However, as will be seen, the encoding of episodic memory is associated with consciousness and can even be voluntarily directed by consciousness. The retrieval or recall of episodic memory requires conscious effort (Munoz-Lopez et al., 2010). Conscious recollection is required to demonstrate episodic memory (Tulving, 1983). Hence both the encoding and recall of episodic memory is associated with consciousness. This is interesting as it can be shown that consciousness itself is also similarly organized in four dimensions at any one time.

Gabriele and Kao (2007) found that the formation of new declarative memories relies on both the hippocampus and parahippocampus. The hippocampus and surrounding structures in the medial temporal lobe of the brain are specifically most important in declarative memory (Eichenbaum, 2001). Eichenbaum (2001) found that the ability to encode, retain, and recall episodic memories is highly dependent on the hippocampus (the hippocampus acts as a memory space). Episodic memory stores context-rich information about everyday events and depends on the hippocampal formation (entorhinal cortex, subiculum, presubiculum, parasubiculum, hippocampus proper, and dentate gyrus-Munoz-Lopez et al., 2010). Lesions that include the hippocampus, amygdala, and adjacent entorhinal and posterior parahippocampal cortices impair episodic memory in humans, with intact perception and working memory (Mishkin, 1978; Munoz-Lopez et al., 2010; Scoville & Milner, 1957). The perirhinal cortex receives most of its input from visual areas of the inferior temporal cortex (Suzuki & Amaral, 1994a) and sends this information to the entorhinal cortex (Insausti et al., 1987a; Suzuki & Amaral, 1994b), from where the information accesses the hippocampus (Witter & Amaral,1991). The inferior temporal cortex is closely connected to the inferior parietal cortex, probably the most important area of multimodal processing, which is involved in attention, perception, working memory, and conscious experience (Sieb, 2004, 2007, 2011, 2013). There are return projections from the hippocampus to the entorhinal cortex primarily, but also directly to the perirhinal cortex (Insausti & Munoz, 2001; Lavenex et al., 2002; Witter & Wouterlood, 2002; Munoz & Insausti, 2005). These connections may be critical for the consolidation of episodic memory for the long term in the cerebral cortex (Munoz-Lopez et al., 2010). These studies indicate the main flow of information underlying episodic memory. The formation of new episodic memories requires the medial temporal lobe of the brain (which includes the hippocampus). Without the medial temporal lobe, one is able to form new procedural memories, but cannot remember (episodic memories) the events in which they arose. The prefrontal cortex (particularly in the left hemisphere) is also involved in the formation of new episodic memories. Patients with damage to the prefrontal cortex can learn new information, but tend to do so in a disordered fashion (Janowsky et al., 1989). The lateral prefrontal cortex may be essential for remembering the contextual details of episodic memory (Gabriele & Kao, 2007). The prefrontal cortex may help organize information for more efficient storage or underlie semantic strategies which enhance encoding, such as thinking about the meaning of the material or rehearsing it in working memory (Gabrieli et al., 1998). There is disagreement on the length of time episodic memories are stored in the hippocampus. Some believe episodic memories always rely on the hippocampus. Others believe the hippocampus only stores episodic memories for a short time, after which they are consolidated to the neocortex. The latter view is strengthened by evidence that neurogenesis in the adult hippocampus may ease the removal of old memories and increase the efficiency of forming new memories (Deisserroth et al., 2004). The amygdala is involved in strengthening memory of emotionally-arousing stimuli, but is not required for memory of emotionally-neutral stimuli (Adolphs et al., 1997; Babinsky et al., 1993). It is much easier to remember an emotionally-charged event than one that is not emotionally-charged.

A central feature of episodic memory is that it is a combination of information from different sensory modalities (olfactory, somatosensory, auditory, and visual). Episodic memory is multimodal. Studies show that although some sensory-specific information reaches the medial temporal cortex directly (primarily olfactory and visual, but also to a lesser extent, somatosensory and auditory), the great bulk of incoming information originates from polymodal areas of the neocortex (Mohedano-Moriano et al., 2007), such as the inferior parietal cortex, providing episodic memory with rich complex contextual information. Episodic memory is based on context and its retrieval often depends on context.

Tulving (1983, 2002) describes episodic memory as a temporally-dated record of a person's experiences. Episodic memories include times, places, associated emotions, and other who, what, when, where, and why information that can be explicitly stated (Schacter et al., 2011). It allows one to figuratively travel back in time to experience again events that took place. All this information forms a context for the memory. Conway (2009) attributes nine properties to episodic memory: (1) it is a summary record of sensory-perceptual-conceptual-affective processing, (2) it retains patterns of activation/inhibition over long periods, (3) it is often represented in the form of (visual) images, (4) it always has a perspective (field or observer, frame of reference), (5) it represents short time slices of experience, (6) it is represented on a temporal dimension roughly in order of occurrence, (7) it is subject to rapid forgetting, (8) it makes autobiographical remembering specific, and (9) it is recollectively experienced when accessed. These properties arise because episodic memory utilizes a discrete type of information processing and is associated with consciousness. The third property may be a consequence of the four-dimensional organization of episodic memory.

In recent years, our understanding of episodic memory has undergone a radical revision, with a transition from a system directed exclusively to the past to a system also oriented to the future. Studies indicate there is a link between the retrieval of personal past events and the construction of future imaginary events. Schacter and Addis (2007a,b, 2009) highlight the role of episodic memory in the construction of future events. This adaptive function of episodic memory extracts components of the past, then assembles and recombines them to remember the past and to imagine the future (Hardt et al., 2010; Howe, 2011; Schacter et al., 2011). Hassabis and Maguire (Hassabis et al., 2007a,b; Hassabis & Maguire, 2007, 2009) argue for a process of "scene construction" that links memory and imagination. Since episodic memory appears to have a four-dimensional organization, this reconstruction process may be even more intricate and fascinating than "imagined".

Since episodic memory encoding and recall is associated with consciousness, the next few sections describe the nature of consciousness and conscious experiences and how they are associated with episodic memory. How the four-dimensional organization of consciousness and episodic memory evolves is then discussed. This entails examination of Einstein's special and general theories of relativity, spacetime, and the concept of invariant spacetime intervals. Components of episodic memory (sensory memory, working memory, short term memory, LTP, LTD, STDP, homeostatic synaptic regulation and plasticity, and long term memory) are subsequently described in great detail.

CONSCIOUSNESS

Consciousness is a state of awareness known as the waking state. The encoding and recall of episodic memory is associated with consciousness (episodic memory encoding and recall occurs only when one is awake or conscious). Consciousness arises and is maintained by a particular system in the brain called the reticular activating system. The reticular activating system and the cerebral cortex are necessary parts of a system that maintains consciousness (Hannaman, 2005; Young, 2009). The reticular activating system originates in the upper brainstem reticular core and projects through the intralaminar thalamic nuclei to the cerebral cortex to mediate arousal, attention, and consciousness, i.e., the waking state (Evans, 2003; Hannaman, 2005; Kinomura et al., 1996; Reiner, 1995; Steriade, 1995, 1996; Young, 2009).

The noradrenergic locus coeruleus, the mesencephalic reticular formation, the cholinergic nucleus basalis of Meynert, the dorsal hypothalamus, the intralaminar thalamic nuclei, and the tegmentum make up the reticular activating system (Afifi & Bergman, 1986; Best, 2010; Hannaman, 2005; Kinomura et al., 1996; Rang, 2003; Steriade et al., 1990; Thierry et al., 1990; Vertes, 1990). Bilateral damage to the intralaminar thalamic nuclei produces lethargy or somnolence (Steriade, 1996). Damage to the reticular activating system results in coma (Daltrozzo et al., 2009, 2010; Liversedge & Hirsch, 2010; Young, 2009). Partial destruction or inhibition of the reticular activating system results in a partial loss of consciousness known as an acute confusion state or delirium (Rull, 2011). The function of the reticular activating system is modulated by its adrenergic and cholinergic components (Burlet et al., 2002; Evans, 2003; Garcia-Rill, 1997; Garcia-Rill et al., 2007; Reiner, 1995). Drugs or activities which enhance adrenergic and cholinergic action in the brain also enhance episodic memory (Apud et al., 2007; Dunbar et al., 2007; Gron et al., 2005).

Information Processing

Discrete and continuous processing are two distinct types of information processing that occur in organized systems. Discrete processing and continuous processing are differentiated by their very different physical mechanics and behavior. Each type has its advantages and disadvantages. One type appears to be involved in the production of consciousness, while the other type is involved in the production of conscious experiences and episodic memories.

Continuous Processing

Continuous processing is constant, consistent, and uninterrupted (Henthorn, 2013; Perry et al., 1997; Ricketts, 2013). There is a seamless flow from input to output. Continuous processing may be so fast as to appear almost immediate or instantaneous. This is exemplified by the extremely fast processing of word processors and the regulation of human vital functions (breathing, heart rate, etc.). Continuous processing is sensitive, accurate, stereotyped, consistent, predictable, and reliable. Thermometers, thermostats, barometers, clocks, watches, engines, machines, human reflexes, and production lines use continuous processing. Continuous processing is context-independent. It occurs in a stereotyped predictable manner regardless of the surrounding conditions. Continuous processing is not labor-intensive. Human vital functions are regulated continuously, without interruption or fatigue, throughout life. Clocks and watches operate continuously and accurately for years, without maintenance. Continuous processing is inflexible and invariable. It does not give rise to anything new or novel. Nonlinearity has no place in continuous processing, i.e., the output is always proportional to the input and precisely as predicted. Negative feedback may be utilized in continuous processing to regulate rate of flow, detect errors, assess trends, and determine rate of change of errors (Henthorn, 2013). Continuous processing may vary in level.

Consciousness arises relatively instantly or immediately upon awaking from sleep or anesthesia. The generation of consciousness is constant, continuous, consistent, reliable, uninterrupted, accurate, stereotyped, inflexible, invariable, free from interference, and predictable. It is context-independent. Small perturbations (psychoactive drugs and chemicals, anesthetics, toxins, fatigue, hypnosis, meditation, sleep, diseases) can change the level of consciousness. The production of consciousness is not labor-intensive. The reticular activating system is relatively non-specific and context-independent, as any type of sensory stimulation may generate consciousness. The reticular activating system therefore appears to be a continuous processing system operating to generate consciousness (the waking state). Consciousness appears to set the stage for conscious experience and episodic memory.

DISCRETE PROCESSING

Discrete processing occurs in stages (Dawson, 2013) and is characterized by distinct separate unit production (Edwards, 2013). Discrete processing is characterized by production of a series of distinct separate identifiable states. Processing occurs over distinct (discrete) intervals or periods of time, called integration periods (episodes, epochs, or packets). Each integration period represents a separate distinct (discrete) identifiable state of the system. Integration periods may be of varying lengths of time and occur serially in a given system. Discrete processing is slower and more labor-intensive than continuous processing. Appreciable interference may occur, as input might backup or accumulate, because of the interpolation of integration periods. Information can interact and be manipulated over integration periods. Nonlinearity therefore is an integral part of the processing. This makes discrete processing variable, flexible, and unpredictable. Information is integrated in new or novel ways, resulting in the emergence of new or novel results. Hence the results of discrete processing might be said to emerge. In a given system, one period of processing ends before another period of processing begins. Thus discrete processing occurs in a sequential or serial manner, one period of processing succeeding another. The serial nature of discrete processing would be expected to slow processing in general and create a bottleneck. One period of processing may also prime subsequent periods of processing.

Episodic memory, as its name implies, consists of a series of distinct (discrete) separate identifiable explicit states (episodes). Episodic memories are self-generated, arise rapidly in recurrent systems, take time to form (integration periods), have a prolonged duration, are seamless, structurally-complex, ineffable, transparent, bounded, unified and coherent, informative, serial, limited in capacity, subject to interference, new or novel, variable, flexible, project outwards, explicit, gain access to other systems, and arise from attention. All these properties suggest that episodic memories develop in a discrete processing system. Conscious experiences, which make up the content of consciousness, have the same properties (Baars, 1988; Koch, 1998; Libet, 1983, 1985; Sieb, 2004, 2007, 2011, 2013; Tononi & Edelman, 1998), as episodic memories. Both conscious experiences and episodic memories therefore appear to arise from a similar type of discrete processing.

Neglect

Neglect syndrome may arise with damage to the inferior parietal cortex (Joseph, 1990, 2000; Mattingley, 1999; Prinz, 2003, Lecture at *Towards a Science of Consciousness* Conference, 2012). Neglect syndrome is a deficit in awareness (conscious experience) of one side of space that occurs as a result of brain damage (Husain, 2008; Unsworth, 2007). Neglect appears to be caused by an inability to pay proper attention to sensory stimuli (Husain, 2008; Unsworth, 2007). Neglect is most profound with damage to the right cerebral hemisphere with the deficit to the left side of space (Husain, 2008; Unsworth, 2007). Left hemisphere damage may produce a less prolonged and long-lasting neglect of right space (Husain, 2008; Kim et al., 1999). Neglect is classically associated with lesions of the right posterior parietal cortex, particularly of the right temporoparietal junction and inferior parietal cortex (Husain, 2008; Mattingley, 1999; Vallar & Perani, 1986). Isolated lesions of the right frontal lobe have also been associated with neglect of left space (Husain & Kennard, 1996).

Severe left unilateral inattention and neglect is seen with massive lesions to the right frontal lobe of the brain (Joseph, 1988a,b, 1990, 2000). Subcortical lesions, without involvement of the overlying cortex, can lead to neglect (Hillis et al., 2002; Hillis et al., 2005). Damage to white matter pathways connecting posterior and frontal cortical areas has also been implicated in neglect (Bartolomeo et al., 2007; Thiebaut de Schotten et al., 2005). Loss of dreaming (a type of conscious experience) can occur with parietal cortex damage (Solms,M., In *"What Are Dreams?"*, KSPS Nova TV Program, 2009). It is also interesting to note that episodic memory does not occur on the affected side during neglect. Thus the neglect syndrome suggests that the connections between the inferior parietal cortex and frontal cortex are quite important in attention, conscious experience, and episodic memory.

Feedback and Conscious Experiences

An increasing number of investigations have found that feedback connectivity between the frontal lobes and parietal lobes of the brain is required to generate conscious experience (Boehler et al., 2008; Crick & Koch, 1995; Fahrenfort et al., 2007; Lamme & Roelfsema, 2000; Ro et al., 2003; Tononi & Koch, 2008). Crick and Koch (2003) and Hudetz and coworkers (2006) determined that conscious experience depends on neuronal coalitions spanning the posterior sensory cortices and the executive areas of the prefrontal cortex. Laplant and coworkers (2005) demonstrated a reduction in frontal activation in anesthetized animals that was normally present in waking animals. Demertzi and coworkers (2013) and Lapitskaya and coworkers (2013) found that external sensory awareness is an emergent property dependent on cortical-cortical interconnections in a widespread frontal-parietal network. Activity in this network was reduced during anesthesia, coma, unresponsive wakefulness syndrome (vegetative state), and minimally conscious states. Imas and coworkers (2005) found recurrent feedback from anterior to posterior brain areas was suppressed earlier in anesthetized animals than feed forward information transfer. It has been found that anesthetics suppress mainly parietal cortex feedback connectivity, but not parietal cortex feed forward connectivity (Avidan et al., 2011; Ku et al., 2011; Lee et al., 2009). This is reversed upon awakening. Mashour (Consciousness in the operating room, Lecture at

TASOC 2012 Conference) concluded that the parietal cortex is affected much more than frontal areas by anesthetics, that conscious experience is likely a cortical-cortical feedback interaction, and that conscious experience may be governed by this feedback interaction.

PERCEPTION AND ATTENTION

Perception and focal attention are necessary for conscious experience (Rowlands, 2002; Velmans, 1999). The inferior parietal cortex receives highly processed input from somesthetic, visual, and auditory association cortex, the frontal lobes, and other higher-order assimilation areas throughout the neocortex (Bogarts, 1997; Bruce et al., 1986; Burton & Jones, 1976; John, 2009; Jones & Powell, 1970; Joseph, 1990, 2000; Petrides & Pandya, 2007; Seltzer & Pandya, 1978; Zeki, 1974). The inferior parietal cortex can simultaneously analyze visual, auditory, and somesthetic information, respond to visual stimuli of any size, shape, or form, and has neurons with visual receptive properties that span almost the entire visual field (Bruce et al., 1982, 1986; Hyvarinen & Shelepin, 1979; Joseph, 1990, 2000). The inferior parietal cortex is involved in the creation and assimilation of cross-modal associations (auditory, visual, and somesthetic equivalents of objects, events, ideas, actions, feelings-Joseph, 1990, 2000). This is perception. Episodic memory is also multimodal in composition and the hippocampal formation involved in the encoding of episodic memory receives primary input from the inferior parietal cortex. Damage to the left inferior parietal cortex may cause various disorders that involve perceptual deficit (which may appear to be a memory deficit)-anomia, object and finger agnosia, acalculia, most forms of apraxia, agraphia, left-right disorientation, pure word blindness, alexia, conductive aphasia, disruption of sequential-grammatical relationships, disruption of spatial-sequential ability, impairment of visual-spatial functioning, impairment of temporal-sequencing ability, disruption of the performance of skilled temporal-sequential motor acts, and the impairment of logic and grammar (Benson et al., 1973; Heilman, 1973, 1979; Hechaen & Albert, 1978; Joseph, 1990, 2000; Luria, 1980; Sauguet et al., 1971; Straub & Geschwind, 1983; Vignolo, 1983). Dressing apraxia and constructional apraxia may occur with right inferior parietal cortex damage (Hier et al., 1983; Mehta et al., 1987; Joseph, 1988a, 1990, 2000). The frontal lobes, especially the right frontal lobe, and the parietal lobes have great attention capacity (Desmedt, 1977; Dimond, 1976, 1979; Hielman & Van Den Abell, 1979, 1980; Joseph, 1982, 1986a, 1988a, b; Tucker, 1981). The frontal cortex and the inferior parietal cortex therefore appear to be involved in perception and focal attention, the two processes necessary for the formation of conscious experiences (and episodic memories).

It appears from the preceding sections that both conscious experience and episodic memory depend on the inferior parietal cortex. The inferior parietal cortex accesses the hippocampus via the perirhinal cortex and entorhinal cortex in the encoding of episodic memory and the hippocampus accesses the entorhinal cortex and perirhinal cortex via reverse connections during episodic memory recall. The feedback interaction between the inferior parietal cortex and the prefrontal cortex is involved in the production of conscious experience as well as in the encoding of episodic memory. Hence one has an anatomical and functional connection between the production of conscious experience and episodic memory.

FOCAL ATTENTION AND WORKING MEMORY

Prinz (2003; Attention as the mechanism of consciousness, Lecture at *TASOC* 2012 Conference) found that awareness (conscious experience) does not arise without attention and develops over a time delay. Prinz suggested this time delay may be the time necessary for attention to act. Joseph (2000) and Prinz found the inferior parietal cortex is involved in attention, perception, and working memory. Working memory also operates in the dorsolateral prefrontal cortex (Fuster, 1997, 2000, 2008; Goldman-Rakic, 1990, 1995). Working memory is the holding in mind of a number of pieces of information and its manipulation for a short period of time so that it can be utilized at a later time for goal-directed responses. Functionally, working memory is the provisional retention of perceptual information for prospective action, a type of focal attention whereby perception is reorganized and rerepresented, becoming explicit, functional, and conscious (Luck & Vogel, 1997; Todd & Marois, 2004; Vogel & Machizawa, 2004). Working memory is a cognitive process that is accompanied by consciousness (Fuster, personal communication, 2013). Working memory represents the two components (perception, focal attention) necessary for the production of conscious experience. Working memory may be formed and maintained by reentry (Fuster, 1997; 2008). Reentry refers to reciprocal interaction and occurs throughout the brain (Tonelli & Edelman, 1998). Consciousness is impaired if the reentry mechanisms of the brain are disrupted (Tonelli & Edelman, 1998). Positive feedback is a type of reentry. Positive feedback occurring in prefrontal-inferior parietal cortex connections may be a type of working memory (focal attention) whereby perception is reorganized and rerepresented, becoming explicit, functional, and conscious. The reorganization and rerepresentation of perception occurring in working memory is not entirely predictable and probably arises in a nonlinear fashion typical of complex systems like the brain (Gershenson & Heylighen, 2004). Positive feedback (and nonlinear emergence) in prefrontal-inferior parietal cortex connections therefore may be a type of focal attention (working memory) which produces conscious experience. Conscious experience is in effect arising from focal attention and perception. Any disruption in the positive feedback interaction between the prefrontal cortex and inferior parietal cortex would be tantamount to a disorder of attention (resulting in neglect). As will be seen, working memory is also involved in the encoding of episodic memory.

Hence working memory appears to be the functional connection between episodic memory and conscious experience, since working memory is involved in the formation of both conscious experience and episodic memory.

POSITIVE FEEDBACK, NONLINEAR EMERGENCE, CONSCIOUS EXPERIENCE, EPISODIC MEMORY

Positive feedback and nonlinear emergence work in conjunction as a basic physical mechanism responsible for a vast number of natural phenomena (Scott, 1996, 1999, 2000, personal communication; Sieb, 2004, 2007, 2011, 2013). A positive feedback cycle arises in a system when part of the output of the system feeds back as an additional excitatory input into the system, producing more output, which feeds back as more input, producing more output, and so on. Positive feedback is positive loop gain around a feedback loop (Zeigler et al.,

2000; Zuckerman & Jefferies, 1996). Positive feedback results in further change in the same direction. Positive feedback will increase the function of a system and the system will grow indefinitely, resulting in instability of the system (Amos & Amos, 2002; Graf, 1999; Zeigler et al., 2000; Zuckerman & Jeffery, 1996). When the loop gain is positive and above 1, there will typically be exponential growth, increasing oscillation, and divergence from equilibrium (Zeigler et al., 2000). System parameters may rapidly progress to extreme values, damage or destruction of the system may occur, or the system may latch onto a new stable state. Some positive feedback cycles produce exponential growth, but all produce unlimited growth and instability. Negative feedback results in change in the opposite direction. Negative feedback will decrease the activity of a system, act to stabilize the system, and bring it back into equilibrium. Negative feedback has a modulating influence. Positive feedback may be reduced, cancelled, or controlled by negative feedback. Negative influences therefore will inhibit or dampen the effects of positive feedback and result in a balanced equilibrium stable state of the system (Scott, 1996, 1999, 2000, personal communication; Sieb, 2004, 2007, 2011, 2013). Because many of the positive and negative influences are nonlinear (are not proportional to the output), the new stable states are said to be nonlinear emergent states.

Phenomena Arising from Positive Feedback and Nonlinear Emergence

The prototypical example of a phenomenon arising from positive feedback and nonlinear emergence is a candle flame (Scott, 1996). When a candle is lit, wax vapor burns, and heat (and light) is dissipated. Some of the dissipated heat feeds back to melt more candle wax, producing more wax vapor, which burns, producing more heat, some of which feeds back, and so on, producing a positive feedback cycle. Nonlinear inhibitory influences (loss of heat, cross-sectional area of the candle, density of the candle wax, wind, temperature, humidity, etc.) act to inhibit or dampen the positive feedback and result in the formation of a stable equilibrium state (a balanced dissipation of heat and light, a candle flame). The candle flame is a new phenomenon (has new properties) that emerge from positive feedback and nonlinear emergence and cannot be precisely predicted in advance. The candle flame is an explicit (physical) state that can affect other physical systems and processes (it can burn fingers, ignite inflammable substances). A wave also may be described in terms of positive feedback and nonlinear emergence (Scott, 1996; Sieb, 2004, 2007, 2011, 2013). When a rock is thrown into a pond, it displaces water molecules. Some adjacent water molecules move in to occupy the space previously occupied by the displaced molecules to further displace the water molecules, some of which feedback, and so on, producing a positive feedback cycle. Nonlinear inhibitory influences (weight and density of water molecules, currents, temperature, wind, obstacles, depth, other waves, etc.) act to inhibit or dampen the positive feedback resulting in the emergence of a new balanced stable state (a wave), which propagates across the water. Waves are new explicit states that can affect other physical systems and processes (boats, swimmers, other waves). A business may be described in terms of positive feedback and nonlinear emergence. An inventory of goods or services is sold. Sale of this inventory provides funds for the purchase of more inventory or services, which are sold, providing more inventory or services, and so on, producing a positive feedback cycle.

Negative influences (size and type of inventory or services, number of employees, number of customers, hours of operation, damage, sickness, expenses, etc.) inhibit the positive feedback and result in a balanced equilibrium state (a specific restaurant, company, railroad, department store, particular business). Stock markets, nations, nerve impulses, cells, groups, families, political parties, and many other phenomena may be described in terms of positive feedback and nonlinear emergence.

Explicit states developed from positive feedback and nonlinear emergence have the following properties (Sieb, 2004, 2007, 2011, 2013): they are self-generated, separate and distinct, take time to form, have new properties, persist for a period of time, are explicit, can affect other physical systems and processes, are bounded, have content, have limited capacity, are variable, flexible, seamless, ineffable, transparent, serial in production, prone to interference, unified and coherent, and project outwards. Nonlinear explicit states have the same properties as conscious experiences and episodic memories. Positive feedback and nonlinear emergence results in distinct separate identifiable states. In fact, positive feedback and nonlinear emergence is a quintessential example of discrete processing. Conscious experiences and episodic memories have all the same properties as states arising from positive feedback and nonlinear emergence. This suggests that conscious experiences and episodic memories may arise from positive feedback and nonlinear emergence. Since both arise from working memory and working memory has been described as a positive feedback-nonlinear emergent process, working memory appears to be a positive feedback-nonlinear emergence process giving rise to conscious experience and episodic memory.

Organization of Consciousness and Explicit Memory

Conscious experiences make up the content of consciousness. These conscious experiences have a particular spatiotemporal organization in consciousness at any one time. Some conscious experiences occur simultaneously, but in different locations, while other conscious experiences occur in the same locations, but at different times. Thus space and time dimensions are introduced into the organization of consciousness. Consciousness therefore appears to be organized in four dimensions (three dimensions of space, one dimension of time). A similar organization of conscious experiences in space and time occurs during episodic memory recall (some conscious experiences occur simultaneously, but in different locations; while other conscious experiences occur at the same location, but at different times). One therefore has a four-dimensional organization of consciousness at any one time and of episodic memory recall. Episodic memory therefore appears to be encoded, retained, and recalled in four dimensions. Einstein's special and general theories of relativity explain this four-dimensional organization quite well.

Special Relativity

Einstein's special theory of relativity depends on frames of reference (observational perspectives in space which may be described using coordinate systems). Special relativity is based on two postulates (Einstein, 2001): (1) the laws of physics (in their simplest form) are invariant (identical) in all inertial frames of reference (frames of reference in uniform transitory motion); and (2) the speed of light is the same for all observers, regardless of the motion of the light source. The invariance (Lorentz invariance) of postulates (1) and (2) is captured by the Lorentz transformation, a mathematical expression utilized to transform the coordinates of one inertial frame of reference into the coordinates of another and to account for the independence of reference frame of the speed of light. The Lorentz transformation describes how measurements of space and time by different observers are related. It reflects that observers moving at different speeds may measure different distances, elapsed times, and orderings of events. The Lorentz transformation predicts relativistic interval conservation and the constancy of the speed of light. Einstein also found that all frames of reference are in inertial motion and that there is no absolute state of rest from which the motion of other inertial frames may be measured (Rindler, 1977; Taylor & Wheeler, 1992).

Special relativity implies a wide range of consequences (which have been experimentally verified). These include length contraction, time dilation, relativistic mass, mass-energy equivalence ($E=mc^2$, where c is the speed of light in a vacuum), a universal speed limit (c), and relativity of simultaneity (Roberts & Schleif, 2007). Combined with other laws of physics, the two postulates of special relativity predict the equivalence of mass and energy (Einstein, 2001; Feynman, 1998). Special relativity has replaced the notion of an absolute universal time with the notion of time that is dependent on reference frame and spatial position. Events that occur at the same time for one observer could occur at different times for another. Rather than an invariant time interval between two events, there is an invariant spacetime interval, which takes into account not only space, but time as well. Time and space cannot be defined separately from one another, but are interwoven into a single continuum (the spacetime continuum). Special relativity is restricted to flat spacetime and its geometry known as Minkowski spacetime (Minkowski, 1909). Galilean relativity (the laws of motion are the same in all inertial systems) is considered an approximation of special relativity valid at low speeds. Special relativity is considered an approximation of general relativity valid in weak gravitational fields.

General Relativity

Einstein's general relativity is a geometric theory of gravitation. General relativity is based on curved spacetime and its geometry known as semi-Riemann spacetime. According to general relativity, gravity is not a force per se, as in Newtonian physics, but is the curvature (distortion) of spacetime produced by the mass-energy contained in that spacetime (any physics text). At the core of general relativity are the Einstein Field Equations, which describe the relationship between the curvature of four-dimensional spacetime and the energy-momentum contained in that spacetime (Wald, 1984; any physics text).

Objects in free fall (not subject to any external forces) move along geodesics (lines in curved spacetime). A body in free fall actually moves freely along a curved path or trajectory (geodesic). This has been repeatedly demonstrated in experiments in which the instantaneous vectors (a vector is a geometric entity having magnitude and direction) calculated along the trajectory of a body in free fall vary with the trajectory. Free fall is equivalent to pure (inertial) motion within the curved geometry of spacetime (Poisson, 2004).

Because general relativity (the Einstein Field Equations) is constructed using tensors (tensors express relations between vectors, scalars, or other tensors and so are dimensionless), it exhibits general covariance (its laws take on the same form in all frames of reference-Wald, 1984). Hence general relativity satisfies the more basic principle of relativity (the laws of physics are the same for all observers-Guilini, 2006). Locally, the spacetime of general relativity is Minkowskian and the Einstein Equivalence Principle holds (the laws of special relativity are a good approximation-Guilini, 2006; Mermin, 2005; Rindler, 2001).

A locally Lorentz invariant frame of reference that abides by the laws of special relativity can be defined at sufficiently small scales in the curved spacetime of general relativity.

SPACETIME

Spacetime is any mathematical model that combines space and time into a single continuum. Mathematically, spacetime is a manifold (a description at small scale using a coordinate system). Dimensions are the independent components of a coordinate grid system used to locate points or positions in a defined space. In a Minkowski (flat spacetime) manifold and a semi-Riemann (curved spacetime) manifold, there are three dimensions of space and one dimension of time (four dimensions).

Occurrences in spacetime are known as events. An event is the fundamental entity of observed physical reality represented by three coordinates of place and one coordinate of time in the spacetime continuum postulated by the theory of relativity (Merriam-Webster). Events represent the basic elements of four-dimensional spacetime. An observer chooses a convenient metrical coordinate system when describing physical phenomena (events) occurring in spacetime. Coordinates specify where and when events occur. In spacetime, an event has a unique position at a unique time (specified by four coordinates). Spacetime may be viewed as a union of all events (like a line is a union of all its points), formally organized into a manifold. Spacetime is independent of any observer (Matolcsi, 1994).

SPACETIME INTERVALS

A line joining two events in spacetime is the shortest distance between the two events. Such lines in flat (Minkowski) spacetime are called world lines. In general relativity, the notion of a geodesic generalizes the notion of a world line to curved spacetime. A world line or geodesic between two events in spacetime is called a spacetime interval. A spacetime interval is the separation of two events in four-dimensional spacetime. Since spacetime has three dimensions of space and one dimension of time, a spacetime interval has both spatial and temporal components. Spacetime intervals are invariant (identical) in all frames of reference because of the symmetry of the laws of physics.

The unification of space and time is exemplified by selecting a metric (a measure) such that all four dimensions of spacetime are measured in units of distance (Petkov, 2010). Thus an event is represented as (xo,x1,x2,x3)= (ct,x,y,z) in the Lorentz Metric or as (x1,x2,x3,x4)=(x,y,z,ict) in the Minkowski Metric. In three-dimensional (Euclidian) space, the separation between two objects is measured by the distance between them. In four-dimensional spacetime, the separation between two events is measured by the invariant spacetime interval between the two events. This may be formulated as follows: the spacetime interval "s" is defined as the difference between the space coordinates "Δr" of two events minus the product of "c" and the difference between the time coordinates "Δt" of the two events (in practice the square is utilized as the sign of the spacetime interval "s" is indefinite-positive, negative, or zero).

(spacetime interval) $s^2=\Delta r^2-c^2\Delta t^2$

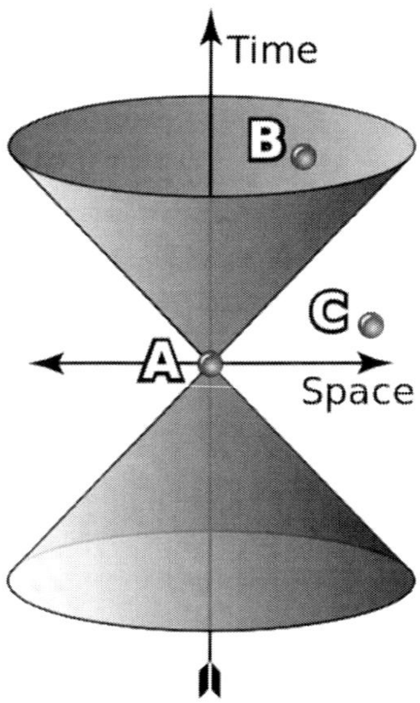

Figure 1. Spacetime interval of different time and events.

There are three types of spacetime interval which define the spatiotemporal organization and relationships of events in spacetime (Figures 1 and 2). In a time-like spacetime interval (Figure 1), the difference in the time coordinates of two events (A and B) is greater than the difference in the space coordinates of the two events ($c^2\Delta t^2>\Delta r^2$). Hence the spacetime interval is negative ($s^2<0$). Time-like spacetime interval separation occurs inside the light cones of Figures 1 and 2. Time-like spacetime separation creates reference frames where two events (A and B) may occur at the same location in space, but are separated only by occurring at different times (they are separated by time, not by space). Time-like separation between events allows signals or information to travel between the events at less than the speed of light. Hence for an event A, there exists a set of events that can either influence or be influenced by A, by signals or information that does not need to travel faster than the speed of light. A cause or effect relationship could exist between the two events (one event could be the cause or the effect of the other event). If B causes A, B exists in the past history (in the past light cone) of A (Figures 1 and 2). If B is caused by A, B exists in the future (the future light cone) of A (Figures 1 and 2). In a light-like spacetime interval (Figures 1 and 2), the difference in the time coordinates of two events is equal to the difference in the space coordinates of the two events ($\Delta r^2=c^2\Delta t^2$). The spacetime interval equals zero ($s^2=0$). Events following one another at light speed have light-like separation. Light-like spacetime intervals define the boundaries of the light cones of Figures 1 and 2. All the events that follow an event at light-like separations define a future light cone (Figures 1 and 2) for that event. All the events that precede an event at light-like separations define a second inverted past light cone (Figure 1 and 2) for that event. Events (A and B) occurring inside a light cone have time-like separation. Events occurring on a light cone have light-like separation (such as for a photon travelling through space). Events occurring outside a light cone have space-like separation (A and C). In a space-like spacetime interval (Figure 1), the difference in space coordinates of two events (A and C) is greater than the difference in time coordinates of the two events ($\Delta r^2 > c^2 \Delta t^2$). Space-like spacetime intervals are positive ($s^2>0$). Space-like spacetime separation creates reference frames in which two events (A and C) may occur simultaneously, but in different locations (they are separated only by space, not by time). There is not enough time between the two events so that one can send signals or information to the other at less than the speed of light. Hence the two events cannot influence one another (they cannot have a cause or effect relationship). Because the two events occur outside the light cone, signals or information must travel faster than the speed of light in order for one event to influence the other (this violates special relativity). A paradox of causality could exist if it was possible for a cause or effect relationship to exist between two events with space-like separation. For example, if A was the effect and C the cause, there could be a frame of reference where the effect would precede the cause. This would require faster than light signals be sent back into one's past. If A was the cause and C the effect, signals from A would still have to travel faster than the speed of light to influence C. Therefore a cause or effect relationship cannot exist between two events with space-like separation.

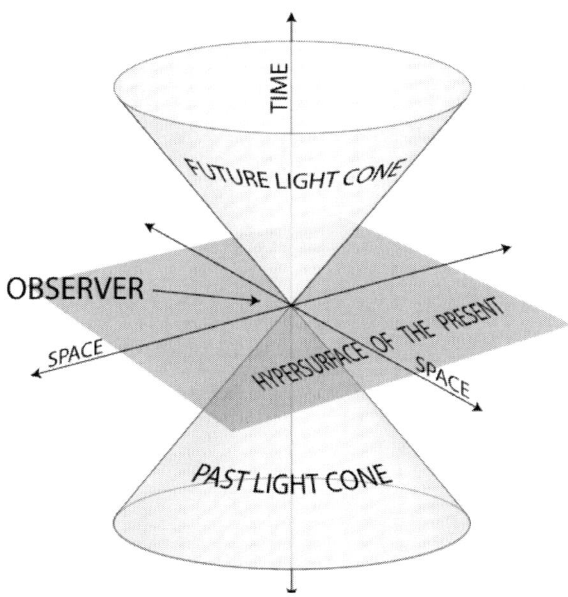

Figure 2. Spacetime intervals boundaries.

Some important points should be noted from the above. Spacetime intervals are invariant (identical) in all frames of reference (for all observers), even though the observers may be travelling at different speeds or have different vantage points. Events with time-like spacetime separation may be separated only by time, not by space. These events may occur at the same location, but at different times. A cause or effect relationship may exist between two events with time-like separation because signals or information may be transferred between them at less than the speed of light. Events with light-like spacetime separation are separated at light speed. Events with space-like spacetime separation may be separated only by space, not by time. These events may occur simultaneously, but at different locations. These events cannot bear a cause or effect relationship to one another because there is not enough time for signals or information to be transferred between them at less than the speed of light. Spacetime intervals determine the organization and relationships of events.

Spacetime Intervals and the Organization of Consciousness and Episodic Memory

Spacetime intervals appear to determine the organization of consciousness and episodic memory. Conscious experiences make up the content of consciousness. At any one time, some conscious experiences occur simultaneously in consciousness, but in different locations. These conscious experiences may have space-like spacetime interval separation. Other conscious experiences occur at the same location in consciousness, but at different times. These conscious experiences may have time-like spacetime interval separation. Spacetime intervals therefore appear to define the organization and relationships of conscious experiences in consciousness at any one time. Consciousness therefore has a four-dimensional organization. Similarly, conscious experiences make up the content of episodic memory recall. During episodic memory recall, some conscious experiences occur simultaneously, but in different locations (space-like separation), while other conscious experiences occur at the same location, but at different times (time-like separation). Spacetime intervals therefore appear to define the organization and relationships of episodic memory recall. Episodic memory therefore appears to have a four-dimensional organization.

Episodic Memory

In general, there are three main stages to memory function. First, information must be encoded or registered into memory. Second, this information must be retained in some way. Third, the stored information may be retrieved or recalled. Four types of memory appear to operate simultaneously and sequentially in the establishment of episodic memory. These are sensory memory, working memory, short term memory, and long term memory.

Sensory Memory

Sensory memory is a fast-decaying register of sensory information. It is a very brief (<1000 msec) pre-categorical high-capacity memory which occurs after stimulus offset (Sperling, 1960; Dick, 1974). Sensory memory is a part of the brain's memory system. It is known as iconic memory in the visual system and echoic memory in the auditory system. Sensory memory is highly colored by the sensory modality in which it occurs. This type of memory decays faster than one can report its details, so it is difficult to determine all the information that it contains.

Iconic memory provides a smooth flow of visual information to the brain which can be extracted over an extended period of time by working memory for consolidation into more stable forms. Iconic memory consists of three components (Coltheart, 1980; DeLollo, 1980; Loftus et al., 1992). Neural persistence is the neural activity that persists in the visual system after stimulus offset. It includes activity arising in photoreceptors, ganglion cells, and simple cells. Visual persistence arises from neural persistence and is the visible phenomena that can be seen after stimulus offset (glow of a light bulb after it is turned off, path of a torch when it is moved) lasting 100-300 msec (Haber, 1970; Irwin & Thomas, 2008; Levick & Zack, 1970; Long, 1980; Nicolic et al., 2009). Visible persistence is sensitive to the physical properties of the stimulus and is inversely related to stimulus duration and luminosity (Coltheart, 1980). Information persistence is the non-visible information about a stimulus that persists after stimulus offset, such as spatial location and other abstract characteristics of the stimulus (Irwin & Yeomans, 1986). Information persistence is proportional to stimulus duration and is the key component of iconic memory contributing to working memory (Coltheart,1980; Green,2007; Irwin & Yeomans, 1986). Information persistence could carry information about stimulus spacetime separation, such as whether stimuli occur simultaneously at different locations or whether stimuli occur at the same location but at different times (spacetime intervals), that might determine the organization of working memory.

WORKING MEMORY

Working memory is the holding in mind for a brief period of time of multiple pieces of information and its manipulation (Baddeley, 2003; Cowan, 2008; Fuster, 1997; Schacter, 2009). Working memory is not prolonged by rehearsal. Functionally, working memory is the provisional retention of perceptual information for prospective action, a type of focal attention whereby perception is reorganized and rerepresented, becoming explicit, functional, and conscious (Luck & Vogel, 1997; Schacter, 2009, 2011; Todd & Marois, 2004; Vogel & Machizawa, 2004). As already discussed, working memory is the nonlinear emergence of a number of explicit stable states (conscious experiences) from the positive feedback interaction between the inferior parietal cortex and the prefrontal cortex.

Cowan (2001) has proposed that working memory has a capacity of about four chunks in young adults (fewer in children and older adults), depending on the category (digits, letters, words, etc.) and features (short words, long words, simple words, complex words, etc.) of the chunks (Hulme et al., 1995). Chunking is the separation of information into meaningful groups called chunks so that all of the information is more easily remembered. The ideal size for a chunk is a series of three letters or numbers (like telephone numbers). Working memory enables the completion of goal-directed actions (voluntary intentional actions). It involves the holding of information over a short period of time (a delay) so that it can be utilized at a later time to perform a response (delayed response). Delayed response paradigms are therefore often used to study working memory (Fuster, 1997, 2000, 2008). Studies indicate the prefrontal cortex, inferior parietal cortex, anterior cingulate cortex, thalamus, and parts of the basal ganglia (caudate, globus pallidus) are crucial for working memory function (Ashby et al., 2005; Benton, 1991; Fuster, 1997, 2000, 2008; Goldman-Rakic, 1990, 1995; Joseph, 2000; Prinz, 2003). Goldman-Rakic (1995) showed that the dorsolateral prefrontal cortex interconnects with all these brain regions and that neuronal microcircuits within the prefrontal cortex are able to maintain information in working memory through recurrent excitatory glutamate networks of pyramidal cells that continue to fire throughout a delay (working memory) period. These microcircuits are tuned by lateral inhibition from GABAergic interneurons (Rao et al., 2000). The neuromodulatory arousal systems markedly alter prefrontal cortex working memory function; for example, either too little or too much dopamine or norepinephrine impairs prefrontal cortex network firing (Arnsten et al., 2010) and working memory performance (Robbins & Arnsten, 2009). A review of numerous studies shows areas of activation during working memory tasks scattered over a large part of the cortex (Smith & Jonides, 1999). There is a tendency for spatial working memory tasks to recruit more right-hemisphere areas and for verbal and object working memory tasks to recruit more left-hemisphere areas. The activation during verbal working memory tasks can be broken down into one component reflecting maintenance in the left posterior parietal cortex and a component reflecting subvocal rehearsal in the left frontal cortex (Smith et al., 1998). There is an emerging consensus that most working memory tasks recruit a network of prefrontal and parietal cortical areas. As discussed previously, a prefrontal cortex-parietal cortex network is recruited during the formation of conscious experiences. Hence working memory appears to be associated with the formation of conscious experience. Honey and coworkers (2002) has shown that during a working memory task the connectivity between these areas increases. Montaghy (2006) has demonstrated that these areas are necessary for working memory and are not simply activated accidentally during working memory tasks. The prefrontal cortex has been found to be active in a variety of tasks that require executive functions (Kane & Engle, 2002). This has led some researchers to argue that the role of the prefrontal cortex in working memory is in controlling attention, selecting strategies, and manipulating information, but not in maintenance. The maintenance function is attributed to more posterior areas of the brain (Curtis & D'Esposito, 2003; Postle, 2006).

Research suggests a close link between working memory capacity and the ability to control the information from the environment that can be selectively enhanced or ignored (Fukuda & Vogel, 2009). The goal-directing of attention is driven by "top-down" signals from the prefrontal cortex that bias processing in posterior cortical areas (Desimone & Duncan, 1995) and saliency capture by "bottom-up" control from subcortical structures and the primary sensory cortices (Yantis & Jonides, 1990). The ability to override saliency capture differs greatly between individuals and this difference closely links to their working memory capacity. The greater a person's working memory capacity, the greater their ability to resist saliency capture (Fukuda & Vogel, 2009). The limited ability to override saliency capture is likely to result in the unnecessary storage of information in working memory (Fukuda & Vogel, 2009), suggesting not only that having a poor working memory affects attention, but that it can also limit the capacity of working memory even further.

There is a link between working memory, attention, learning, cognitive development, cognitive function, and memory. The greater is the capacity of working memory, the better are these functions (Alloway & Alloway, 2010; Andrews & Halford, 2002; Berry et al., 2009; Case, 1985; Cowan & Alloway, 2008; Kail, 2007; Klingberg et al., 2002; Jarrold & Bayliss, 2007; Zanto & Gazzaley, 2009). These studies show that working memory capacity is a better predictor of academic and occupational success than IQ.

Conscious experience and episodic memory appear to be related by working memory. They both arise from working memory and both have the same four-dimensional organization and relationships. Consciousness is also required for episodic memory recall. Working memory appears to be responsible for the basic four-dimensional organization and relationships of consciousness and episodic memory recall. As previously described, working memory is a positive feedback interaction between the inferior parietal cortex and the prefrontal cortex which produces and maintains a number of nonlinear explicit states. Working memory is the holding in mind and manipulation of a number of pieces of information for a short period of time. Hence working memory is the holding in mind and manipulation of a number of nonlinear explicit states (conscious experiences) developed from positive feedback and nonlinear emergence. Each nonlinear explicit state has a spatiotemporal location in working memory. Working memory therefore consists of a four-dimensional organization of a number of nonlinear emergent explicit stable states (conscious experiences) which is maintained (and manipulated) for a short period of time and utilized at a later time for some purpose (goal-directed action, memory). The four-dimensional organization and relationships of consciousness and episodic memory is thus developed from working memory. Working memory is a link between consciousness and episodic memory.

SHORT TERM MEMORY

Thanks to the hippocampus, individuals are able to store and retain new memories and retrieve them from pieces of the original (Serino & Riva, 2014). It is generally agreed that the hippocampus plays an important role in the formation of new memories about experienced events (Eichenbaum & Cohen, 1993; Squire, 1992; Squire & Schacter, 2002). Part of this function is hippocampal detection of novel events, places, and stimuli (VanElzakker et al., 2008). Some regard the hippocampus as part of a larger medial temporal lobe memory system responsible for general declarative memory (Squire, 1992). A distinction may be made between conscious *recollection*, which depends on the hippocampus, and *familiarity*, which depends on portions of the medial temporal cortex (Diana et al., 2007). The encoding of information into short term memory (hours to days to weeks) appears to involve establishment of a pattern of long term potentiation (long term depression may also occur) in the hippocampus, adjacent structures (dentate gyrus, subiculum), and the cerebral cortex (occipital, temporal, parietal cortex). Long-term potentiation (LTP) is a form of neural plasticity, first known to occur in the hippocampus, which is widely believed to be one of the main neural mechanisms by which memory is stored in the brain. Severe anterograde amnesia (an inability to form new memories, to learn or remember new things) occurs with extensive hippocampal damage.

The hippocampus and LTP also appears to be involved in the retention of short term memory. LTP has been shown to occur in the hippocampus and dentate gyrus (Deadwyler et al., 1988; Eccles et al., 1987; Gustafsson & Wigstrom, 1990; Haas & Buzsaki, 1988; Lynch et al., 1990; Lynch et al., 1990; Steward et al., 1988). LTP is a prolonged increase in transmission through pathways subjected to repetitive high frequency stimulation. The effect may last for days to weeks and exhibits many of the characteristics of a learning and memory mechanism. The establishment of patterns of LTP in the hippocampus and related structures is a way of encoding and retaining information for the short term. Hippocampal damage appears to disrupt the retention of short term memories. Hippocampal damage appears to affect not only the encoding and retention of new memories (anterograde amnesia), but also affects access (recall, retrieval) of memories laid down prior to the damage (retrograde amnesia) in a graded manner (the retrograde amnesia effect decreases the further back in time were recorded the memories). This retrograde effect may extend back several years prior to the damage and in some cases memories beyond this period remain unaffected. This sparing effect suggests that consolidation of memories over time involves the transfer of information out of the hippocampus to other brain areas. However, the role of the hippocampus in maintaining older memories is uncertain.

HIPPOCAMPUS

There are four main histological divisions of the hippocampus: CA1, CA2, CA3, and CA4 (Wechsler, 2004). The hippocampus plays an important role in the encoding and retention of short term episodic memory, the consolidation of information from short term memory to long term memory, and spatial navigation. The hippocampus receives highly processed sensory, emotional, and cognitive information from cortical association areas (temporal, parietal, and occipital), the frontal cortex, the amygdaloid nuclei, the septal nuclei, the dentate gyrus, and other structures via the entorhinal cortex (Carlson et al., 1982; Flood & Coleman, 1990; Goldman-Rakic, 1990; Van Hoesen & Hyman, 1990). The superficial layers of the entorhinal cortex provide most of the input to the hippocampus, while the deeper layers receive the most output from the hippocampus. The entorhinal cortex is located in the parahippocampal gyrus and is considered part of the hippocampal region because of its anatomical connections. The entorhinal cortex is strongly and reciprocally connected with many other parts of the cerebral cortex. In addition, the medial septal nucleus, the anterior nuclear complex and nucleus reuniens of the thalamus, the supramammillary nucleus of the hypothalamus, as well as the raphe nuclei (serotonergic) and locus coeruleus (norepinephrinergic) in the brainstem send axons to the entorhinal cortex. The main output pathway of the entorhinal cortex comes from the large stellate pyramidal cells in layer II that "perforate" the subiculum and project densely to the granule cells in the dentate gyrus, less densely to the apical dendrites of CA3, and sparsely to the apical dendrites of CA1. This “perforant path” establishes the entorhinal cortex as the main "interface" between the hippocampus and other parts of the cerebral cortex. Dentate granule cell axons (called mossy fibers) pass on the information from the entorhinal cortex to the proximal apical dendrites of CA3 pyramidal cells. Then, CA3 axons exit from the deep part of the cell body and loop up into the region where the apical dendrites are located, then extend all the way back into the deep layers of the entorhinal cortex (the Schaffer collaterals which complete the reciprocal circuit); field CA1 also sends axons back to the entorhinal cortex, but these are more sparse than the CA3 projection. Within the hippocampus, the flow of information from the entorhinal cortex is largely unidirectional, with signals propagating through a series of tightly packed cell layers, first to the dentate gyrus, then to the CA3 layer, then to the CA1 layer, then to the subiculum, and then out of the hippocampus to the entorhinal cortex, mainly due to collateralization of the CA3 axons. Each of these layers also contains complex intrinsic circuitry and extensive longitudinal connections (Amaral & Lavenex, 2006). Several other connections play important roles in hippocampal function (Amaral & Lavenex, 2006). Besides the output to the entorhinal cortex, additional output pathways go to other cortical areas including the prefrontal cortex, the lateral septal area, and the mammillary body of the hypothalamus. The hippocampus receives modulatory input from the serotonin, norepinephrine, and dopamine systems and from the nucleus reuniens of the thalamus to field CA1. A very important projection comes from the medial septal area, which sends cholinergic and GABAergic fibers to all parts of the hippocampus. The inputs from the septal area play a key role in controlling the physiological state of the hippocampus; destruction of the septal area abolishes the hippocampal theta rhythm and severely impairs certain types of memory (Winson, 1978).

The cortical region adjacent to the hippocampus is known collectively as the parahippocampus (Eichenbaum et al., 2007). It includes the entorhinal cortex and the perirhinal cortex. The perirhinal cortex plays an important role in visual recognition of complex objects. There is also substantial evidence that the perirhinal cortex makes a contribution to episodic memory, which can be distinguished from the contribution of the hippocampus. It is apparent that complete amnesia occurs only when both the hippocampus and the parahippocampus are damaged (Eichenbaum et al., 2007). Various parts of the hippocampal formation are shown to be functionally and anatomically distinct.

The dorsal, ventral, and intermediate regions of the hippocampal formation have varying degrees of place field neurons, have different projections, and serve different functions (Fanselow & Dong, 2009). The dorsal region serves for spatial memory, verbal memory, and learning of conceptual information (Pothuizen et al., 2004). It also has more place field neurons than both the ventral and intermediate regions (Jung et al., 1994). The intermediate region has overlapping characteristics with the ventral and dorsal regions (Fanselow & Dong, 2009). Cenquizca and Swanson (2007) found moderate projections to two primary olfactory cortical areas and prelimbic areas of the medial prefrontal cortex. This region has the smallest number of place field neurons. The ventral region functions in fear conditioning and affective processes. Anagnostaras and coworkers (2002) showed that alterations to the ventral region reduced the amount of information sent to the amygdala by the dorsal and ventral regions, consequentially altering fear conditioning in rats.

Because of its densely packed neural layers, the hippocampus generates some of the largest EEG signals of any brain structure. The hippocampus has two major "modes" of activity named after the EEG patterns associated with them (Buzsaki, 2006). The theta mode appears during states of active, alert behavior (especially locomotion) and also during REM sleep. When cortical desynchronization (the presence of rapid irregular low amplitude EEG activity) occurs (during active searching, orientation, maintained and selective attention, initial stages of learning, discrimination responses, initial exposure to novel stimuli, in other words, during conscious processing), hippocampal theta appears (Buzsaki, 2006; Foreman & Stevens, 1987); that is, theta appears in the hippocampus whenever an animal or person is actively engaged with its environment. The EEG is dominated by regular waves at 3 to 9 Hertz, often continuing for many seconds. These reflect subthreshold membrane potentials which strongly modulate the spiking of hippocampal neurons and synchronise across the hippocampus in a travelling wave pattern (Lubenov & Siapas, 2009). This EEG pattern is known as a theta rhythm. Theta is very obvious in rabbits and rodents and also clearly present in cats and dogs. Whether theta can be seen in primates is a vexing question (Cantero et al.,

2003). In rats, theta is seen mainly when an animal is walking or in some other way actively interacting with its surroundings and during REM sleep. The EEG is dominated by large regular waves (6 to 9 Hertz) and the main groups of neurons (pyramidal cells and granule cells) show sparse population activity, which means that in any short time interval, the great majority of cells are silent, while the small remaining fraction fire at relatively high rates, up to 50 spikes per second, for the most active of them. An active cell typically stays active for half a second to a few seconds. As the animal behaves, the active cells fall silent and new cells become active, but the overall percentage of active cells remains more or less constant. In many situations, cell activity is determined largely by the spatial location of the animal, but other behavioral variables also clearly influence it. The function of theta has not yet been convincingly explained (Buzsaki, 2006). The most popular hypothesis has been to relate it to learning and memory. Theta may affect those aspects of learning and memory that are dependent upon synaptic plasticity (Huerta & Lisman, 1993). It is well established that lesions of the medial septum (the central node of the theta system) cause severe disruptions of memory. However, the medial septum is more than just the controller of theta, it is also the main source of cholinergic projections to the hippocampus (Amaral & Lavenex, 2006). It has not been established that septal lesions exert their effects specifically by eliminating the theta rhythm (Kahana et al., 2001).

The large irregular activity mode appears during slow-wave sleep and also during states of waking immobility, such as resting or eating (Buzsaki et al., 1990). During sleep or waking states when an animal is resting or otherwise not engaged with its surroundings, the hippocampal EEG shows a pattern of irregular slow waves, somewhat larger in amplitude than theta waves, which is occasionally interrupted by large surges called *sharp waves* (Buzsaki, 1986). Sharp waves (randomly timed large deflections of the EEG signal lasting for 25-50 msec) dominate the EEG. They are frequently generated in sets, with sets containing up to 5 or more individual sharp waves and lasting up to 500 msec. The spiking activity of neurons within the hippocampus is highly correlated with sharp wave activity. Most neurons decrease their firing rate between sharp waves; however, during a sharp wave, there is a dramatic increase of firing rate in up to 10% of the hippocampal population. These events are also associated with short-lived high-frequency EEG oscillations called "ripples", with frequencies in the range of 150 to 200 Hertz in rats. Sharp waves are most frequent during sleep when they occur at an average rate of around 1 per second (in rats), but in a very irregular temporal pattern. Sharp waves are less frequent during inactive waking states and are usually smaller.

Sharp waves have been observed in humans and monkeys. The two hippocampal activity modes can be seen in primates as well as rats, with the exception that it has been difficult to see robust theta rhythm in the primate hippocampus. There are, however, qualitatively similar sharp waves and state-dependent changes in neural population activity (Skaggs et al., 2007).

One of the most interesting aspects of sharp waves is that they appear to be associated with memory. Wilson and McNaughton (1994) and numerous later studies reported that when hippocampal place cells have overlapping spatial firing fields (and therefore often fire in near-simultaneity), they tend to show correlated activity during sleep following the behavioral session. This enhancement of correlation, commonly known as *reactivation*, has been found to occur mainly during sharp waves (Jackson et al., 2006). It has been proposed that sharp waves are reactivations of neural activity patterns that were encoded into memory during behavior driven by strengthening of synaptic connections within the hippocampus (Sutherland

& McNaughton, 2000). Buzsáki (2006) and others propose that memories are stored within the hippocampus during behavior and then later transferred to the neocortex during sleep. Sharp waves are suggested to drive Hebbian synaptic changes in the neocortical targets of hippocampal output pathways (Buzsaki, 1989).

Dreaming might be a manifestation of this consolidation process. Forgotten memories could also be reactivated at these times. Thus the hippocampus monitors and modulates cortical activity at various levels of cortical activation and encodes and consolidates episodic memory. The hippocampus, however, docs cease to play a role in the retention of episodic memory after a certain period of time (Squire & Schacter, 2002). That is, after a certain period of time, episodic memory appears to be retained independently of the hippocampus. This appears to occur because of the establishment of more stable and permanent connections in the neocortex. The period of consolidation may last up to five to seven years, as electroshock therapy produces retrograde amnesia which extends back five to seven years, but no further (Mohs, 1991).

Damage to the hippocampus has detrimental effects on short term and long term episodic memory and the production of goal-directed behavior. Hippocampal damage appears to have an overall inhibitory effect on cortical activity (Gray & McNaughton, 2000). Hippocampal damage may produce difficulties in inhibiting behavioral responsiveness, difficulties in inhibiting shifts in attention, perseveration, loss of habituation, impaired divergent thinking and response, impaired ability to learn and remember new information (anterograde amnesia) and to remember old information (retrograde amnesia), and the individual may be overwhelmed, hyperresponsive, distracted, and confused (Joseph, 1990). Attention, learning, memory, and cognitive function are severely impaired. Bilateral destruction of the hippocampus produces severe anterograde amnesia (Ganong, 1988; Joseph, 1990). Iterative hippocampal discharge during induced seizures and the administration of drugs that alter normal hippocampal discharge also produce anterograde amnesia (Ganong, 1988). Alcoholics who have anterograde amnesia (Korsokoff's Syndrome) often have lesions of the mammillary bodies, which are targets of the hippocampus (Ganong, 1988).

Spatial Navigation, Spatial Memory

The hippocampus is involved in spatial navigation and spatial memory. Evidence suggests that the hippocampus is involved in processing and storing spatial information. Studies in animals have shown that there are neurons in the hippocampus called "place cells" which have spatial firing fields. These cells fire when the animal is in a particular location (place) in the environment, regardless of direction of travel. Most of these neurons are at least partially also sensitive to head direction and direction of travel. Different cells fire at different spatial locations, so one can tell where an animal is in an environment by observing which neurons are firing. Place cells have been found in humans finding their way around a virtual reality town (Eckstrom et al., 2003).

The hippocampus may act as a neural representation of the layout of the environment (it is a cognitive map). Spatial navigation involves the ability to orient oneself in space, to follow directions, and to recognize familiar places (spatial memory). It depends on the hippocampus. An intact hippocampus has been shown to be necessary for simple spatial memory tasks, such as finding the way to a hidden goal (Morris et al., 1982). A fully functional hippocampus appears to be necessary for humans to remember where they have been and where they are going. Getting lost and disorientation are common symptoms of hippocampal damage and amnesia. The hippocampi are more active when people are correctly navigating (Maguire et al., 1998). The hippocampus has also been shown to be involved in finding shortcuts and new routes between familiar places (Maguire et al., 2000).

Place Cells

A place cell is a type of pyramidal neuron found within the hippocampus that becomes active when an animal enters a particular location or place in the environment-the place field (Binder, 2009; Fenton et al., 2008; O'Keefe et al., 1998). Studies have shown that many hippocampus neurons have place fields and fire bursts of action potentials when the freely moving animal passes through that particular part of the environment (Ekstrom et al., 2003; Matsumara et al., 1999; Rolls & Xiang, 2006). A place cell has only one, or a few, place fields in a typical small laboratory environment, but more in a larger environment (Fenton et al., 2008). There is no apparent topography in place fields in the hippocampus, neighboring place cells are as likely to have distant fields as neighboring ones (O'Keefe et al., 1998). In a different environment, typically about half the place cells will still have place fields, but these will be for new places, unrelated to their former locations (Muller & Kubie, 1987). Place cell firing is strongly correlated to sensory input and very responsive to spatial surroundings. Place cells tend to fire quickly when an animal enters a new open environment (Bures et al., 1997). Place cells have proven to have the ability to suddenly change their firing pattern from one pattern to another ("re-mapping"). Although place cells do change their firing according to the external environment, they are stabilized to enable the system to resist small changes in sensory input, but respond collectively and coherently to large changes (Jeffery, 2007).

Place cell responses are shown by pyramidal cells in the hippocampus proper and granule cells in the dentate gyrus (Moser et al., 2008). Place cells constitute the majority of neurons in the hippocampus. Inhibitory interneurons, which make up most of the remaining cell population, frequently show significant place-related variations in firing rate that are much weaker than those displayed by pyramidal or granule cells. Place cells are typically almost silent when an animal is moving outside of the place field, but reach sustained rates as high as 40 Hertz when the animal is near the center. Neural activity sampled from 30 to 40 randomly chosen place cells carries enough information to allow an animal's location in an environment to be reconstructed with high confidence. The size of place fields varies in a gradient along the length of the hippocampus, with cells at the dorsal end showing the smallest fields, cells near the center showing larger fields, and cells at the ventral tip fields that cover the entire environment (Moser et al., 2008). In some cases, the firing rate depends not only on place, but also on the direction the animal is moving, the destination toward which it is traveling, and other task-related variables (Smith & Muzimori, 2006).

O'Keefe and Dostrovsky first found the existence of place cells in the hippocampus in 1971. They later demonstrated that place cells would fire whenever a rat was within a certain place in the environment (Moser et al., 2008). O'Keefe (1978) found that place cells fire more rapidly when rats ran past places in the environment, when a new item was added to the environment, or when an item that is usually there is not present. This was one of the first indicators that place cells were related to spatial orientation. It might also be indicative of a temporal orientation (the presence of an item in the same location, but at different times). It was also discovered that place cells fired in different areas of the hippocampus depending on where the rat went and this whole firing network made up the rat's environment (Wilson & McNaughton, 1993). As environments changed, the same place cells would fire, but the relationship and dynamic between firing fields would change (O'Keefe & Conway 1978). Therefore place cells are thought to give humans and animals a guide to the environment they are navigating and their position in that environment. As humans and animals navigate large environments and then arrive at a particular location, there is a notable increase in the place cell firing rate once that specific location has been reached. Place cells fire in different, often widespread, hippocampal locations at the same time (O'Keefe, 1999). This also might indicate that place cells signal temporal orientation. Place cells that fire simultaneously, but in different locations in the hippocampus, and place cells that fire at the same locations in the hippocampus, but at different times, may add a temporal dimension to hippocampal function. Hence place cells may be encoding, not just a three-dimensional cognitive map of the environment, but a four-dimensional cognitive map of the environment.

In an open field environment, firing fields of place cells prove to be similar even when a rat travels in different directions (O'Keefe, 1999). When limitations are placed in the aforementioned environment, firing fields prove to be directional and fire in one direction, but not in another (O'Keefe, 1999). There are both simple place cells with purely location correlates and also complex place cells that increase their firing rate when an animal encounters a particular object or experience (O'Keefe, 1999). Others fire when expectations in a particular location are not met or when novelty is encountered along the path.

The existence of place cells within the hippocampus demonstrates the important role it plays in spatiotemporal adaptation and awareness. There have been recorded increases in firing patterns in open environments and spatial learning and awareness impairments after damage to the hippocampus and the place cells within (Binder, 2009). Place fields are considered to be allocentric rather than egocentric, meaning that they are defined with respect to the outside world rather than the body. By orientation based on the environment, rather than the individual, place fields can work effectively as neural maps of the environment (Jeffery et al., 2003). If this cognitive map is in four dimensions (a neural representation of the four-dimensional layout of the environment) then the hippocampus could participate in the four-dimensional organization of awareness (consciousness) and episodic memory. Several lines of evidence support this hypothesis. It is a frequent observation, that without a fully functional hippocampus, humans may not remember where they have been or how to get where they are going (getting lost is one of the most common symptoms of amnesia-Chui et al., 2004). Studies have shown that an intact hippocampus is required for initial learning and retention in some spatial memory tasks, in particular, ones that require finding the way to a hidden goal (Clarke et al., 2005; Morris et al., 1982; Sutherland et al., 1982; Sutherland et al., 2001). The cognitive map hypothesis has been further advanced by recent discoveries of head direction cells, grid cells, and boundary cells in several parts of the brain that are strongly connected to the hippocampus (Moser et al., 2008; Solstad et al., 2008). People have more active hippocampi when correctly navigating, as tested in computer-simulated "virtual" navigation tasks (Maguire et al., 1998). Also, there is evidence that the hippocampus plays a role in finding shortcuts and new routes between familiar places. A study by Maguire and coworkers (2000) showed that part of the hippocampus is larger in taxi drivers than in the general public and that more experienced drivers have bigger hippocampi. The study found a positive correlation between the length of time an individual had spent as a taxi driver and the volume of the right hippocampus.

Place cells therefore may act collectively as a four-dimensional cognitive map of the environment. Place cells work with other types of neurons in the hippocampus and surrounding regions to perform this kind of processing (Muir & Bilkey, 2001). Head direction neurons increase their firing rates above baseline levels only when the animal's head points in a specific direction. When stimulated these neurons fire at a steady rate (they do not show adaptation), but decrease back to their baseline rates, as the animal's head turns away from the preferred direction (usually about 45° away from this direction-Taube, 2007). These cells are found in many brain areas, including the post-subiculum, retrosplenial cortex, the thalamus (anterior and the lateral dorsal thalamic nuclei), lateral mammillary nucleus, dorsal tegmental nucleus, striatum, and entorhinal cortex (Sargolini et al., 2006). This system is related to the place cell system, which is mostly orientation-invariant and location-specific, while head direction cells are mostly orientation-specific and location-invariant. However, head direction cells do not require a functional hippocampus to show their head direction specificity.

Place cells are a part of a complex circuit that informs awareness and memory (Smith & Mizumori, 2006). Place cell circuits have important implications for memory as they provide a context for such memory (Smith & Mizumori, 2006). Like many other parts of the brain, place cell circuits are dynamic. They are constantly adjusting and remapping to suit the current environment. Place fields are roughly analogous to the receptive fields of sensory neurons in that firing corresponds to a representation of sensory information in the environment (Jones, 2013). Place cells respond to complex stimuli rather than simple individual sensory cues (Jeffery et al., 2003; Moser et al., 2008). Sensory information is processed in various cortical structures upstream of the hippocampus before actually reaching the structure, so that the information received by place cells is a compilation of different stimuli (Jeffery et al., 2003). Sensory information received by place cells can be metric or contextual (Jeffery, 2007). Metric corresponds to where place cells should fire while contextual corresponds to whether or not a place field should fire in a certain environment. Metric sensory information may be any kind of input that might indicate a separation between two points (such as separation in spacetime). Metric signals can be either linear or directional. Directional inputs provide information about the orientation of a place field, whereas linear inputs essentially form a representational grid.

The entorhinal cortex is the main input to the hippocampus. The entorhinal cortex contains head direction cells and grid cells. It seems probable that it is the head direction cell input that directs the orientation of place fields in the hippocampus and the grid cells that direct the grid-like representation of place fields in the hippocampus. Contextual cues allow established place fields to adapt to minor changes in the environment, such as a change in object color, shape, or location. Metric and contextual inputs are processed together in the entorhinal cortex before reaching the hippocampal place cells. Visual and olfactory inputs are examples of sensory inputs that are utilized by place cells. These types of sensory cues can include both metric and contextual information (Jeffery, 2007).

Spatial cues such as geometric boundaries or orienting landmarks are also important metric input. So-called boundary or border cells are found in the entorhinal cortex which appears to provide such input to place cells. Place cells rely on set distal cues rather than cues in the immediate proximal environment (Jeffery, 2007). Movement can also be an important metric cue. The ability of place cells to incorporate new movement information is called *path integration* (Moser et al., 2008). Path integration is largely aided by grid cells in the entorhinal cortex that relay information to place cells in the hippocampus. As will be seen, grid cells may establish a four-dimensional regular grid representation of spacetime in the hippocampus, so that during movement place cells can fire according to a reference grid of their external environment (Jeffery, 2007). Visual inputs can supply important contextual information. A change in color of a specific object can affect whether or not a place field fires in a particular environment (Jeffery, 2007). Although place cells primarily rely on visual input, some studies suggest that olfactory input may also play a role in generating and recalling place fields and orientation during movement (Save et al., 2000; Zhang & Manahan-Vaughn, 2013).

The hippocampus plays an essential role in episodic memory (Rolls, 2013). One important aspect of episodic memory is the context in which the event occurred (Smith & Mizumori, 2006). Hippocampal place cells have been shown to exhibit stable firing patterns even when cues from a location are removed. Additionally, specific place fields begin firing when exposed to signals or a subset of signals from a previous location (Nakazawa et al., 2004). Place cells therefore appear to prime memory by differentiating the context (Smith & Mizumori, 2006). Hence place cells may complete memory patterns (Rolls, 2013). Furthermore, place cells can maintain representation of one location while recalling the neural map of a separate location, effectively differentiating between present experience and past experience (Smith & Mizumori, 2006). Place cells therefore demonstrate both pattern completion and pattern separation (Mozer et al., 2008; Rolls, 2013). Place cells often exhibit reactivation outside their place fields. This reactivation has a much faster time scale than the actual experience and it occurs mostly in the same order in which it was originally experienced, or, more rarely, in reverse. Replay is believed to have a functional role in memory retrieval and memory consolidation. It was also shown that the same sequence of activity may occur before the actual experience. This phenomenon, termed preplay, may have a role in prediction and learning.

GRID CELLS

Grid cells were discovered by Moser and coworkers (2005) in the entorhinal cortex. Grid cells have multiple firing fields, with regular spacing, which tessellate the environment in a hexagonal pattern. The arrangement of the firing fields of grid cells all at equal distances from their neighbors led to a hypothesis that these cells encode a cognitive representation of Euclidian space (Hafting et al., 2005). What makes grid cells especially interesting is that the regularity in grid spacing does not derive from any regularity in the environment or in the sensory input available. Grid cells appear to encode a type of abstract structure inside the brain which is imposed on the environment with no regard for the sensory features of the environment.

Unique properties of grid cells include the following (Hafting et al., 2005):

1. Grid cells have firing fields dispersed over the entire environment (place fields are restricted to certain specific regions of the environment),
2. The firing fields of grid cells are organized into a hexagonal lattice,
3. Firing fields of grid cells are generally equally spaced apart such that the distance from one firing field to all six adjacent firing fields is approximately the same (though when an environment is resized, the field spacing may shrink or expand differently in different directions-Barry et al., 2007),
4. Firing fields of grid cells are equally positioned such that the six neighboring fields are located at approximately 60 degree increments.

Grid cells at similar dorsal-to-ventral entorhinal cortex levels had similar grid spacing and grid orientation, but the phase of the grid (the offset of the grid vertices relative to the x and y axes) appeared to be randomly distributed between cells (Hafting et al., 2005).

A grid cell fires when a freely moving animal traverses a set of small regions which are roughly equal in size and arranged in a periodic triangular array that covers the entire available environment (Hafting et al., 2005). The grid cell periodic firing pattern was anchored to external landmarks, was expressed independently of the configuration of landmarks, was expressed in darkness as well as in the presence of visible landmarks, and occurred independently of changes in the animal's speed and direction. Cells with this firing pattern have been found in all layers of the dorsocaudal medial entorhinal cortex, but cells in different layers tend to differ in other respects. Layer II contains the largest density of pure grid cells, in that they fire equally regardless of the direction in which an animal traverses a grid location. Grid cells from deeper layers are intermingled with conjunctive cells and head direction cells (in layers III, V, and VI there are cells with a grid-like pattern that fire only when the animal is facing a particular direction-Sargolini et al., 2006). Grid cells that lie next to one another usually show the same grid spacing and orientation, but their grid vertices are displaced from one another by apparently random offsets. Cells recorded from a distance from one another frequently show different grid spacing. Cells that are located more ventrally (that is, farther from the dorsal border of the entorhinal cortex) generally have larger firing fields at each grid vertex and correspondingly greater spacing between the grid vertices (Hafting et al., 2005). Grid cell activity does not require visual input, since grid patterns remain unchanged when all the lights in an environment are turned off. When visual cues are present, however, they exert strong control over the alignment of the grids (rotating a cue card on the wall of a cylinder causes grid patterns to rotate by the same amount). Grid patterns appear on the first entrance of an animal into a novel environment and usually remain stable thereafter. When an animal is moved into a completely different environment, grid cells maintain their grid spacing and the grids of neighboring cells maintain their relative offsets. The interactions of grid cells with place cells are interesting. When a rat is moved to a different environment, the activity patterns of hippocampal place cells usually show *"complete remapping"* — that is, the pattern of place fields reorganizes in a way that bears no detectable resemblance to the pattern in the original environment (Muller and Kubie, 1987). If the features of an environment are altered less radically, however, the place field pattern may show a lesser degree of change, referred to as *"rate remapping"*, in which many cells alter their firing rates, but the majority of cells retain place fields in the same locations as before. Fyhn and coworkers (2007) examined this phenomenon using simultaneous recordings of hippocampal and entorhinal cells and found that in situations where the hippocampus shows rate remapping, grid cells show unaltered firing patterns, whereas when the hippocampus shows complete remapping, grid cell firing patterns show unpredictable shifts and rotations.

Neural activity in nearly every part of the hippocampal system is modulated by the theta rhythm. The entorhinal cortex is no exception. Like the hippocampus, it receives cholinergic and GABAergic input from the medial septal area, the central controller of theta. Grid cells, like hippocampal place cells, show strong theta modulation (Hafting et al., 2005). Grid cells from layer II of the medial entorhinal cortex also resemble hippocampal place cells in that they show phase precession—that is, their spike activity advances from late to early phases of the theta cycle as an animal passes through a grid vertex. Most grid cells from layer III do not show precession, but their spike activity is largely confined to half of the theta cycle. The grid cell phase precession is not derived from the hippocampus because it continues to appear in animals whose hippocampus has been inactivated by an agonist of GABA (Hafting et al., 2008).

Boundary Cells

Boundary cells (border cells, boundary vector cells) are neurons found in the hippocampal formation that respond to the presence of an environmental boundary at a particular distance and direction from an animal. The firing fields of place cells, which characteristically respond only in a circumscribed area of an animal's environment, tended to fire in corresponding locations when the shape and size of the environment was altered (Burgess et al., 2000; Hartley et al., 2000; O'Keefe & Burgess, 1996). Boundary cells relied on inputs sensitive to the geometry of the environment (environmental boundaries at particular distances and allocentric directions) to determine where a given place cell would fire in environments of different shapes and sizes. Cells were identified with these characteristics in the subiculum (Barry et al., 2006; Lever et al., 2009), entorhinal cortex (Savelli et al., 2008; Solstad et al., 2008), and pre- and para-subiculum (Boccara et al., 2010). In medial entorhinal cortex border/boundary cells comprise about 10% of the local population, being intermingled with grid cells and head direction cells.

Long Term Potentiation (LTP)

Since the time of Ramon y Cajal (1894), scientists have speculated that the brain stores memory by altering the strength of connections between neurons that are simultaneously active. This idea was formalized by Donald Hebb in 1948, but for many years thereafter attempts to find a brain mechanism for such changes failed. In 1973, Bliss and Lømo described a phenomenon in the rabbit hippocampus that appeared to meet Hebb's specifications: a change in synaptic responsiveness induced by brief strong activation and lasting for hours to days or longer. This phenomenon was referred to as long term potentiation (LTP). As a candidate mechanism for memory, LTP has since been studied intensively.

The hippocampus is particularly favorable for studying LTP because of its densely packed and sharply defined layers of neurons, but similar types of activity-dependent synaptic change have been observed in many other brain areas (Cooke & Bliss, 2006). The best-studied form of LTP occurs at synapses that terminate on dendritic spines and use the neurotransmitter glutamate. Several of the major pathways within the hippocampus fit this description and exhibit LTP (Malenka & Bear, 2004). The synaptic changes in LTP depend on a special type of glutamate receptor, the NMDA receptor, which has the special property of allowing calcium to enter the postsynaptic spine only when presynaptic activation and postsynaptic depolarization occur at the same time (Nakazawa et al., 2004). LTP appears to occur because of increased Ca++ release in the synapses of active pathways. Drugs that interfere with NMDA receptors block LTP and have major effects on some types of memory, especially spatial memory. Transgenic mice, which have a disabled LTP mechanism, also generally show severe memory deficits (Nakazawa et al., 2004).

ACTIVITY DRIVEN SYNAPTIC PLASTICITY

Activity driven changes occur in the synapses between neurons which influence their further activity. This is known as activity driven synaptic plasticity. These changes may be inhibitory or excitatory. Learning and memory are hypothesized to occur via activity driven changes in the connections between neurons. Certain mechanisms normally operate at the neural synapse level to regulate firing according to the average level of activation. A mechanism that normally operates at the neural synapse level to regulate postsynaptic firing according to the average level of synaptic activity is known as synaptic (spike) timing dependent plasticity (STDP). STDP depends on the temporal occurrence of presynaptic and postsynaptic spikes. When one neuron repeatedly fires after another neuron fires, changes occur in the connections between the two neurons which facilitate further such firing. The firing of the two neurons is said to be correlated. The firing of the postsynaptic neuron is in effect regulated by the firing of the presynaptic neuron. This regulation consists of changes in the connections between the two neurons so that the connections are either strengthened or weakened. A synapse is strengthened if the firing of one neuron is repeatedly followed by the firing of the other neuron. A synapse is weakened if a neuron repeatedly fails to fire the other neuron. STDP refers to the observation that the precise timing of presynaptic and postsynaptic spikes significantly affects the magnitude and sign of synaptic response. STDP follows Donald Hebb's famous postulate for neural learning: "two neurons that fire together, wire together". Hebb's postulate may also be interpreted in another way: "if presynaptic activity repeatedly produces postsynaptic activity, then changes occur in the synapses that facilitate further postsynaptic activation by the presynaptic activation." Hence STDP is also known as Hebbian plasticity. Hebb's postulate was later extended by others so that when the presynaptic input failed to produce postsynaptic output, then the synapses were depressed. In the archetypical form of STDP, a neural connection is strengthened if the presynaptic neuron fires a few milliseconds before the postsynaptic neuron fires and weakened if the reverse temporal situation occurs, the postsynaptic neuron fires before the presynaptic neuron. The time windows for the production of potentiation and depression in this model of STDP are very different with the time window for the production of depression (up to negative 100 msec) being much broader than the time window for potentiation (positive 1-10 msec). If the synaptic connection is strengthened (STDP is called LTP). If the synaptic connection is weakened (STDP is called long term depression-LTD). STDP may be considered a type of learning and memory.

STDP provides a powerful mechanism for the modification of neural circuits. It has been implicated in cortical map development, the sharpening of sensory receptive fields, working memory, associative learning, and the stabilization of neural activity (Watt & Desai, 2010). However, evidence suggests that the spike timing case for the production of STDP is an oversimplification (Shouval et al., 2010; Watt & Desai, 2010). STDP does not develop in a simple graded additive (linear) fashion, as would occur if it arose simply from the incremental summed effects of individual spike pairs. A simple spike timing rule does not explain many of the nonlinearities that occur in the development of activity driven synaptic plasticity (Shouval et al., 2010; Watt & Desai, 2010). Many forms of STDP occur in many different synapse types. Also, if the presynaptic input and/or the postsynaptic output consist of spike trains (more than one spike), then the changes that occur in the synapses are much more complex because of nonlinear interactions among the multiple spikes (Froemke et al., 2010). In the particular case of high frequency input, LTP occurs regardless of the temporal order of the firings. Hence temporal order is not the most important determinant of STDP (Buchanan & Mellor, 2010; Shouval et al., 2010). Nonlinearities arise in the production of activity driven synaptic plasticity as a result of stimulation frequency, total number of presynaptic and postsynaptic spikes, the history of spike activity, and the location and spatial pattern of activity on and across neurons. STDP arises in an all or none manner over minutes. It does not arise in a graded incremental manner. The induction of synaptic plasticity involves a number of receptor-generated second messengers (like calcium) and the kinases, phosphatases, and other downstream targets they activate (Shouval et al., 2010). All these molecular factors are ignored in a simple plasticity or learning rule involving the temporal order of spike pairs. A study of the cellular processes involved in synaptic plasticity provides a better rule for synaptic plasticity and learning by accounting for much of the nonlinearity that arises in the development of such plasticity. Calcium accounts for much of the complexity of the induction of a number of types of synaptic plasticity (Shouval et al., 2010). The critical trigger for STDP is actually the influx of Ca++ through NMDA receptors in postsynaptic membranes where the peak local postsynaptic Ca++ concentration achieved is crucial in determining CAMKII and PPI activation and consequently the direction and magnitude of STDP (Buchanan & Mellor, 2010). It is the amount of postsynaptic depolarization that is achieved at a synapse that determines the magnitude of the Ca++ influx. Postsynaptic depolarization is responsible for removing the Mg++ block from NMDA receptors and allowing the influx of Ca++. Hence the greater the postsynaptic depolarization, the greater the influx of Ca++. The size of the local increase in Ca++ concentration determines the magnitude and sign of STDP. The amount of postsynaptic depolarization achieved depends on the EPSP (excitatory presynaptic potential) amplitude, the dendritic membrane potential and excitability, the presence of inhibitory synaptic transmission, action potential back-propagation, and the frequency of stimulation (Buchanan & Mellor, 2010). If the total postsynaptic depolarization is below a certain threshold, then LTD occurs and the synapse is weakened (the postsynaptic neuron fires less strongly to the same presynaptic stimulation or it takes stronger presynaptic stimulation to fire the postsynaptic neuron). In other words, the connection weight is decreased. If the postsynaptic depolarization is above the threshold, then LTP occurs and the synapse is strengthened (the postsynaptic neuron fires more strongly to the same presynaptic input or it takes less presynaptic stimulation to fire the postsynaptic neuron). The connection weight is increased. Hence LTP and LTD are determined by the amount of postsynaptic depolarization achieved and the calcium dynamics at synapses. It is a certain threshold of postsynaptic depolarization which determines the switch over point from LTD to LTP. One therefore has the potential to have a sliding threshold of postsynaptic depolarization controlling the production of LTD or LTP. This is very important in controlling and stabilizing synaptic activity and maintaining synaptic activity at an optimal level. The threshold may be changed (regulated) by a mechanism which increases or decreases the overall excitability of the neurons involved. This mechanism is sensitive to the activity level across synapses and is very important in maintaining the stability and function of synapses when the activity levels become high or low or fluctuate. Thus, if the activity level across a synapse is increased, then an increase in the threshold (by decreasing the excitability of the presynaptic and/or postsynaptic neuron) determining the production of LTD or LTP would lead to a dampening and stabilization of the postsynaptic firing. Conversely, if the activity level across a synapse is decreased, then a decrease in the threshold (by increasing the excitability of the presynaptic and/or postsynaptic neuron) for the production of LTD or LTP would lead to an increase and stabilization of postsynaptic firing. If the excitability of a neuron was determined by the activity level (which it seems to be), then it is the activity level which controls the sliding threshold for producing LTD or LTP. Hence the activity level regulates the output. How the excitability of neurons is controlled by their activity level will be discussed in the next section (homeostatic intrinsic plasticity).

HOMEOSTATIC SYNAPTIC REGULATION

Runaway potentiation may occur in the hippocampus from high frequency input, leading to the destabilization of neuron and neural network activity and the loss of effective information transfer and function (Watt & Desai, 2010). Other presynaptic inputs to postsynaptic neurons, which were not originally correlated with the postsynaptic firing, will also become correlated with the postsynaptic firing and LTP will occur in these connections as well. Consequently, the runaway potentiation will spread from neuron to neuron and throughout the neural network, leading to ever increasing excitation, destabilization of neural network activity, a loss of effective information transfer and function, and pathology (like in some seizures). Since such a situation does not normally occur in the brain, some mechanisms must operate to prevent such instability and pathology from occurring. In order to maintain the stability and functionality of neurons and neural circuits in the face of high frequency stimulation (or fluctuating activation), changes in synaptic strength must be regulated and coordinated across multiple synapses and with other neuronal properties. This is known as homeostatic synaptic regulation. Homeostatic synaptic regulation helps balance high and low activity levels of a neuron or neural network in order to achieve balanced and stable functionality of the neuron or neural network. This is necessary for the neuron or neural network to function optimally.

Homeostatic synaptic regulatory mechanisms combat runaway potentiation, maintain the optimal dynamic functional range during periods of high or low activity levels and during periods of intense synaptogenesis and neural growth, and face the different stability issues arising during different developmental stages. In complicated neural circuits, homeostatic synaptic regulation encompasses different processes acting on different timescales and in different ways. There are homeostatic synaptic mechanisms that regulate a neuron's excitability, synaptic drive (input), or spiking output. STDP does not work alone. Homeostatic forms of neuronal plasticity complement and work in conjunction with STDP and forms of developmental plasticity to constrain synaptic strength and/or neuronal properties. Homeostatic plasticity mechanisms control the total synaptic strength of a neuron, modulate the intrinsic excitability of a neuron as a function of its average activity, and control the ability of synapses to undergo Hebbian modification according to their history of use (Watt & Desai, 2010). Homeostatic plasticity mechanisms serve to regulate a neuron's output to keep it relatively stable and at an optimal level. In other words, a neuron's activity is kept in the optimal dynamic range, so that the neuron can continue to transfer information and function in an effective manner. Major forms of homeostatic synaptic plasticity that function is this manner are synaptic scaling, homeostatic intrinsic plasticity, metaplasticity, and stability mechanisms intrinsic to STDP (Watt & Desai, 2010).

SYNAPTIC SCALING

Synaptic scaling directly regulates the synaptic strength of neurons to promote cellular stability. Synaptic scaling is bidirectional. The synaptic strength of neurons is scaled down if general activity levels are increased and scaled up if general activity levels are decreased. This tends to keep the synaptic strength and activity of neurons within the optimal dynamic functional range. Synaptic scaling occurs gradually over hours and days. It appears to occur globally over all of a neuron's synapses, but a local type of synaptic scaling has also been described (Watt & Desai, 2010). This is in contrast to STDP which occurs in only a fraction of a neuron's synapses (those that are active). Synaptic scaling also occurs so that differences in the relative strengths of various inputs of a neuron are maintained. This is important for learning and memory, as these are thought to be encoded as different synaptic strengths of neurons. The firing rate of a neuron appears to be an important determinant of synaptic scaling. Studies indicate that at the neuronal level the input that a neuron receives is titrated to keep its output stable and within an optimal range (Watt & Desai, 2010). The synaptic input to a neuron was scaled up when its firing rate was reduced to bring the firing rate of the neuron back to optimal control levels. The synaptic input to a neuron was scaled down when its firing rate was increased to bring the firing rate of the neuron back to optimal levels. In synaptic scaling, changes in postsynaptic activity are precisely balanced with presynaptic changes in release dynamics. Synaptic scaling may involve presynaptic and/or postsynaptic changes and depends on calcium dynamics and a variety of soluble second messengers (BDNF-brain derived neurotrophic factor, calmodulin activated kinases, etc.), maybe the same molecules as are involved in producing STDP (Watt & Desai, 2010). Synaptic scaling may involve changes in both AMPA and NMDA currents (Watt & Desai, 2010). Since both synaptic scaling and STDP result from changes in Ca++ dynamics and the same second messengers, the two forms of plasticity may be coordinated and complement each other. The changes for synaptic scaling and STDP occur at different time scales, however. Synaptic scaling is much slower than STDP.

HOMEOSTATIC INTRINSIC PLASTICITY

A neuron's intrinsic electrical properties arise largely as a result of the expression of various voltage and calcium dependent ion channels (Watt & Desai, 2010). The expression of these channels at any one time is very complex, varies throughout life, and is under the influence of the activity level. Changes in the intrinsic properties of a neuron result in an overall shift in cellular excitability. This is known as homeostatic intrinsic plasticity. This plasticity may help neurons and neural circuits maintain appropriate levels of electrical activity under changing conditions of input. Thus the gain and/or threshold (excitability) of individual neurons for firing may be adjusted or matched to the average input received. A neuron's output (firing rate) therefore has a relation to its input. If the average input is too low, the neuron will hardly ever fire, because of the spike threshold. If the average input is too high, the neuron's firing will saturate, because there is a physical limit to how fast a neuron can fire. Between these two extremes is the sensitive region in which the neuron's output really does reflect its input. Thus there is a robust strategy for maintaining neuronal firing rate stability. When there are large or long-lasting fluctuations in average input, changes occur in the intrinsic properties (excitability) of the neurons, so that the sensitive region of the neurons' firing corresponds to the distribution of inputs received. The dynamic range of a neuron's firing is therefore preserved. In other words, neurons respond to low or high activity conditions by becoming more or less sensitive to whatever input they receive. Evidence for such a mechanism has been found in the neocortex, hippocampus, and other parts of the brain (Watt & Desai, 2010). As for synaptic scaling, the signal for homeostatic intrinsic plasticity seems to be calcium influx due to depolarization and spike firing (Watt & Desai, 2010). Under physiological conditions, intracellular calcium levels track average firing rates quite well. Thus one has a robust self-regulatory activity driven signaling control of the firing of individual neurons and of large neural networks. One also has the coordination of homeostatic intrinsic plasticity, STDP, and other homeostatic plasticity mechanisms, as all depend on calcium levels to some extent. Other extracellular signals may also play a role in homeostatic intrinsic plasticity, like in synaptic scaling. The relationship of homeostatic intrinsic plasticity and synaptic scaling is unclear. These two forms of homeostatic plasticity may act together in parallel, singly at different stages of development, or synergistically. Neither appears to be sufficient alone (Watt & Desai, 2010). It appears like homeostatic intrinsic plasticity affects STDP. When the average activity of a neuron is low, an increase in intrinsic excitability (a decrease in threshold) would make the subsequent production of synaptic potentiation more likely. If the average activity of a neuron is high, a decrease in intrinsic excitability (an increase in threshold) would make the subsequent production of synaptic potentiation less likely and the production of synaptic depression more likely. This plasticity of plasticity is known as metaplasticity and is a type of homeostatic adaptation.

HISTORY

From a network point of view, the most stable type of homeostatic regulation may be embedded in the rules of Hebbian plasticity (STDP). One way of doing this is to make the capacity of synapses to undergo Hebbian plasticity depend on their history of activity (their history of use). This is a form of metaplasticity. LTP occurs if the postsynaptic depolarization is above a certain threshold, while LTD occurs if the postsynaptic depolarization is below the threshold. By moving the threshold for producing LTD and LTP as a slow function of the average activity level, the capacity for undergoing Hebbian plasticity can be varied over time. In other words, the capacity of synapses to undergo Hebbian plasticity may be varied according to the history of activity level. One consequently has a sliding threshold controlling neural firing which is dependent on the history of use of the synapses. This history effect stabilizes synaptic dynamics. Many experiments have demonstrated the existence of a threshold that moves based on prior activity (Watt & Desai, 2010). This sliding of the threshold has been shown to be a cell wide property and not restricted to specific synapses. The cause of this shift may be activity driven changes in NMDA receptor subunit composition, synaptic scaling of NMDA conductance, and intrinsic homeostatic plasticity (Watt and Desai, 2010). STDP therefore is tempered by homeostatic synaptic plasticity mechanisms which naturally change synaptic weights so that neurons can continue to effectively transfer information. Homeostatic plasticity mechanisms stabilize synapses. Homeostatic regulation may encompass regulating and stabilizing the firing of individual neurons or of large neural networks, preserving the uniform distribution of synaptic weights across neurons and neural networks, and ensuring that the dynamic range of response properties matches that of inputs.

SHORT TERM EPISODIC MEMORY

Working memory appears to access the hippocampus via the entorhinal cortex (perforant path). A reverberatory circuit of activation has been shown to be set-up between the entorhinal cortex and the hippocampus during conscious processing (Deadwyler et al., 1988). Such a circuit could provide prolonged repetitive high frequency stimulation adept at producing LTP in place cell assemblies of the hippocampus. An allocentric representation of the environment appears to be encoded and stored in the hippocampus (Ekstrom et al., 2003; O'Keefe & Dostrovsky, 1971; O'Keefe & Nadel, 1978; Ono et al., 1993; Serino & Riva, 2014). Neurons in the CA3 region of the hippocampus, receiving inputs from the entorhinal cortex, form an *allocentric view-point dependent representation;* whereas neurons in the CA1 region of the hippocampus, receiving inputs from CA3 via Schaffer collaterals, rapidly encode an *allocentric view-point independent representation* (Robertson et al., 1998; Rolls, 2007). It is possible to distinguish between the allocentric representation formed by CA3 and the allocentric representation formed by CA1 (Behrendt, 2013). Hippocampal place cells play a crucial role in the establishment and maintenance of both representations (Serino & Riva, 2014). The encoding of activity of both the CA3 and CA1 regions in sync is necessary for the encoding of coherent short term episodic memory. Hippocampal place cells play a critical role in retrieving a coherent episodic memory because of the reciprocal connectivity between the hippocampus and neocortical regions. When prompted by a retrieval cue, the full episodic memory can be retrieved through the process of pattern completion (Burgess et al., 2001; Byrne et al., 2007). Episodic recall (including the simulation of future or imaginary events) may be traceable to retrieval cues (encoded in neocortical areas and introduced to the hippocampus via the entorhinal cortex) inducing the reinstatement of neuronal assemblies in

CA3 and CA1 (Serino & Riva, 2014). Serino and Riva (2014) suggest that for effective episodic retrieval, it is crucial that the allocentric view-point independent representation in CA1 be synced with the allocentric view-point dependent representation in CA3. In Alzheimer's disease a break in the "mental frame syncing" between these two allocentric representations may occur, caused by damage to the hippocampus, and may contribute significantly to the early impairment of episodic memory (Serino & Riva,2014).

LONG TERM MEMORY

The storage of episodic memories for a long period of time (long term memory) involves the establishment of more stable and permanent neuronal connections in pathways previously subjected to LTP. This involves additional protein synthesis and structural change (like phosphorylation of proteins). This probably occurs most prominently in the cerebral cortex (parietal, temporal, and occipital cortex), where long term episodic memories appear to be stored. Patients with lesions in the parietal and occipital cortex may have markedly reduced memory spans, but intact learning abilities (Mohs, 1991). The hippocampus may be involved in changing neuronal connections for three months or more after the initial learning. It may take years (sometimes 5-7 years) for long term episodic memories to be fully consolidated (Mohs, 1991). The hippocampus appears to be responsible for the consolidation of short term episodic memory into long term episodic memory. Hence the hippocampus is necessary for not only the formation of new memories, but also for long term retention of those memories. In the absence of the hippocampus, the formation and retention of new memories and the storage and retrieval of old memories are severely impaired (producing anterograde and retrograde amnesia respectively). Such individuals have very short attention spans, are unable to remember items just seen, have graded retrograde amnesia, and have impaired spatial location and navigation ability. In Alzheimer's disease, the hippocampus may be one of the first structures affected, with memory impairments and disorientation the first symptoms appearing. Damage to the hippocampus and the production of these symptoms may result from anoxia, encephalitis, medial temporal lobe epilepsy, alcoholism, and taking certain drugs.

AMYGDALA

It is interesting to note that the hippocampus receives input from the amygdaloid nuclei and orbitofrontal cortex. These areas play a critical role in the expression of emotions (Sieb, 2013). The amygdala also mediates the strengthening effect that emotion has on episodic memory. If one is aroused or emotional in any way, memory of experiences is greatly increased (Sieb, 2013). The amygdala has been implicated in the acquisition and expression of classical (Pavlovian) fear conditioning (a form of conditioning of emotional responses) and LTP appears to be involved (Amunts et al., 2005; Blair et al., 2001; Helmuth, 2003; Marin,1999; Pare et al., 2002; Paton et al., 2005; Sapolsky, 2003). The amygdala is also involved in appetitive (positive) conditioning (Blair, 2008; Paton et al., 2005). The amygdala plays a primary role in the formation and storage of memories associated with emotional events and it appears to be the role of the amygdala to regulate memory consolidation in other brain regions (Marin, 1999). The amygdala has many connections with the hippocampus (more going to the hippocampus than received from it-Ben Best, The amygdala and the emotions, *Anatomical Basis of Mind*, Ch.9, see website) and emotions can facilitate the encoding of memories (Gasbarri & Tomaz, 2012). Increased amygdala activation immediately after an event increases the strength and retention of memory of that event (Asari et al., 2010; Hutcherson et al., 2008; Nathan et al., 2006; Pare et al., 2002). The more emotionally-arousing the event, the greater is the activation of the amygdala (especially the basolateral nuclei) and the greater the strength and retention of the memory of the event. Injections of cortisol and adrenaline and stimulation of the amygdaloid nuclei facilitate memory. Excessive or prolonged stress (with prolonged cortisol release) may hinder memory. Patients with amygdaloid damage are no longer able to better remember emotionally-charged words than non-emotionally-charged words.

Factors Affecting Episodic Memory

A number of factors affect episodic memory. Among these are the effect of spacing rehearsal over an extended period of time, cramming, the Zeigarnik effect (people remember interrupted or uncompleted tasks better than completed ones), use of certain odors, size of the knowledge base (children may be able to remember things better than adults because they have a smaller knowledge base), retroactive interference (learning new information causes one to forget old information), proactive interference (learning one piece of information makes it harder to remember similar new information), emotion (or stimulation of the amygdaloid nuclei), increasing the supply of oxygen to the brain, staying intellectually active, keeping physically active, socializing, reducing stress, getting regular sleep, avoiding depression and emotional instability, and observing good nutrition.

Conclusion

Episodic memory is a record of personal four-dimensional past experiences. This is evident during episodic memory recall. Episodic memory recall entails the retrieval of a particular four-dimensional organization of conscious experiences: some conscious experiences occur simultaneously, but in different locations; while other conscious experiences occur at the same location, but at different times. Such organization indicates that both space and time must be considered together in the organization of episodic memory recall. Because conscious experiences make up the content of episodic memory recall, episodic memory recall requires consciousness (conscious recollection is required to demonstrate episodic memory), as conscious experiences make up the content of consciousness. It is interesting to note that consciousness at any one time has a similar four-dimensional organization of conscious experiences (some conscious experiences occur simultaneously, but in different locations; while other conscious experiences occur at the same location, but at different times).

Conscious experiences which occur simultaneously, but in different locations in episodic memory recall and consciousness, are reminiscent of space-like spacetime interval separation of events in four-dimensional spacetime, as described by Einstein's special theory of relativity. Similarly, conscious experiences occurring in the same location, but at different times in episodic memory recall and consciousness are reminiscent of time-like spacetime interval separation of events in four-dimensional spacetime, also as described by Einstein's special theory of relativity. Hence the organization and relationships of episodic memory recall and consciousness at any one time may be described utilizing the concept of invariant spacetime intervals as predicted by Einstein's special theory of relativity. This suggests that episodic memory is encoded in four dimensions in order for its recall to have a four-dimensional organization. Episodic memory appears to be encoded in four dimensions via the prefrontal cortex-inferior parietal cortex-inferior temporal cortex-perirhinal cortex-entorhinal cortex-hippocampus network. Working memory appears to access the hippocampus via this network for the four-dimensional encoding of episodic memory in the hippocampus. Working memory consists of the positive feedback interaction between the prefrontal cortex and inferior parietal cortex. Conscious experiences appear to arise in a nonlinear emergent manner from this positive feedback interaction (working memory). Working memory is the holding in mind of a small number of pieces of information for a short period of time and their manipulation. Positive feedback and nonlinear emergence results in the formation of explicit stable states (conscious experiences). The pieces of information that are maintained in working memory for a short period of time thus appear to be a number of explicit stable states or conscious experiences. Consciousness, at any one time, and episodic memory recall have the same basic four-dimensional organization of conscious experiences. The organization of both consciousness and episodic memory arises from input from the prefrontal cortex and inferior parietal cortex (from working memory). Hence the four-dimensional organization of consciousness and episodic memory must come from working memory. The information persistence component of sensory memory (such as iconic memory) is the main contributor to working memory. Information persistence is a brief representation of abstract properties of sensory stimuli, such as spatial location, temporal occurrence, and spatiotemporal separation. Information persistence may guide the organization of conscious experiences in working memory via the prefrontal cortex and/or inferior parietal cortex. Sensory memory, working memory, short term memory, and long term memory operate together in the development of episodic memory. The encoding and retention of episodic memory involves the establishment of LTP in place cell assemblies of the hippocampus by working memory. This encoding and retention of episodic memory involves STDP and other mechanisms of homeostatic intrinsic plasticity working together to mould, stabilize, and optimize neural connections. Working memory accesses the hippocampus via the prefrontal cortex-inferior parietal cortex-inferior temporal cortex-perirhinal cortex-entorhinal cortex perforant path to encode an allocentric four-dimensional viewpoint dependent representation in the CA3 region of the hippocampus. Grid cells, head direction cells, and boundary cells of the entorhinal cortex are responsible for the encoding of this representation in four dimensions in place cell assemblies of the CA3 region of the hippocampus. This encoding is a static four-dimensional neural (cognitive) map of the environment. The CA3 region then sets up an allocentric viewpoint independent representation in place cell assemblies of region CA1 of the hippocampus. This is a representation of the dynamic relationships arising from CA3. The representations in CA3 and CA1 in sync make up the complete episodic memory and must be reactivated in sync for episodic memory recall, imagination, and future ideations.

Consciousness and episodic memory are linked by working memory. They are both organized by working memory. Episodic memory is associated with consciousness via working memory. Consciousness and episodic memory recall are frames of reference (observational perspectives). Hence they both have a subjective or perspective-like aspect. Their four-dimensional organization gives them a "picture" or "image-like" quality. The simultaneous production of conscious experiences at different locations in consciousness and episodic memory recall provides a four-dimensional awareness of our surroundings and of our self in relation to those surroundings (a context). The production of conscious experiences at the same location, but at different times in consciousness and episodic memory recall, provides us with a four-dimensional awareness of dynamic changes in our surroundings and of cause and effect relationships occurring in those surroundings. Syncing of these two types of awareness enable the more effective production of appropriate voluntary (adaptive or goal-directed) actions. Because of four-dimensional episodic memory, we can more effectively look toward the future and use our imagination. This greatly improves our success and survival in our complex societies. These processes would be much less effective if they were not four-dimensional.

The utilization of relativistic principles in human consciousness and memory offer a different avenue of support for the validity of Einstein's theories of relativity. This suggests that our brain may create and remember its own four-dimensional reality.

References

Adolphs, R., Cahill, L., Schul, R., Babinsky, R., (1997). Impaired declarative memory for emotional material following bilateral amygdala damage in humans. *Learning and Memory* 4, 291-300.

Afifi, A.K., Bergman, R.A., (1986). *Basic Neuroscience.* Baltimore/Munich:Urban & Schwarzenberg.

Alloway, T. P., Alloway, R. G., (2010). Investigating the predictive roles of working memory and IQ in academic attainment. *Journal of Experimental Child Psychology* 80(2), 606–621.

Amaral, D., Lavenex, P., (2006). Ch 3. Hippocampal Neuroanatomy. In: P. Andersen, R. Morris, D. Amaral, T. Bliss, & J.O'Keefe, *The Hippocampus Book.* Oxford University Press.

Amos, S.W., Amos, R.W., (2002). *Newnes Dictionary of Electronics*, 4th Ed., Newness, 276.

Amunts, K., Kedo, O., Kindler, M., Pieperhoff, P., Mohlberg, H., Shah, N., Habel, U., Schneider, F., Zilles, K., (2005). Cytoarchitectonic mapping of the human amygdala, hippocampal region and entorhinal cortex: Intersubject variability and probability maps. *Anatomy and Embryology (Berlin)* 210(5-6), 343-352.

Anagnostaras, S.G., Gale, G.D., Fanselow, M.S., (2002). The hippocampus and Pavlovian fear conditioning: Reply to Bast et al. *Hippocampus* 12(12), 561–565.

Andrews, G., Halford, G.S., (2002). A cognitive complexity metric applied to cognitive development. *Cognitive Psychology* 45(2), 153–219.

Apud, J.A., Mattay, V., Chen, J., et al., (2007). Tolcapone improves cognition and cortical information processing in normal human subjects. *Neuropsychopharmacology* 32(5), 1011–1020.

Arnsten, A.F.T., Paspalas, C.D., Gamo, N.J., Wang, M., (2010). Dynamic network connectivity: A new form of neuroplasticity. *Trends in Cognitive Science* 14(8), 365–375.

Asari, T. Konishi, S., Jimura, K., Chikazoe, J., Nakamura, N., Miyashita, Y., (2010). Amygdala enlargement associated with unique perception. *Cortex* 46(1), 94-99.

Ashby F.G., Ell, S.W., Valentin, V.V., Casale, M.B., (2005). FROST: a distributed neurocomputational model of working memory maintenance. *Journal of Cognitive Neuroscience* 17(11), 1728–1743.

Avidan, M.S., et al., (2001). Prevention of intraoperative awareness in a high-risk surgical population. *The New England Journal of Medicine* 865, 591-600.

Baars, B., (1988). *A Cognitive Theory of Consciousness*, Cambridge: Cambridge University Press.

Babinsky, R., Calabrese, P., Durwen, H., Markowitsch, H., Brechtelsbuauer, D., (1993). The possible contribution of the amygdala to memory. *Behavioural. Neurology* 6, 167-170.

Baddeley, A., (2003). Working memory: looking back and looking forward. *Nature Reviews Neuroscience* 4(10), 829–839.

Barry, C., Lever, C., Hayman, R., Hartley, T., Burton, S., O'Keefe, J., Jeffery, K., Burgess, N., (2006). The boundary vector model of place cell firing and spatial memory. *Reviews in the Neurosciences* 17(1–2), 71– 97.

Bartolomeo, P., Thiebaut de Schotten, M., Doricchi, F., (2007). Left unilateral neglect as a disconnection syndrome. *CerebralCortex* 17, 2479-2490.

Behrendt, R.P., (2013). Conscious experience and episodic memory: Hippocampus at the crossroads. *Frontiers Psychology* 4, 304.

Benson, D.F., Sheremata, W.A., Bouchard, R., Seggarram, J., Price, D., Geschwind, N., (1973). Conduction aphasia. *Archives of Neurobiology* 28, 339-346.

Benton, A.L., (1991). The prefrontal region: Its early history. In: H.S. Levin, H.M. Eisenberg, & A.L. Benton (Eds.), *Frontal lobe function and dysfunction* 19. New York: Oxford University Press.

Berry, A.S., Zanto, T.P., Rutman, A.M., Clapp, W.C., Gazzaley, A., (2009). Practice-related improvement in working memory is modulated by changes in processing external influences. *Journal of Neurophysiology* 102(3), 1779–1789.

Best, B., (2004). The amygdala and the emotions. *Anatomical Basis of Mind*, Chapter 9, website.

Best, B., (2010). *Brain Neurotransmitters*, Chapter 10, Website.

Binder, M.D., (2009). *Encyclopedia of Neuroscience* 3166. Springer.

Blair, H.T., Schafe, G.E., Bauer, E.P., Rodrigues, S.M., Ledoux, J.E., (2001). Synaptic plasticity in the lateral amygdala: A cellular hypothesis for fear conditioning. *Learning and Memory* 8, 229-242.

Blair, R.J.R., (2008). The amygdala and ventromedial prefrontal cortex: Functional contribution and dysfunction in psychopathy. *Philosophical Transactions of the Royal Society* 363(1503), 2557-2565.

Bliss T, Lømo T., (1973). Long-lasting potentiation of synaptic transmission in the dentate area of the anaesthetized rabbit following stimulation of the perforant path. J. Physiol. 232 (2), 331–356.

Boccara, C.N., Sargolini, F., Thoresen, V.Y.H., Solstad, T., Witter, M.P., Moser, E.I., Moser, M.B. (2010). Grid cells in pre- and parasubiculum. *Nature Neuroscience* 13(8), 987–994.

Boehler, C.N., Schoenfeld, M.A., Heinze, H.J., Hopf, J.M., (2008). Rapid recurrent processing gates awareness in primary visual cortex. *Proceedings of National Academy of Science USA* 105, 8742-8747.

Bogarts, B., (1997). The temporolimbic system theory of positive schizophrenic symptoms. *Schizophrenia Bulletin* 23.

Bruce, C.J., Desimone, R., Gross, C.G., (1982). Visual properties of neurons in a polysensory area in superior temporal sulcus of the macaque. *Journal of Neurophysiology* 46, 369-384.

Bruce, C.J., Desimone, R., Gross, C.G., (1986). Both striate and superior colliculus contribute to visual properties of neurons in superior polysensory area of macaque monkey. *Journal of Neurophysiology* 58, 1057-1076.

Buchanan, K.A., Mellor, J.R., (2010). The activity requirements for spike timing-dependent plasticity in the hippocampus. *Frontiers in Synaptic Neuroscience* 2, 11.

Bures, J., Fenton, A.A., Kaminsky, Y., Zinyuk, L., (1997). Place cells and place navigation. *Proceedings of the National Academy of Sciences* 94(1), 343–350.

Burgess, N., Becker, S., King, J.A., O'Keefe, J., (2001). Memory for events and their spatial context: Models and experiments. *Philosophical Transactions Royal Society Lond. B Biological Science* 356, 1493–1503.

Burgess, N., Jackson, A., Hartley, T., O'Keefe, J., (2000). Predictions derived from modelling the hippocampal role in navigation. *Biological Cybernetics* 83(3), 301–312.

Burlet, S., Tyler, C.J., Leonard, C.S., (2002). Direct and indirect excitation of laterodorsal tegmental neurons by hypocretin/orexin peptides: Implication for wakefulness and narcolepsy. *Journal of Neuroscience* 22(7), 2862-2872.

Burton, H., Jones, E.G., (1976). The posterior thalamic region and its cortical projection in New World and Old World Monkeys. *Journal of Comparative Neurology* 168, 249-302.

Buzsáki, G., (1986). Hippocampal sharp waves: Their origin and significance. *Brain Research* 398(2), 242– 252.

Buzsáki, G., (1989). Two-stage model of memory trace formation: A role for "noisy" brain states. *Neuroscience* 31(3), 551–570.

Buzsáki, G., (2006). *Rhythms of the Brain*. Oxford University Press.

Buzsáki, G., Chen, L.S., Gage, F.H., (1990). Spatial organization of physiological activity in the hippocampal region: Relevance to memory formation. *Progress in Brain Research* 83, 257–268.

Byrne, P., Becker, S., Burgess, N., (2007). Remembering the past and imagining the future: A neural model of spatial memory and imagery. *Psychological Review* 114, 340–375.

Cajal, R.S. y, (1894). The Croonian Lecture: La fine structure des centres nerveux. *Proceedings of the Royal Society* 55(331–335), 444–468.

Cantero, J.L., Atienza, M., Stickgold, R., Kahana, M.J., Madsen, J.R., Kocsis, B., (2003). Sleep-dependent theta oscillations in the hippocampus and neocortex. *Journal of Neuroscience* 23(34), 10897–10903.

Carlsen, J., DeOlmos, J., Heimer, L., (1982). Tracing of two neuron pathways in the olfactory system by the aid of transneuronal degeneration: Projection to the amygdaloid body and hippocampal formation. *Journal of Comparative Neurology* 208, 196-208.

Case, R., (1985). *Intellectual development. Birth to adulthood.* New York: Academic Press.

Cenquizca, L.A., Swanson, L.W., (2007). Spatial organization of direct hippocampal field CA1 axonal projections to the rest of the cerebral cortex. *Brain Research* 56(12), 1–26.

Chiu, Y.C., Algase, D., Whall, A., et al., (2004). Getting lost: Directed attention and executive functions in early Alzheimer's disease patients. *Dementia Geriatric Cognitive Disorders* 17(3), 174–80.

Clark, R.E., Broadbent, N.J., Squire, L.R., (2005). Hippocampus and remote spatial memory in rats. *Hippocampus* 15(2), 260–272.

Conway, M.A., (2009). Episodic Memory. *Neuropsychologia 47*, 2305-2306.

Coltheart, M., (1980). Iconic memory and visible persistence. *Perception & Psychophysics* 27(3), 183–228.

Cooke, S.F., Bliss, T.V., (2006). Plasticity in the human central nervous system. *Brain* 129 (Pt 7), 1659–1673.

Cowan, N., (2001). The magical number 4 in short-term memory: A reconsideration of mental storage capacity. *Behavioral and Brain Sciences* 24, 87–185.

Cowan, N., (2008). What are the differences between long-term, short-term, and working memory? *Progress in Brain Research* 169(169), 323–338.

Cowan, N., Alloway, T.P., (2008). The development of working memory. In: N. Cowan (Ed). *Development of Memory in Childhood*, 2nd edition, pp. 303–342. Hove, England: Psychology Press.

Crick, F., Koch, C., (1995). Are we aware of neural activity in primary visual cortex? *Nature* 375, 121-123.

Crick, F., Koch, C., (2003). A framework for consciousness. *Nature Neuroscience* 6(2), 119-126.

Curtis, C.E., D'Esposito, M., (2003). Persistent activity in the prefrontal cortex during working memory. *Trends in Cognitive Sciences* 7(9), 415–423.

Daltrozzo, J., Wioland, N., Mutschler, V., Lutun, P., Jaeger, A., Calon, B., Meyer, A., Pottecher, T., Lang, S., Kotchoubey, B., (2009). Cortical information processing in coma. *Cognitive and Behavioral Neurology* 22(1), 53-62.

Daltrozzo, J., Wioland, N., Mutschler, V., Lutun, P., Jaeger, A., Calon, B., Meyer, A., Pottecher, T., Lang, S., Kotchoubey, B., (2010). Electrodermal response in coma and other low response patients. *Neuroscience Letters* 475(1), 44-47.

Dawson, M.W.R., (2013). *Blackwell's Dictionary of Cognitive Science.*

Deadwyler, S.A., Hampson, R.E., Foster, T.C., Marlow, G., (1988). The functional significance of long-term potentiation: Relation to sensory processing by hippocampal circuits. In: P.W. Landfield & S.A. Deadwyler (Eds.), Long-term potentiation: From biophysics to behavior, *Neurology and Neurobiology* 35, 499-534.

Deisseroth, K., Singla, S., Toda, H., Monje, M., Palmer, T.D., Malenka, R.C., (2004). Excitation-neurogenesis coupling in adult neural stem/progenitor cells. *Neuron* 42(4), 535–552.

Demertzi, A., Soddu, A., Laureys, S., (2013). Consciousness supporting networks. *Current Opinion in Neurobiology*, 23(2), 239-244.

Desmedt, J.E., (1977). Active touch exploration of extrapersonal space elicits specific electrogenesis in the right cerebral hemisphere of intact right handed man. *Proceedings of the National Academy of Sciences* 74, 4037-4040.

Desimone, R., Duncan, J., (1995). Neural mechanisms of selective visual attention. *Annual Review of Neuroscience* 18, 193–222.

Diana, R.A., Yonelinas, A.P., Ranganath, C., (2007). Imaging recollection and familiarity in the medial temporal lobe: A three-component model. *Trends in Cognitive Science* 11(9), 379–86.

Dick, A.O., (1974). Iconic memory and its relation to perceptual processing and other memory mechanisms. *Perception & Psychophysics* 16(3), 575–596.

Di Lollo, V., (1980). Temporal integration in visual memory. *Journal of Experimental Psychology: General* 109, 75–97.

Dimond, S.J., (1976). Depletion of attentional capacity after total commissurotomy in man. *Brain* 99, 347-356.

Dimond, S.J., (1979). Tactual and auditory vigilance in split-brain man. *Journal of Neurology, Neurosurgery and Psychiatry* 42, 70-74.

Dunbar, G., Boeijinga, P.H., Demazières, A., et al., (2007). Effects of TC-1734 (AZD3480), a selective neuronal nicotinic receptor agonist, on cognitive performance and the EEG of young healthy male volunteers. *Psychopharmacology (Berl.)* 191(4), 919–929.

Eccles, J.C., McGeer, F.L., McGeer, E.G., (1987). *Molecular Neurobiology of the Mammalian Brain*, Plenum Press: New York.

Ekstrom, E.D., Kahana, M.J., Caplan, J.B., Fields, T.A., Isham, E.A., Newman, E.L., Fried, I., (2003). Cellular networks underlying human spatial navigation. *Nature* 425, 184-188.

Edwards, J.D., (2013). World product data management-discrete guide, Release A9.3 E21777-04. *Oracle Corporation.*

Eichenbaum, H., (2001). The hippocampus and declarative memory: Cognitive mechanisms and neural codes. *Behavioural Brain Research* 127, 199-207.

Eichenbaum, H., Cohen, N.J., (1993). *Memory, Amnesia, and the Hippocampal System.* MIT Press.

Eichenbaum, H., Yonelinas, A.P., Ranganath, C., (2007). The medial temporal lobe and recognition memory. *Annual Review of Neuroscience* 30, 123–152.

Einstein, A., (2001). *Relativity: The Special and the General Theory* (Reprint of 1920 translation by Robert W. Lawson ed.). Routledge. p.48.

Evans, B.M., (2003). Sleep, consciousness and the spontaneous and evoked electrical activity of the brain. Is there a cortical integrating mechanism? *Neurophysiologie Clinique* 33, 1-10.

Fahrenfort, J.J., Scholte, H.S., Lamme, V.A., (2007). Masking disrupts reentrant processing in human visual cortex. *Journal of Cognitive Neuroscience* 19, 1488-1497.

Fanselow, M.S., Dong, H.-W., (2009). Are the dorsal and ventral hippocampus functionally distinct structures? *Neuron* 65(1), 7–19.

Fenton, A.A., Kao, H.Y., Neymotin, S.A., Olypher, A, Vayntrub, Y., Lytton, W.W., Ludvig, N., (2008). Unmasking the CA1 ensemble place code by exposures to small and large environments: More place cells and multiple, irregularly arranged, and expanded place fields in the larger space. *Journal of Neuroscience* 28(44), 11250–11262

Feynman, R.P., (1998). *Six Not-so-easy Pieces: Einstein's relativity, symmetry, and space-time* (Reprint of 1995 ed.). Basic Books. p.68.

Flood, D.G., Coleman, P.D., (1990). Hippocampal plasticity in normal aging and decreased plasticity in Alzheimer's Disease. In: J. Storm-Mathisen, J. Zimmer, & O.T.Ottersen (Eds.), Understanding the brain through the hippocampus, *Progress in Brain Research* 83, 435-444.

Foreman, N., Stevens, R., (1987). Relationships between the superior colliculus and hippocampus: Neural and Behavioral considerations. *Behavioral and Brain Sciences* 10, 101-152.

Froemke, R.C., Debanne, D., Bi, G.-Q., (2010). Temporal modulation of spike-timing-dependent plasticity. *Frontiers in Synaptic Neuroscience* 2(19), 1-16.

Fukuda, K., Vogel, E.K., (2009). Human variation in overriding attentional capture. *The Journal of Neuroscience* 29(27), 8726–8733.

Fuster, J.M., (1997). *The Prefrontal Cortex. Anatomy, Physiology, and Neuropsychology of the Frontal Lobe*, 3rd ed, New York: Lippincott-Raven.

Fuster, J.M., (2000). Cross-modal and cross-temporal association in neurons of frontal cortex. *Nature* 405, 347-351.

Fuster, J.M., (2008). *The Prefrontal Cortex*, 45th ed, London: Academic Press.

Fyhn, M., Hafting, T., Treves, A., Moser, M.B., Moser, E.I., (2007). Hippocampal remapping and grid realignment in entorhinal cortex. *Nature* 446(7132), 190–194.

Gabrieli, J., Kao, Y., (2007). Development of the declarative memory system in the human brain. *Nature Neuroscience* 10, 1198-1205.

Gabrieli, J.D., Poldrack, R.A., Desmond, J.E., (1998). The role of left prefrontal cortex in language and memory. *Proceedings National Academy of Science U.S.A.* 95(3), 906–913.

Ganong, W.F., (1988). *Review of Medical Physiology*, Lange: Los Altos, California.

Garcia-Rill, E., (1997). Disorders of the reticular activating system. *Medical Hypotheses* 49(5), 379-387.

Garcia-Rill, E., Heister, D.S., Ye, M., Charlesworth, A., Hayar, P., (2007). Electrical coupling: Novel mechanism for sleep-wake control. *Sleep* 30(11), 1405-1414.

Gasbarri, A., Tomaz, C. (2012). Memory and motivational/emotional processes. *Frontiers of Behavioral Neuroscience*, 6, 71.

Gershenson, C., Heylighen, F., (2004). How can we think the complex. In K.Richardson (Ed.), *Managing the Complex Volume One: Philosophy, Theory and Application*, Image Age Publishers.

Giulini, D., (2006). Algebraic and geometric structures in special relativity. In: J.Ehlers & C.Lämmerzahl (Eds.), *Special Relativity—Will it Survive the Next 101 Years?* Springer, 45–111.

Goldman-Rakic, P.S., (1990). Cellular and circuit basis of working memory in prefrontal cortex of nonhuman primates. In H.B.M.Uylings, C.G.Van Eden, J.P.E.DeBruin, M.A.Corner, & M.G.P.Feenstra (Eds.), The prefrontal cortex, *Progress in Brain Research* 85, 325-336.

Goldman-Rakic, P.S., (1995). Cellular basis of working memory. *Neuron* 14, 447–485.

Graf, R.F., (1999). *Modern Dictionary of Electronics*, 7th Ed., Newness, 276.

Graf, P., Schacter, D.L., (1985). Implicit and explicit memory for new associations in normal and amnesic subjects. *Journal of Experimen-tal Psychology: Learning, Memory, and Cognition* 11, 501-518.

Gray, J.A., McNaughton, N., (2000). *The Neuropsychology of Anxiety: An Enquiry into the Functions of the Septal-Hippocampal System*, Oxford University Press.

Greene, E., (2007). Information persistence in the integration of partial cues for object recognition. *Perception & Psychophysics* 69(5), 772–784.

Grön, G., Kirstein, M., Thielscher, A., Riepe, M.W., Spitzer, M., (2005). Cholinergic enhancement of episodic memory in healthy young adults. *Psychopharmacology (Berl.)* 182(1), 170–179.

Gustafsson, B., Wigstrom, H., (1990). Long-term potentiation in the hippocampus CA1 region: Its induction and early temporal development. In: J.Storm-Mathisen, J.Zimmer, & O.P.Ottersen (Eds.), Understanding the brain through the hippocampus, *Progress in Brain Research* 83, 223-232.

Haas, H.L., Buzsaki, G., (1988). *Synaptic Plasticity in the hippocampus*, New York:Springer-Verlag.

Haber, R., Standing, L., (1970). Direct measures of visual short-term visual storage. *Quarterly Journal of Experimental Psychology* 21, 216–229.

Hafting, T., Fyhn, M., Bonnevie, T., Moser, M.B., Moser, E.I., (2008). Hippocampus-independent phase precession in entorhinal grid cells. *Nature* 453(7199), 1248–1252.

Hafting, T., Fyhn, M., Molden, S., Moser, M., Moser, E., (2005). Microstructure of a spatial map in the entorhinal cortex. *Nature* 436(7052), 801–806.

Hannaman, R.A., (2005). *Medstudy Internal Medicine Review Core Curriculum: Neurology 11th Ed. Medstudy*, (11-1)-(11-2).

Hardt, O., Einarsson, E.O., Nader, K., (2010). A bridge over troubled water: Reconsolidation as a link between cognitive and neuroscientific memory research traditions. *Annual Review of Psychology* 61, 141–167.

Hartley, T., Burgess, N., Lever, C., Cacucci, F., O'Keefe, J., (2000). Modeling place fields in terms of the cortical inputs to the hippocampus. *Hippocampus* 10(4), 369–379.

Hassabis, D., Kumaran, D., Maguire, E.A., (2007a). Using imagination to understand the neural basis of episodic memory. *Journal of Neuroscience* 27, 14365–14374.

Hassabis, D., Kumaran, D., Vann, S.D., Maguire, E.A., (2007b). Patients with hippocampal amnesia cannot imagine new experiences. *Proceedings National of Academic Science U S A* 104, 1726–1731.

Hassabis, D., Maguire, E.A., (2007). Deconstructing episodic memory with construction. *Trends in Cognitive Science* 11(7), 299–306.

Hassabis, D., Maguire, E.A., (2009). The construction system of the brain. *Philosophical Transactions Royal Society Lond. B Biological Science* 364, 1263–1271.

Hebb, D.O., (1949). *Organization of Behavior: A Neuropsychological Theory*. New York:John Wiley.

Hecaen, H., Albert, M.L., (1978). *Human Neuropsychology*. New York:John Wiley & Sons.

Heilman, K.M., (1973). Ideational apraxia. *Brain* 96, 861-864.

Heilman, K.M., Van Den Abell, T., (1979). Right hemispheric dominance for mediating cerebral activation. *Neuropsychologia* 17, 315-322.

Heilman, K.M., Van Den Abell, T., (1980). Right hemisphere dominance for attention. The mechanism underlying hemispheric asymmetries of inattention (neglect). *Neurology (New York)* 30, 327-330.

Helmuth, L., (2003). Fear and trembling in the amygdala. *Science* 300, 568-569.

Henthorn, T., (2013). What is continuous processing? *Wise Geek, Conjecture Corporation.*

Hier, D.B., Mondlock, J., Caplan, L.R., (1983). Behavioral abnormalities after right hemisphere stroke. *Neurology (New York)* 33, 337-344.

Hillis, A.E., Newhart, M., Kessler, J., Barker, P.B., Herskovits, E.H., Degaonkar, M., (2005). Anatomy of spatial attention: Insights from perfusion imaging and hemispatial neglect in acute stroke. *Journal of Neuroscience* 25, 3161- 3167.

Hillis, A.E., Wityk, R.J., Barker, P.B., Beauchamp, N.J., Gailloud, P., Murphy, K., Cooper, O., Mitter, E.J., (2002). Subcortical aphasia and neglect in acute stroke: The role of cortical hypoperfusion. *Brain* 125, 1094-1104.

Honey, G.D., Fu, C.H., Kim, J., et al., (2002). Effects of verbal working memory load on corticocortical connectivity modeled by path analysis of functional magnetic resonance imaging data. *NeuroImage* 17(2), 573–82.

Howard, M.W., Kahana, M.J., (2002). When does semantic similarity help episodic retrieval. *Journal of Memory and Language* 46, 85–96.

Howe, M.L., (2011). The adaptive nature of memory and its illusions. *Current Directions Psychological Science* 20, 312–315.

Hutcherson, C.A., Seppala, E.M., Gross, J.J., (2008). Loving-kindness meditation increases social-connectedness. *Emotion* 8(5), 720-724.

Hudetz, A.G., (2006). Suppressing consciousness: Mechanisms of general anesthesia. *Perioperative Medicine And Pain* 25(4), 196-204.

Huerta, P.T., Lisman, J.E., (1993). Heightened synaptic plasticity of hippocampal CA1 neurons during a cholinergically induced rhythmic state. *Nature* 364(6439), 723–725.

Hulme, C., Roodenrys, S,, Brown, G., Mercer, R., (1995). The role of long-term memory mechanisms in memory span. *British Journal of Psychology* 86(4), 527–536.

Husain, M., (2008). Hemineglect. *Scholarpedia* 3(2), 3681.

Husain, M., Kennard, C., (1996). Visual neglect associated with frontal lobe infarction. *Journal of Neurology* 243, 652-657.

Hyvarinen, J., Shelepin, Y., (1979). Distribution of visual and somatic functions in the parietal association cortex of the monkey. *Brain Research* 169, 561-564.

Imas, O.A., Ropella, K.M., Ward, B.D., Wood, J.D., Hudetz, A.G. (2005). Volatile anesthetics disrupt frontal-posterior recurrent information transfer at gamma frequencies in rats. *Neuroscience Letters* 387(3), 145-150.

Insausti, R., Amaral, D.G., Cowan, W.M., (1987). The entorhinal cortex of the monkey, II. Cortical afferents. *Journal Comparative Neurology* 264, 356–395.

Insausti, R., Munoz, M., (2001). Cortical projections of the non-entorhinal hippocampal formation in the cynomolgus monkey (*Macaca fascicularis*). *European Journal Neuroscience* 14, 435–451.

Irwin, D., Thomas, L., (2008). Neural basis of sensory memory. In: S.Luck & A.Hollingworth. *Visual Memory*. New York:Oxford University Press. 32–35.

Irwin, D., Yeomans, J., (1986). Sensory registration and informational persistence. *Journal of Experimental Psychology: Human Perception and Performance* 12(3), 343–360.

Jackson, J.C., Johnson, A., Redish, A.D., (2006). Hippocampal sharp waves and reactivation during awake states depend on repeated sequential experience. *Journal of Neuroscience* 26(48), 12415–12426.

Janowsky, J.S., Shimamura A.P., Squire, L.R., (1989). Source memory impairment in patients with frontal lobe lesions. *Neuropsychologia* 27(8), 1043–1056.

Jarrold, C., Bayliss, D.M., (2007). Variation in working memory due to typical and atypical development. In: A.R.A. Conway, C.Jarrold, M.J.Kane, A.Miyake. & J.N.Towse (Eds.), *Variation in working memory* (pp. 137–161). New York:Oxford University Press.

Jeffery, K., (2007). Integration of sensory inputs to place cells: What, where, why, and how? *Hippocampus* 17(9), 775–785.

Jeffery, K., Anderson, M., Hayman, R., Chakraborty, S., (2003). A proposed architecture for the neural representation of spatial context. *Neuroscience and Behavioral Reviews* 28, 201–218.

John, P.J., (2009). Frontal-temporal dysfunction in schizophrenia: A selective review. *Indian Journal Of Psychiatry* 51(3), 180-190.

Jones, E.G., Powell, T.P.S., (1970). An anatomical study of converging sensory pathways within the cerebral cortex of the monkey. *Brain* 93, 793-820.

Jones, R., (2013). Forming memories, one neuron at a time. *Knowing Neurons.*

Joseph, R., (1982). The neuropsychology of development: Hemispheric laterality, limbic language, and the origin of thought. *Journal of Clinical Psychology* 44, 3-33.

Joseph, R., (1986). Confabulation and delusional denial: Frontal lobe and lateralized influences. *Journal of Clinical Psychology* 42, 845-860.

Joseph, R., (1988a). The right cerebral hemisphere: Emotion music, visual-spatial skills, body-image, dreams, and awareness. *Journal of Clinical Psychology* 44, 630-673.

Joseph, R., (1988b). Dual mental functioning in a "split-brain" patient. *Journal of Clinical Psychology* 44, 770-779.

Joseph, R., (1990). *Neuropsychology, Neuropsychiatry, and Behavioral Neurology*, New York:Plenum Press.

Joseph, R., (2000). *Neuropsychiatry, Neuropsychology, Clinical Neuroscience*, New York:Academic Press.

Jung, M.W., Wiener, S.L., McNaughton, B.L., (1994). Comparison of spatial firing characteristics of units in dorsal and ventral hippocampus of the rat. *Journal of Neuroscience* 14 (12), 7347–7356.

Kahana, M.J., Seelig, D., Madsen, J.R., (2001). Theta returns. *Current Opinion in Neurobiology* 11(6), 739–744.

Kail, R., (2007). Longitudinal evidence that increases in processing speed and working memory enhance children's reasoning. *Psychological Science* 18(4), 312–313.

Kane, M.J., Engle, R.W., (2002). The role of prefrontal cortex in working-memory capacity, executive attention, and general fluid intelligence: An individual-differences perspective. *Psychonomic Bulletin & Review* 9(4), 637–671.

Kim, M., Na, D.L., Kim, G.M., Adair, J.C., Lee, K.H., Heilman, K.M., (1999). Ipsilesional neglect: Behavioral and anatomical features. *Journal of Neurology, Neurosurgery & Psychiatry* 67, 35-38.

Kinomura, S., Larsson, J., Gulyas, B., Roland, P.E., (1996). Activation by attention of the human reticular formation and thalamic intralaminar nuclei. *Science* 271(5248), 512-515.

Klingberg, T., Forssberg, H., Westerberg, H., (2002). Training of working memory in children with ADHD. *Journal of Clinical and Experimental Neuropsychology* 24(6), 781–91.

Koch, C., (1998). The neuroanatomy of visual consciousness. In: H.H.Jasper, L.Descarries, V.F.Costelluchi, & S.Rossignol (Eds.), *Advances in Neurology, Consciousness at the Frontiers of Neuroscience* 77, 229-239, New York:Lippincott-Raven.

Ku, S., Lee, U., Noh, G., Jun, I., Mashour, G.A., (2011). Preferential inhibition of frontal-to-parietal feedback connectivity is a neurophysiologic correlate of general anesthesia in surgical patients. *PLoS/One* 6(10).

Lamme, V. A., Roelfsema, P. R., (2000). The distinct modes of vision offered by feedforward and recurrent processing. *Trends of Neuroscience* 23, 571-579.

Lapitskaya, N., Gosseries, O., De Pasqua, V., Pederson, A.R., Nielsen, J.F., de Noordhout, A.M., Laureys, S. (2013). Abnormal corticospinal excitability in patients with disorders of consciousness. *Brain Stimulation* preprint.

Laplante, F., Morin, Y., Quirion, R., Vaucher, E., (2005). Acetylcholine release is elicited in the visual cortex, but not in the prefrontal cortex, by patterned visual stimulation: A dual in vivo microdialysis study with functional correlates in the rat. *Neuroscience* 132(2), 501-510.

Lavenex, P., Suzuki, W.A., Amaral, D.G., (2002). Perirhinal and parahippocampal cortices of the macaque monkey, projections to the neocortex. *Journal Comparative Neurology* 447, 394–420.

Lee, U., Kim, S., Nok, G., Choi, B., Hwang, E., Mashour, G.A., (2009). The directionality and functional organization of frontalparietal connectivity during consciousness and anesthesia in humans. *Consciousness and Cognition* 18(4), 1069-1078.

Lever, C., Burton, S., Jeewajee, A., O'Keefe, J., Burgess, N., (2009). Boundary vector cells in the subiculum of the hippocampal formation. *Journal of Neuroscience* 29(31), 9771–9777.

Levick, W., Zacks, J., (1970). Responses of cat retinal ganglion cells to brief flashes of light. *Journal of Physiology* 206(3), 677–700.

Libet, B., et al., (1983). Time of conscious intention to act in relation to onset of cerebral activity (readiness-potential). The unconscious initiation of a freely voluntary act. *Brain* 106, 623-642.

Libet, B., (1985). Unconscious cerebral initiation and the role of conscious will in voluntary action. *Behavioral Brain Science* 8, 529-566.

Liversedge, T., Hirsch, N., (2010). Coma. *Anesthesia & Intensive Case Medicine* 11(9), 337-339.

Loftus, G., Bursey, T., Senders, J., (1992). On the time course of perceptual information that results from a brief visual presentation. *Journal of Experimental Psychology* 54, 535–554.

Long, G., (1980). Iconic Memory: A review and critique of the study of short-term visual storage. *Psychological Bulletin* 88(3), 785–820.

Lubenov, E.V., Siapas, A.G., (2009). Hippocampal theta oscillations are travelling waves. *Nature* 459(7246), 534.

Luck, S.J., Vogel, E.K., (1997). The capacity of visual working memory for features and conjunctions. *Nature* 390, 279-281.

Luria, A.R., (1980). *Higher Cognitive Functions in Man*, New York:Basic Books.

Lynch, G., Kessler, M., Arai, A., Larson, J., (1990). The nature and causes of hippocampal long-term potentiation. In: J.Storm-Mathisen, J.Zimmer, & O.P.Ottersen (Eds.), Understanding the brain through the hippocampus. *Progress in Brain Research* 83, 233-250.

Lynch, M.A., Errington, M.L., Clements, M.P., Bliss, T.V.P., Redini-Del Negro, C., Laroche, S., (1990). Increases in glutamate release and phosphoinositide metabolism associated with long-term potentiation and classical conditioning. In: J.Storm-Mathisen, J.Zimmer, & O.P.Ottersen (Eds.), Understanding the brain through the hippocampus. *Progress in Brain Research* 83, 251-256.

Maguire, E.A., Burgess, N., Donnett, J.G., Frackowiak, R.S.J., Frith, C.D., O'Keefe, J., (1998). Knowing where and getting there: A human navigation network. *Science* 280(5365), 21–24.

Maguire, E.A., Gadian, D.G., Johnsrude, J.D., Good, C.D., Ashburner, J., Frackowiak, R.S., Frith, C.D., (2000). Navigatgion-related structural change in the hippocampi of taxi drivers. *PNAS* 97(8), 4398–403.

Malenka, R.C., Bear, M.F., (2004). LTP and LTD: An embarrassment of riches. *Neuron* 44(1), 5–21.

Maren, S., (1999). Long-term potentiation in the amygdale: A mechanism for emotional learning and memory. *Trends in Neuroscience* 22(12), 561-567.

Matolcsi, T., (1994). *Spacetime Without Reference Frames*. Budapest:Akadémiai Kiadó.

Matsumura, N., Nishijo, H., Tamura, R., Eifuku, S., Endo, S., Ono, T. (1999). Spatial and task-dependent neuronal responses during real and virtual translocation in the monkey hipppocampal formation. *Journal of Neuroscience* 19(6), 2381–2393.

Mattingley, J.B., (1999). Attention, consciousness, and the damaged brain: Insights from parietal neglect and extinction. *Psyche* 5, 14.

Mehta, Z., Newcombe, F., Damasio, H., (1987). A left hemisphere contribution to visiospatial processing. *Cortex* 23, 447-451.

Mermin, N.D., (2005). *It's About Time. Understanding Einstein's Relativity*, Princeton University Press. Meunier, M., Bachevalier, J., Mishkin, M., and Murray, E.A. (1993). Effects on visual recognition of combined and separate ablations of the entorhinal and perirhinal cortex in rhesus monkeys. *J. Neurosci.* 13, 5418–5432.

Minkowski, H., (1909). Raum und zeit. *Physikalische Zeitschrift* 10, 75–88.

Mishkin, M., (1978). Memory in monkeys severely impaired by combined but not by separate removal of amygdala and hippocampus. *Nature* 273, 297–298.

Mohedano-Moriano, A., Pro-Sistiaga, P., Arroyo-Jimenez, M.M., Artacho-Perula, E., Insausti, A.M., Marcos, P., Cebada-Sanchez, S., Martinez-Ruiz, J., Munoz, M., Blaizot, X., Martinez-Marcos, A., Amaral, D.G., Insausti, R., (2007). Topographical and laminar distribution of cortical input to the monkey entorhinal cortex. *Journal of Anatomy* 211, 250–260.

Mohs, R.C., (1991). Learning and memory. In: K.Davis, H.Klar, & J.T.Coyle (Eds.), *Foundations of Modern Psychiatry*, 11-18. Toronto: W.R. Sanders.

Morris, R.G.M., Garrud, P., Rawlins, J.N.P., O'Keefe, J., (1982). Place navigation impaired in rats with hippocampal lesions. *Nature* 297, 681-683.

Moser, E., Kropff, E., Moser, May-Britt, (2008). Place cells, grid cells, and the brain's spatial representation system. *Annual Review of Neuroscience* 31, 69–89.

Mottaghy, F.M., (2006). Interfering with working memory in humans. *Neuroscience* 139(1), 85–90.

Muir, G., Bilkey, D.K., (2001). Instability in the place field location of hippocampal place cells after lesions centered on the perirhinal cortex. *The Journal of Neuroscience* 21(11), 4016–4025.

Muller, R.U., Kubie, J.L., (1987). The effects of changes in the environment on the spatial firing of hippocampal complex-spike cells. *The Journal of Neuroscience: The official journal of the Society for Neuroscience* 7(7), 1951–1968.

Munoz, M., Insausti, R., (2005). Cortical efferents of the entorhinal cortex and the adjacent parahippocampal region in the monkey (*Macaca fascicularis*). *European Journal of Neuroscience* 22, 1368–1388.

Munoz-Lopez, M.M., Mohedano-Moriano, A. Insausti, R., (2010). Anatomical pathways for auditory memory in primates. *Frontiers Neuroanatomy* 4, 129.

Nakazawa, K., McHugh, T., Wilson, M., Tonegawa, S., (2004). NMDA receptors, place cells and hippocampal spatial memory. *Nature Reviews* 5, 368–369.

Nathan, P., Phan, L., Fitzgerald, D., Tancer, M., (2006). Studying brain activity could aid diagnosis of social phobia. *Science Daily*, http://wwn.sciencedaily.com/releases/2006/01/060118205940.htm

Nikolić, D., Häusler, S., Singer, W., Maass, W., (2009). Distributed fading memory for stimulus properties in the primary visual cortex. In: J.D.Victor (Ed.), *PLoS Biology* 7(12), e1000260.

O'Keefe, J., (1978). *The Hippocampus as a Cognitive Map*. Oxford: Claredon Press.

O'Keefe, J., (1999). Do hippocampal pyramidal cells signal non-spatial as well as spatial information. *Hippocampus*, 9(4), 352-363.

O'Keefe, J., Burgess, N., (1996). Geometric determinants of the place fields of hippocampal neurons. *Nature* 381(6581), 425–428.

O'Keefe, J., Burgess, N., Donnett, J.G., Jeffery, K.J., Maguire, E.A., (1998). Place cells, navigational accuracy, and the human hippocampus. *Philosophical Transactions of the Royal Society B: Biological Sciences* 353(1373), 1333–1340.

O'Keefe, J., Conway, D.H., (1978). Hippocampal place units in the freely moving rat: Why they fire where they fire. *Experimental Brain Research* 31(4).

O'Keefe, J., Dostrovsky, J., (1971). The hippocampus as a spatial map. Preliminary evidence from unit activity in the freely-moving rat. *Brain Research* 34(1), 171–175.

O'Keefe, J., Nadel, L., (1978). *The Hippocampus as a Cognitive Map*. Oxford University Press.

Ono, T., Nakamura, K., Nishijo, H., Eifuku, S., (1993). Monkey hippocampal neurons related to spatial and nonspatial functions. *Journal of Neurophysiology* 70, 1516–1529.

Pare, D., Collins, D.R., Pelletier, J.G., (2002). Amygdala oscillations and the consolidation of emotional memories. *Trends in Cognitive Sciences* 6(7), 306-314.

Paton, J.J., Belova, M.A., Morrison, S.E., Salzman, C.D., (2006). The primate amygdala represents the positive and negative value of visual stimuli during learning. *Nature* 439, 865-870.

Perry, R.H., Chilton, C.H., Green, C.W., (1997). *Perry's Chemical Engineers' Handbook (7th Ed.),* New York:McGraw-Hill.

Petkov, V., (2010). *Minkowski Spacetime: A Hundred Years Later*. Springer. p. 70.

Petrides, M., Pandya, D.N., (2007). Efferent association pathways from the rostral prefrontal cortex in the macaque monkey. *Journal of Neuroscience* 27, 11573-11586.

Poisson, E., (2004), The motion of point particles in curved spacetime. *Living Rev. Relativity* 7, doi:10.12942/lrr-2004-6

Postle, B.R., (2006). Working memory as an emergent property of the mind and brain. *Neuroscience* 139(1), 23–38.

Pothuizen, H.H., Zhang, W.N., Jogen-Relo, A.L., Feldon, J., Yee, B.K., (2004). Dissociation of function between the dorsal and the ventral hippocampus in spatial learning abilities of the rat: A within-subject, within-task comparison of reference and working spatial memory. *European Journal of Neuroscience* 19(3), 705–712.

Prinz, J.J., (2003). *A Neurofunctional Theory of Consciousness*. On his personal website.

Rang, H.P., (2003). *Pharmacology*, 474,476,480,483. Edinburgh: Churchill Livingstone.

Rao, S.G., Williams, G.V., Goldman-Rakic, P.S., (2000). Destruction and creation of spatial tuning by disinhibition: GABA(A) blockade of prefrontal cortical neurons engaged by working memory. *Journal of Neuroscience* 20(1), 485–494.

Reiner, P.B., (1995) Are mesencephalic cholinergic neurons either necessary or sufficient components of the ascending reticular formation? *Seminars In The Neurosciences* 7(5), 355-359.

Ricketts, J.A., (2013). How a blast furnace works. *American Iron and Steel Institute*.

Rindler, W., (1977). Essential Relativity. Birkhäuser, Ro, T., Breitmeyer, B., Burton, P., Singhal, N. S., Lane, D., (2003). Feedback contributions to visual awareness in human occipital cortex. *Current Biology* 13(12), 1038-1041.

Robbins, T.W., Arnsten, A.F. (2009). The neuropsychopharmacology of fronto-executive function: monoaminergic modulation. *Annual Review of Neuroscience* 32, 267–287.

Roberts, T., Schleif, S. (2007). What is the experimental basis of Special Relativity? *Usenet Physics FAQ*.

Robertson, R.G., Rolls, E.T., Georges-Francois, P., (1998). Spatial view cells in the primate hippocampus: Effects of removal of view details. *Journal of Neurophysiology*, 79, 1145–1156.

Rolls, E.T., (2007). An attractor network in the hippocampus: Theory and neurophysiology. *Learning and Memory*, 14, 714–731.

Rolls, E.T., (2013). The mechanisms for pattern completion and pattern separation in the hippocampus. *Frontiers in Systems Neuroscience* 7, 74.

Rolls, E.T., Xiang, J.Z., (2006). Spatial view cells in the primate hippocampus and memory recall. *Review of Neuroscience* 17(1–2), 175–200.

Rowlands, M., (2002). Two dogmas of consciousness. *Journal of Consciousness Studies* 9, 158-180.

Rull, G., (2011). Delirium. *Patient.co.uk*. Patient.co.uk/doctor/Acute-Confusional-state.htm.

Sapolsky, R.M., (2003). Stress and plasticity in the limbic system. *Neurochemical Research* 28(11), 1735-1742.

Sargolini, F., Fyhn, M., Hafting, T., McNaughton, B.L., Witter, M.P., Moser, M.-B., Moser, E.I., (2006). Conjunctive representation of position, direction, and velocity in entorhinal cortex. *Science* 312(5774), 758–762.

Save, E., Nerad, L., Poucet, B., (2000). Contribution of multiple sensory information to place field stability in hippocampal place cells. *Hippocampus* 10(1), 64–76.

Savelli, F., Yoganarasimha, D., Knierim, J.J., (2008). Influence of boundary removal on the spatial representations of the medial entorhinal cortex. *Hippocampus* 18(12), 1270–1282.

Sauguet, J., Benton, A.L., Hecaen, H., (1971). Disturbances of the body schema in relation to language impairment and hemispheric locus of lesion. *Journal of Neurology, Neurosurgery and Psychiatry* 34, 496-501.

Schacter, D.L., (2009-2011). *PYCHOLOGY*. Catherine Woods. pp. 226.

Schacter, D.L., Addis, D.R., (2007a). The cognitive neuroscience of constructive memory: Remembering the past and imagining the future. *Philosophical Transactions Royal Society London B Biological Science* 362, 773–786.

Schacter, D.L., Addis, D.R., (2007b). On the constructive episodic simulation of past and future events. *Behavioral Brain Science* 30, 331–332.

Schacter, D.L., Addis, D.R., (2009). Remembering the past to imagine the future: A cognitive neuroscience perspective. *Milestones of Psychology* 21, 108–112.

Schacter, D.L., Gilbert, D.T., Wegner, D.M., (2011). Semantic and episodic memory. *Psychology*, 2nd Ed., 240-241. New York:Worth Incorporated.

Scott, A.C., (1996). The hierarchical emergence of consciousness. In: S.R.Hameroff, A.W.Kaszniak, & A.C.Scott (Eds.), *Toward A Science Of Consciousness*, The First Tucson Discussions And Debates, 659- 672, Cambridge, MA:MIT Press.

Scott, A.C., (1999). *Nonlinear Science: Emergence and Dynamics of Coherent Structures*, Oxford:Oxford University Press.

Scott, A.C., (2000). *Modern Science of the Mind*, University of Arizona Course.

Scoville, W.B., Milner, B., (1957). Loss of recent memory after bilateral hippocampal lesions. *J. Neurol. Neurosurg. Psychiatry* 20 11–21.

Seltzer, B., Pandya, D.N., (1978). Afferent cortical connections and architectonics of the superior temporal sulcus and surround cortex in the rhesus monkey. *Brain Research* 149, 1-24.

Serino, S., Riva, G., (2014). What is the role of spatial processing in the decline of episodic memory in Alzheimer's disease? The "mental frame syncing" hypothesis. *Frontiers of Aging Neuroscience* 6, 33.

Shouval, H.Z., Wang, S.S.-H., Wittenberg, G.M., (2010). Spike timing dependent plasticity: A consequence of more fundamental learning rules. *Frontiers in* Computational *Neuroscience* 4(19), 1-13.

Sieb, R.A., (2004). The emergence of consciousness. *Medical Hypotheses* 63(5), 900-904.

Sieb, R.A., (2007). Consciousness and voluntary action. In: S.K.Turrini (Ed.), *Consciousness and Learning Research*, 165-199, New York:Nova Science Publishers, Inc.

Sieb, R.A., (2011). Consciousness and adaptive behavior. *Activitas Nervosa Superior* 53(N.1-2), 21-26.

Sieb, R.A., (2013). The emergence of emotion. *Activitas Nervosa Superior* 55(4), 115-145.

Skaggs, W.E., McNaughton, B.L., Permenter, M., et al., (2007). EEG sharp waves and sparse ensemble unit activity in the macaque hippocampus. *Journal of Neurophysiology* 98(2), 898–910.

Smith, D., Mizumori, S., (2006). Hippocampal place cells, context, and episodic memory. *Hippocampus* 16(9), 716–729.

Smith, E.E., Jonides, J., (1999). Storage and executive processes in the frontal lobes. *Science* 283(5408), 1657–1661.

Smith, E.E., Jonides, J., Marshuetz, C., Koeppe, R.A., (1998). Components of verbal working memory: Evidence from neuroimaging. *Proceedings of the National Academy of Sciences of the United States of America* 95(3), 876–882.

Solstad, T., Boccara, C.N., Kropff, E., Moser, M.B., Moser, E.I., (2008). Representation of geometric borders in the entorhinal cortex. Science 322(5909), 1865–1868.

Sperling, G., (1960). The information available in brief visual presentations. *Psychological Monographs* 74, 1– 29.

Squire, L.R., (1992). Memory and the hippocampus: A synthesis from findings with rats, monkeys, and humans. *Psychological Reviews* 99(2), 195–231.

Squire, L.R., Schacter, D.L., (2002). *The Neuropsychology of Memory*. Guilford Press.

Steriade, M. (1995). Neuromodulatory systems of the thalamus and neocortex. *Seminars in the Neurosciences*, 7(5), 361-370.

Steriade, M., (1996). Arousal revisiting the reticular activating system. *Science* 272(5257), 225-226.

Steriade, M., Jones, E.B., Llinus, R.R., (1990). *Thalamic Oscillations and Signalling*, New York:John Wiley & Sons, Inc.

Steward, O., White, G., Korol, D., Levy, W.B., (1988). Cellular events underlying long-term potentiation and depression in hippocampal pathways: Temporal and spatial constraints. In: P.W.Landfield & S.A.Deadwyler (Eds.), Long-term potentiation from biophysics to behavior, *Neurology and Neurobiology* 35, 139-166.

Strub, R.L., Geschwind, N., (1983). Localization in Gerstmann syndrome. In: A.Kertesz (Ed.), *Localization in Neuropsychology* 173-190, New York:Academic Press.

Sutherland, R.J., Kolb, B., Whishaw, I.Q., (1982). Spatial mapping - definitive disruption by hippocampal or medial frontal cortical damage in the rat. *Neuroscience Letters* 31(3), 271–276.

Sutherland, G.R., McNaughton, B., (2000). Memory trace reactivation in hippocampal and neocortical neuronal ensembles. *Current Opinion in Neurobiology* 10(2), 180–186.

Sutherland, R.J., Weisend, M.P., Mumby, D., et al., (2001). Retrograde amnesia after hippocampal damage: Recent vs. remote memories in two tasks. *Hippocampus* 11(1), 27–42.

Suzuki, W.A., Amaral, D.G., (1994a). Topographic organization of the reciprocal connections between the monkey entorhinal cortex and the perirhinal and parahippocampal cortices. *Journal of Neuroscience* 14, 1856–1877.

Suzuki, W.A., Amaral, D.G., (1994b). Perirhinal and parahippocampal cortices of the macaque monkey, cortical afferents. *Journal Comparative Neurology* 350, 497–533.

Taube, J.S., (2007). The head direction signal: Origins and sensory-motor integration. *Annual Review of Neuroscience* 30, 181–207.

Taylor, E.F., Wheeler, J.A., (1992). *Spacetime Physics: Introduction to Special Relativity*. W. H. Freeman.

Thiebaut de Schotten, M., Urbanski, M., Duffau, H., Volla, E., Levi, R., Dubois, B., Bartolomeo, P., (2005). Direct evidence for a parietal-frontal pathway subserving spatial awareness in humans. *Science* 309, 2226-2228.

Theirry, A.M., Godbout, R., Mantz, J., Glowinski, J. (1990). Influence of the ascending monaminergic systems on the activity of the rat prefrontal cortex. In: H.B.M.Uylings, C.G.Van Eden, J.P.C.DeBruin, M.A.Corner, & M.G.P.Feenstra (Eds.), *The Prefrontal Cortex, Progress in Brain Research* 85, 357-366.

Todd, J.J., Marois, R., (2004). Capacity limit of visual short-term memory in human posterior parietal cortex. *Nature* 428, 751-754.

Tononi, G., Edelman, G., (1998). Consciousness and the integration of information in the brain. In: H.H.Jasper, L.Descarries, V.F.Costelluchi, & S.Rossignol (Eds.), *Advances in Neurology, Consciousness At The Frontiers Of Neuroscience* 77, 245-280. New York:Lippincott-Raven.

Tononi, G., Koch, C., (2008). The neural correlates of consciousness: An update. *Annals of The New York Academy of Science* 1124, 239-261.

Tucker, D.M., (1981). Lateral brain, function, emotion, and conceptualization. *Psychological Bulletin* 89, 19- 46.

Tulving, E., (1972). Episodic and semantic memory. In: E.Tulving & W.Donaldson (Eds.), *Organization of Memory,* 381–403. New York: Academic Press.

Tulving, E., (1983). *Elements of Episodic Memory.* New York:Oxford University Press.

Tulving, E., (1985). Memory and consciousness. *Canadian Psychology* 26, 1–12.

Tulving, E., (2001). Episodic memory and common sense: How far apart? Philosophical Transactions Royal Society Lond. B Biological Science 356, 1505–1515.

Tulving, E., (2002). Episodic memory: From mind to brain. *Annual Review of Psychology* 53, 1–25.

Tulving, E., Schacter, D.L., (1990). Priming and human memory systems. *Science* 247(4940), 301–306.

Tulving, E., Markowitsch, H.J., (1998). Episodic and declarative memory: Role of the hippocampus. *Hippocampus* 8(3), 198–204.

Ullman, M.T., (2004). Contributions of memory circuits to language: The declarative/procedural model. *Cognition* 92, 231–270.

VanElzakker, M.B., Fevurly, R.D., 7Breindel, T., Spencer, R.L., (2008). Environmental novelty is associated with a selective increase in Fos expression in the output elements of the hippocampal formation and the perirhinal cortex. *Learning & Memory* 15(12), 899–908.

Van Hoesen, G.W., Hyman, B.T., (1990). Hippocampal formation: Anatomy and the patterns of pathology in Alzheimer's Disease. In: J.Storm-Mathisen, J.Zimmer, & O.P.Ottersen (Eds.), Understanding the brain through the hippocampus, *Progress in Brain Research* 83, 445-458.

Velmans, M., (1999). When perception becomes conscious. *British Journal Of Psychology* 90(4), 543-566.

Vertes, R.P., (1990). Fundamentals of brainstem anatomy: A behavioral perspective. In: W.P.Klemm & R.P.Vertes (Eds.), *Brainstem Mechanisms Of Behavior* 33-104, New York:John Wiley & Sons, Inc.

Vignolo, L.A., (1983). Modality-specific disorders of written language. In: A.Kertesz (Ed.), *Localization in Neuropsychology*, New York:Academic Press.

Vogel, E.K., Machizawa, M.G., (2004). Neural activity predicts individual differences in visual working memory capacity. *Nature* 428, 748-751.

Wald, R.M., (1984), *General Relativity*, University of Chicago Press.

Watt, A.J., Desai, N.S., (2010). Homeostatic plasticity and STDP: Keeping a neuron's cool in a fluctuating world. *Frontiers in Synaptic Neuroscience*, 2, 5.

Wechsler, R.T., Morss, A.M., Wustoff, C.J., Caughey, A.B., (2004). Blueprints, notes, & cases: Neurosceince, p. 37. Oxford:Blackwell Publishing.

Wilson, M.A., McNaughton, B.L., (1994). Reactivation of hippocampal ensemble memories during sleep. *Science* 265(5172), 676–679.

Winson, J., (1978). Loss of hippocampal theta rhythm results in spatial memory deficit in the rat. *Science* 201(4351), 160–163.

Witter, M.P., Amaral, D.G., (1991). Entorhinal cortex of the monkey, V. Projections to the dentate gyrus, hippocampus, and subicular complex. *Journal Comparative Neurology* 307, 437–459.

Witter, M., Wouterlood, F., (2002). The Parahippocampal Region, Organization and Role in Cognitive Function. Oxford:Oxford University Press.

Yantis, S., Jonides, J., (1990). Abrupt visual onsets and selective attention: Voluntary versus automatic allocation. *Journal of Experimental Psychology. Human Perception and Performance* 16(1), 121–134.

Young, G.M., (2009). Coma. *Annals of The New York Academy of Sciences* 1157, 32-47.

Zanto, T.P., Gazzaley, A., (2009). Neural suppression of irrelevant information underlies optimal working memory performance. *The Journal of Neuroscience* 29(10), 3059–3066.

Zeigler, B.P., Praehofer, H., Kim, T.G., (2000). *Theory of Modelling and Simulation: Integrating Discreet Event and Continuous Complex Dynamic Systems*, 55. New York: Academic Press.

Zeki, S.M., (1974). Functional organization of a visual area in the posterior bank of the superior temporal sulcus of the Rhesus monkey. *Journal of Physiology* 236, 549-573.

Zhang, S., Manahan-Vaughn, D., (2013). Spatial olfactory learning contributes to place field formation in the hippocampus. *Cerebral Cortex*, doi:10.1093/cercor/bht239.

Zuckerman, B., Jefferson, D., (1996). *Human Population and Environmental Crisis*, Jones & Barlett Learning, 42.

INDEX

D

E

F

G

H

I

K

L

M

N

O

P

Q

R

S

T

U

V

W

Y